UNITARY REPRESENTATIONS OF THE POINCARÉ GROUP AND RELATIVISTIC WAVE EQUATIONS

UNITARY REPRESENTATIONS OF THE POINCARÉ GROUP AND RELATIVISTIC WAVE EQUATIONS

Y. OHNUKI
(Nagoya University)

Translated by
S. KITAKADO
T. SUGIYAMA
(NAGOYA)

World Scientific
Singapore • New Jersey • Hong Kong

Published by

World Scientific Publishing Co. Pte. Ltd.
P.O. Box 128, Farrer Road, Singapore 9128

U. S. A. office: World Scientific Publishing Co., Inc.
687 Hartwell Street, Teaneck NJ 07666, USA

Library of Congress Cataloging-in-Publication data is available.

UNITARY REPRESENTATIONS OF THE POINCARÉ GROUP AND RELATIVISTIC WAVE EQUATIONS

ISBN 9971-50-250-X

Printed in Singapore by JBW Printers & Binders Pte. Ltd.

Preface to the English Edition

This book was written in 1975 and published in Japanese in September of 1976. Since that time a lot of new developments have been made in the theory of elementary particles as seen for example in superstring models. In spite of this the role of the Poincaré group is still very fundamental in understanding relativistic particles. At least in this sense the publication of an English version could be worthwhile for students who are learning the theory of elementary particles. Though nothing new has been added, some trivial errors in the original book have been corrected through the translation.

I would like to thank Prof. K. K. Phua who strongly recommended the publication of the English version. I am also extremely grateful to Drs. T. Sugiyama and S. Kitakado who have kindly carried out the laborious work of translation into English.

January, 1988 Yoshio Ohnuki

Preface to the Original Edition

This book is based on my lectures for graduate students of physics, and was written with a considerable amount of addition. It is intended also for undergraduate students and readers in other fields.

An attempt to unify quantum mechanics and special relativity was made by Dirac toward the end of the 1920's, and the famous Dirac equation for an electron, together with relativistic wave equations for particles of finite mass and of arbitrary spin, was discovered. The studies along this line, also including massless particles, have been developed by many authors thereafter. However, the waves obeying this kind of equations are different from those of the first Dirac's idea, and do not represent the probability amplitudes by themselves. For example, the probability wave function for a photon is neither the electric nor magnetic field which satisfies the Maxwell's equations. In a sense, the state vector of a photon lies behind the Maxwell's equations. Other particles are also in similar situations. In this sense, the so-called relativistic wave equations must be regarded only as indirect representations for the description of one-particle probability waves, and the forms of equations themselves have a direct connection to relativistic matter waves, i.e., the equations of field theory.

On the other hand, the idea in which quantum states of relativistic particles are formulated directly without the use of wave equation, was proposed by Wigner (1938). From the point of view in which states of a free particle are given by unitary irreducible representations of the Poincaré group, i.e., the group formed by translations and Lorentz transformations in the

Minkowski space, Wigner considered the problem of representations of the group, and revealed its structure. Although this laborious work, which was written as a mathematical paper on the representation of group, does not put much emphasis on concrete relations to physics, the presented argument has a significant meaning in understanding the relativistic picture of particles in quantum mechanics. As a matter of fact, if we get all unitary irreducible representations of the Poincaré group, we do have a complete knowledge of every free particle states and their behavior. It is a surprising fact that a simple framework as the Poincaré group, when unified with quantum theory, fixes our possible picture of particles so severely and without any exception. As a natural consequence it must be recognized that Wigner's idea had great influences on basic understanding of subsequent relativistic quantum mechanics or elementary particle theories.

Nevertheless, to the author's knowledge, there is no textbook discussing the details of the relation between physics and unitary representations of the Poincaré group. Thus it may be meaningful to publish a book emphasizing this problem. At least to the author, this was one of the motivations of writing this text.

The text is divided into three parts. The first five chapters cover the theory of unitary representations of the Poincaré group and their concrete contents, the connection to covariant formalism is considered in Chapters 6 and 7, and Chapter 8 is devoted to discussions of relations with quantum field theory. This last chapter has the smallest number of pages because it restricts itself only to the fundamental problems. Applications of field theory to the elementary particle theories are beyond the scope of this book, and are expected to be found in other appropriate textbooks.

Needless to say, this is not a book of pure mathematics. So it does not take account of generality or exactness which are not considered to be necessary from a physical point of view, rather it takes care of developing the argument as specifically as possible. Consequently, it involves somewhat tedious calculations and slightly complicated handling of expressions. This is unavoidable because, in natural science, understanding the specific cases seems to be very important in any abstract argument. I have not chosen such an easy-going way of writing as rearranging contents collected from many literatures. Because each literature has its own limitations, any effort of arrangement will not lead to a self-contained textbook. In addition, there were several problems which were not covered by the existing literatures. Therefore, I argued to some extent in my own way, and derived each formula by myself. So I am afraid of unexpected errors or misunderstandings.

A few sections in the primary plan had to be omitted because of lack of space. But the whole volume was written carefully so that a self-consistent style was kept till the end. I would like to accept reader's comments as to whether the aim of the author has been accomplished, and hope to have an opportunity to improve the text by taking away a superfluity and making up an insufficiency. Further, notations used in the text follow the author's private usage. Of course, this is not so essential, and readers should understand them in their own styles.

Professor Mikio Namiki, who invited me to write this book, read the manuscript and made valuable comments. Mr. Hiromi Katayama of Iwanami publishing Co. took on the management during the preparation of this book. I would like to express my sincere gratitude to these people.

March 1976 Yoshio Ohnuki

Contents

Chapter 1

Introduction

§1.1 Transformation and Invariance

Let us denote the state vectors (or simply states), which describe a system according to the rules of quantum mechanics, as $|A\rangle$, $|B\rangle, \ldots$. For an arbitrary complex number $c, c|A\rangle$ is also a state vector, and the well known superposition principle for the state vectors indicates that

$$|C\rangle = |A\rangle + |B\rangle \tag{1.1.1}$$

is also a state of the system. An inner product $\langle A|B\rangle$ is defined for arbitrary states $|A\rangle$ and $|B\rangle$. The product is a complex number fixed by these two states, and satisfies

$$\langle A|B\rangle^* = \langle B|A\rangle\,, \tag{1.1.2}$$

$$\langle A|c|B\rangle = c\langle A|B\rangle\,, \tag{1.1.3}$$

$$\langle A|A\rangle \geq 0\,, \tag{1.1.4}$$

and for $|C\rangle$ of (1.1.1),

$$\langle D|C\rangle = \langle D|A\rangle + \langle D|B\rangle\,. \tag{1.1.5}$$

Here the symbol * denotes complex conjugate, and the equality in (1.1.4) stands only for $|A\rangle = 0$. The norm of a state $|A\rangle$ is given by $\langle A|A\rangle$.*)

*) We use this definition in this text. In mathematics, however, the norm is usually defined by $\sqrt{\langle A|A\rangle}$.

Another important property of the state vectors is that the probability of getting a particular result in an observation is expressed in terms of them. That is, when a physical quantity is observed in a state $|A\rangle$ of the system, the transition probability from $|A\rangle$ into an eigenstate $|B\rangle$ is given by

$$P_{BA} = |\langle B|A\rangle|^2 , \tag{1.1.6}$$

where $|A\rangle$ and $|B\rangle$ are both normalized ($\langle A|A\rangle = \langle B|B\rangle = 1$).

Assuming these properties for state vectors, let us consider transformations of the states. To do this, let state vectors $|A'\rangle, |B'\rangle, \ldots$ correspond to $|A\rangle, |B\rangle, \ldots$ respectively, and consider replacements

$$\left.\begin{array}{l} |A\rangle \longrightarrow |A'\rangle , \\ |B\rangle \longrightarrow |B'\rangle , \\ \cdots\cdots\cdots\cdots . \end{array}\right\} \tag{1.1.7}$$

Needless to say, the transformed states must satisfy (1.1.2)–(1.1.5). Further requirement of invariance of the theory under the transformation leads, at least, to invariances of the superposition principle and of the definition of probability.*) In other words, if $|C\rangle$ is given by (1.1.1) with arbitrary $|A\rangle$, and $|B\rangle$, then

$$|C'\rangle = |A'\rangle + |B'\rangle , \tag{1.1.8}$$

$$|\langle B|A\rangle|^2 = |\langle B'|A'\rangle|^2 , \tag{1.1.9}$$

must be satisfied. At this stage, either the relation

$$\langle B|A\rangle = \langle B'|A'\rangle \tag{1.1.10}$$

or

$$\langle B|A\rangle = \langle A'|B'\rangle \tag{1.1.11}$$

can be derived in the following way.

Let $|A\rangle = |B\rangle$ in (1.1.9), then (1.1.4) gives $\langle A|A\rangle = \langle A'|A'\rangle$. Since this relation holds for any $|A\rangle$, setting $\langle C|C\rangle = \langle C'|C'\rangle$ and substituting (1.1.1) and (1.1.8) for $|C\rangle$ and $|C'\rangle$, we get

$$\mathrm{Re}\,\langle B|A\rangle = \mathrm{Re}\,\langle B'|A'\rangle , \tag{1.1.12}$$

*) This may be a definition of the invariance in a broader sense.

where Re $\langle B \mid A \rangle$ is the real part of $\langle B \mid A \rangle$, and Im $\langle B \mid A \rangle$ in the following equation is the imaginary part. Substituting (1.1.12) into (1.1.9) we immediately get

$$\mathrm{Im} \langle B \mid A \rangle = \mathrm{Im} \langle B' \mid A' \rangle \tag{1.1.13}$$

or

$$\mathrm{Im} \langle B \mid A \rangle = -\mathrm{Im} \langle B' \mid A' \rangle \ . \tag{1.1.14}$$

Equations (1.1.12), (1.1.13) and (1.1.14) are nothing but Eqs. (1.1.10) and (1.1.11).

Therefore, if an operator D, which relates $\mid A \rangle, \mid B \rangle, \ldots$ to $\mid A' \rangle, \mid B' \rangle, \ldots$ is introduced by

$$\left.\begin{array}{l} \mid A' \rangle = D \mid A \rangle \ , \\ \mid B' \rangle = D \mid B \rangle \ , \\ \ldots\ldots\ldots\ldots \ , \end{array}\right\} \tag{1.1.15}$$

then D is restricted to a unitary operator in the case of (1.1.10) by the invariance of the inner product, and to an anti-unitary operator in the case of (1.1.11).

A transformation of state vectors may be considered as a result of some external action to the system that causes the transition of the states, and it may also be considered to be caused by a transformation of the framework describing the system, e.g., a change of the direction of the coordinate axis.*)

In either case, however, if a set of transformations $\{a, b, c, \ldots)$ leaves the theory invariant, the state vectors are transformed by the corresponding unitary or anti-unitary operators $D(a), D(b), D(c), \ldots$. In particular, when $\{a, b, c, \ldots\}$ forms a group, and when $ab = c$, let us assume the following relation:

$$D(a)D(b) = D(c) \ . \tag{1.1.16}$$

Then we call $D(a), D(b), \ldots$ representations of the group G, and refer that the state vectors transform according to the group G. If no change of state vectors is considered as one of the transformation, i.e., the identity transformation, its representation can be expressed by 1.**) This is, of course, a unitary representation. If G is a continuous group and if any element g of

---

*)In order to avoid confusion, we consider here that state vectors are transformed, but dynamical variables are not.

**)This does not mean that there exists no other representation of the unit element of G than 1. Because there is a case where $D(e) = 1$ does not hold for e given by a successive application of transformations on the identity. The so-called multi-valued representation belongs to this case (cf. §2.1).

G is obtained by successive multiplication of infinitesimal transformations on the identity, $D(g)$ is written as a product of corresponding representations of the infinitesimal transformations. Hence, according to the continuity of representation, we can conclude that $D(g)$ is always unitary. This means that an anti-unitary representation is possible only in a case with some discontinuous transformation. In quantum mechanics, the transformation of time reversal is a typical and familiar example of such a transformation (cf. §6.6).

§1.2 Poincaré Group and Free Particles

In the world of elementary particles, the theory of special relativity is necessary to describe the motion of particles, because the motion frequently approaches the light velocity. Quantum mechanics and special relativity are two fundamental prerequisites for the following arguments.

Before entering on the main subject, let us give some remarks on the notation which will be used in the text.

Let $x_\mu = (x_1, x_2, x_3, x_4)$ and $y_\mu = (y_1, y_2, y_3, y_4)$ be 4-vectors in the Lorentz space, and write an inner product $\sum_{\mu=1}^{4} x_\mu y_\mu$ as

$$x_\mu y_\mu = \mathbf{xy} + x_4 y_4 = \mathbf{xy} - x_0 y_0 \ , \tag{1.2.1}$$

where the boldface letter $\mathbf{x}$ denotes a 3-vector whose components are x_i $(i = 1, 2, 3)$, and $\mathbf{xy}$ denotes an inner product of $\mathbf{x}$ and $\mathbf{y}$ defined by

$$\mathbf{xy} = \sum_{i=1}^{3} x_i y_i \ , \tag{1.2.2}$$

further x_4 and y_4 are written as

$$x_4 = i x_0 \ , \quad y_4 = i y_0 \ . \tag{1.2.3}$$

In the case of general 4-dimensional tensor, the index 4 also can be changed to 0 by taking the imaginary factor i outside. The length of a 3-vector $\mathbf{x}$ is written as

$$|\mathbf{x}| = \sqrt{\mathbf{x}^2} \ . \tag{1.2.4}$$

The expression $\mathbf{x} \times \mathbf{y}$ is a vector product of $\mathbf{x}$ and $\mathbf{y}$, and its components are

$$(\mathbf{x} \times \mathbf{y})_i = \sum_{j,k=1}^{3} \varepsilon_{ijk} x_j y_k \ , \tag{1.2.5}$$

where ε_{ijk} is completely antisymmetric with respect to the indices i, j and k, and $\varepsilon_{123} = 1$.

We adopt the natural unit in which the Planck constant $\hbar = h/2\pi$ and the light velocity c are both 1:

$$\hbar = c = 1 . \tag{1.2.6}$$

When x_μ describe the position of a point in 4-dimensional space-time, x_0 indicates the time t. Although in this text both x_0 and t are used to describe the parameter for the time, these are of course the same. For other symbols we shall mention where they appear.

If Lorentz transformed x_μ are written as $x'_\mu = \Lambda_{\mu\nu} x_\nu$, the relation $x'_\mu x'_\mu = x_\mu x_\mu$ leads to

$$\Lambda_{\mu\nu}\Lambda_{\mu\rho} = \delta_{\nu\rho} . \tag{1.2.7}$$

This formula implies that the transposed matrix Λ^{T} of the 4×4 matrix Λ is equal to the inverse matrix Λ^{-1}, and thus it can be rewritten as

$$\Lambda_{\nu\mu}\Lambda_{\rho\mu} = \delta_{\nu\rho} , \tag{1.2.8}$$

where Λ_{ij} $(i, j = 1, 2, 3)$ and Λ_{44} are real, and Λ_{i4} and Λ_{4i} are purely imaginary numbers.

Hereafter, Λ is restricted to such Λ that can be obtained by successive applications of infinitesimal Lorentz transformations on the identity transformation. From (1.2.7) we find that the determinant of Λ satisfies $(\det(\Lambda))^2 = 1$ and that $\Lambda_{44}^2 = 1 - \sum_{i=1}^{3} (\Lambda_{i4})^2 \geq 1$. By continuity of the transformation, Λ is further restricted to satisfy

$$\det(\Lambda) = 1 , \tag{1.2.9}$$

$$\Lambda_{44} \geq 1 . \tag{1.2.10}$$

The Lorentz transformations defined by (1.2.7)–(1.2.10) obviously form a group, which is called the continuous Lorentz group or the orthochronous proper Lorentz group. We call it simply the Lorentz group in the text. Considering the condition (1.2.7), we can easily find that such Λ is characterized by six real parameters, and Λ varies with continuous variation of these parameters.

For the discussions of the world of elementary particles this is, however, not sufficient, and another transformation which translates the coordinates as

$$x'_\mu = x_\mu + a_\mu \tag{1.2.11}$$

has to be introduced. This transformation combined with the previous Lorentz transformation obviously forms a continuous group, which is called the proper Poincaré group or the continuous Poincaré group. We call it the Poincaré group for simplicity hereafter. Taking account of (1.2.11) we find that the Poincaré group is a continuous group characterized by ten real parameters. If the theory is invariant under this transformation, and if the state vectors are transformed according to the Poincaré group, their representations must be unitary according to the argument of the previous section. In other words, the state vectors are transformed according to a unitary representation of the Poincaré group. The following discussion will be focused on the case where the states particularly represent free particles.

At an instant of time sufficiently before or sufficiently after the scattering, we can suppose that particles move separated from each other. Then the interaction between the particles becomes so small that we can neglect it completely, and the particles behave as free particles without getting any effect from each other. As a result, the naming of each particle, the electron or the deuteron for example, becomes possible in such a system. However, it must be noticed that such a naming of a one-particle state is usually based on the invariance of it under the coordinate transformations of the Poincaré group. For example, when there is a free electron, we call the object an electron either in the Lorentz transformed reference frame or in another reference frame which has a different origin. That is, unitary representations of the Poincaré group play a role of connecting various state vectors of the same free particle. In this sense, if we consider a superposition of two state vectors which cannot be connected by any transformation of the Poincaré group, we cannot regard it as a state vector of a free particle. In other words, any two state vectors of a given free particle can be transformed mutually by an appropriate transformation of the Poincaré group. Assuming that a free particle is such an object, let us proceed to the following discussions. However there may remain still a problem whether we consider free particles in a larger transformation group containing discrete transformations such as the space reflection which connects the right-handed frame with the left-handed frame. But in the theory of special relativity, the requirement of the invariance under discrete transformations does not exist. Therefore we consider free particles in the above sense with weak restriction, and will discuss later their behavior under the discrete transformations.

If U is a unitary operator and D is a unitary representation of the Poincaré group $UDU^\dagger$ is also a unitary representation of the group, where the symbol $\dagger$ denotes Hermitian conjugate. Two representations D and $UDU^\dagger$ are called unitary equivalent to each other. For an appropriate U and for any element of the Poincaré group, we assume that $UDU^{\dagger}$ is represented by a matrix

$$UDU^{\dagger} = \begin{pmatrix} D_1 & 0 \\ 0 & D_2 \end{pmatrix} , \tag{1.2.12}$$

where $D_i (i = 1, 2)$ are $d_i \times d_i$ square matrices, and upper right 0 and lower left 0 are $d_1 \times d_2$ and $d_2 \times d_1$ rectangular matrices which consist of only 0. Both D_1 and D_2 are obviously unitary representations of the Poincaré group, but the state vector transforming according to D_1 and the state vector transforming according to D_2 cannot represent different states of the same particle. Because any transformation of the Poincaré group cannot connect them. That is, a state vector of a free particle in the above sense must transform according to a unitary irreducible representation of the Poincaré group, i.e., a representation which is not decomposed into a direct sum of two representations as (1.2.12) by any U. Therefore, if we obtain all unitary irreducible representations of the group, we can determine completely the behavior of every possible free particle at least in the framework of 4-dimensional Minkowski space, because each free particle must transform according to one of the irreducible representations. We shall investigate the properties of the Poincaré group from this standpoint.

Chapter 2
Lorentz Group

§2.1 Double-Valued Representations

Let $L(\Lambda)$ and $T(a)$ be unitary representations of a Lorentz transformation $\Lambda_{\mu\nu}$ and a translation a_μ of coordinates respectively. Then the relations between these transformations are given by

$$L(\Lambda)L(\Lambda') = L(\Lambda\Lambda') , \tag{2.1.1}$$

$$T(a)T(b) = T(a+b) , \tag{2.1.2}$$

$$L(\Lambda)T(a) = T(\Lambda a)L(\Lambda) , \tag{2.1.3}$$

where $\Lambda\Lambda'$ and Λa are brief accounts of $\Lambda_{\mu\rho}\Lambda'_{\rho\nu}$ and $\Lambda_{\mu\nu}a_\nu$ respectively. These notations will be frequently used hereafter. The meanings of (2.1.1) and (2.1.2) are obvious. Equation (2.1.3) implies that the application of a translation a_μ followed by a Lorentz transformation $\Lambda_{\mu\nu}$ is equal to that of the Lorentz transformation $\Lambda_{\mu\nu}$ followed by a translation $\Lambda_{\mu\nu}a_\nu$. The set of these relations is a basis of finding representations of the Poincaré group. Before going on, however, let us consider what kinds of multi-valued representations of the group are possible.

As has been mentioned in the previous section, an element of the Poincaré group has ten parameters, and various transformations are given corresponding to the values of these parameters. The whole set of values of the parameters are called a parameter space. The identity is, of course, a point in this space, and is represeted by e. Let us now consider a closed curve which starts from e, goes around in the space and returns again to e. Each point on the closed curve has the corresponding transformation, and by the continuity of representation the transformation must vary continuously with the continuous movement of the point. Therefore, if the closed curve can be shrunk continuously into the point e, the representations corresponding to the starting point e and to the end point e of the curve are the same. In a simply connected parameter space any closed curve is shrinkable in this way, but in another space with different property there exists a closed curve which cannot be shrunk. In such a case, the representations corresponding to the starting point e and to the end point e are not necessarily the same. As a result, there will appear the so-called multi-valued representations. In order to apply this method to the Poincaré group, we have to investigate the connectedness of the space formed by ten parameters. However, since it is complicated and not intuitive enough to consider the parameter space at a time, we shall investigate it step by step.

An element of the Poincaré group is always expressed in the form LT, which is a product of L and T, with the help of (2.1.1)–(2.1.3). First, consider a subgroup formed by translations characterized by parameters a_1, a_2, a_3 and a_0. Since the parameter space formed by these parameters is a 4-dimensional Euclidean space and is simply connected, if the Poincaré group has any multi-valued representation, it must have its origin in the other subgroup, i.e., the Lorentz group.

Thus it is necessary to investigate the Lorentz group. As a preparation we shall prove that any $\Lambda_{\mu\nu}$ which satisfies (1.2.7)–(1.2.10) can always be written as

$$\Lambda_{\mu\nu} = R_{\mu\rho} B_{\rho\sigma}(\tau) R'_{\sigma\mu} \,, \tag{2.1.4}$$

where R is a spatial rotation satisfying (1.2.7)–(1.2.10), and $R_{i4} = R_{4i} = 0$, $R_{44} = 1$. Similar equations stand for R'. The matrix $B(\tau)$ represents a Lorentz transformation in the direction of the first axis:

$$B(\tau) = \begin{pmatrix} \cosh\tau & 0 & 0 & i\sinh\tau \\ 0 & 1 & 0 & 0 \\ 0 & 0 & 1 & 0 \\ -i\sinh\tau & 0 & 0 & \cosh\tau \end{pmatrix} . \tag{2.1.5}$$

The proof is as follows. Let us introduce a 3-dimensional vector $\mathbf{x} = (i\Lambda_{14}, i\Lambda_{24}, i\Lambda_{34})$. When $\mathbf{x} = 0$, (1.2.7) gives $(\Lambda_{44})^2 = 1$. Substituting this into $\Lambda_{4\mu}\Lambda_{4\mu} = 1$ we have $\Lambda_{4i} = 0$ and $\Lambda_{44} = 1$ with the help of (1.2.10). Hence Λ is a spatial rotation, and (2.1.4) is obtained by putting $B_{\mu\nu} = \delta_{\mu\nu}$. Next, we consider the case $\mathbf{x} \neq 0$. Let us normalize $\mathbf{x}$ as $\mathbf{e}_1 = \mathbf{x}/|\mathbf{x}| = (\alpha_1, \alpha_2, \alpha_3)$, and let $\mathbf{e}_2 = (\beta_1, \beta_2, \beta_3)$ and $\mathbf{e}_3 = (\gamma_1, \gamma_2, \gamma_3)$ be two unit vectors perpendicular to $\mathbf{e}_1$. When $\mathbf{e}_3 = \mathbf{e}_1 \times \mathbf{e}_2$,

$$\bar{R} = \begin{pmatrix} \alpha_1 & \alpha_2 & \alpha_3 & 0 \\ \beta_1 & \beta_2 & \beta_3 & 0 \\ \gamma_1 & \gamma_2 & \gamma_3 & 0 \\ 0 & 0 & 0 & 1 \end{pmatrix} \tag{2.1.6}$$

represents a spatial rotation, and $\bar{\Lambda} = \bar{R}\Lambda$ satisfies (1.2.7)–(1.2.10). Since $\mathbf{e}_1$ is perpendicular to $\mathbf{e}_2$ and $\mathbf{e}_3$, $\bar{\Lambda}$ has a form

$$\bar{\Lambda} = \begin{pmatrix} a_{11} & a_{12} & a_{13} & a_{14} \\ a_{21} & a_{22} & a_{23} & 0 \\ a_{31} & a_{32} & a_{33} & 0 \\ \Lambda_{41} & \Lambda_{42} & \Lambda_{43} & \Lambda_{44} \end{pmatrix} . \tag{2.1.7}$$

Because $\mathbf{f}_1 = (a_{21}, a_{22}, a_{23})$ and $\mathbf{f}_2 = (a_{31}, a_{32}, a_{33})$ are unit vectors perpendicular to each other, setting $\mathbf{f}_3 = \mathbf{f}_1 \times \mathbf{f}_2 = (c_1, c_2, c_3)$ we can define $\bar{R}'$ by

$$\bar{R}' = \begin{pmatrix} c_1 & a_{21} & a_{31} & 0 \\ c_2 & a_{22} & a_{32} & 0 \\ c_3 & a_{23} & a_{33} & 0 \\ 0 & 0 & 0 & 1 \end{pmatrix} \tag{2.1.8}$$

which represents a spatial rotation. Estimating $B = \bar{\Lambda}\bar{R}'$ we have

$$B = \begin{pmatrix} b_{11} & 0 & 0 & b_{14} \\ 0 & 1 & 0 & 0 \\ 0 & 0 & 1 & 0 \\ b_{41} & 0 & 0 & b_{44} \end{pmatrix} \tag{2.1.9}$$

which, of course, satisfies (1.2.7)–(1.2.10). Since we can write $b_{11} = b_{44} = \cosh\tau$, $b_{14} = -b_{41} = i\sinh\tau$, setting $R = \bar{R}^{-1}, R' = \bar{R}'^{-1}$ we obtain (2.1.4) with the help of $B = R^{-1}\Lambda R'^{-1}$.

In this way, every Lorentz transformation is written as a product of two spatial rotations R, R' and a Lorentz transformation $B(\tau)$ in the direction of the first axis. Among them $B(\tau)(-\infty < \tau < \infty)$ forms a group whose parameter space is a line, which is simply connected, and which does not produce the multi-valuedness of the Lorentz transformation. Therefore, the multi-valuedness of the Lorentz group, and consequently that of the Poincaré group depend on the multi-valuedness of R and R', i.e., the rotation group. If we accept an usual argument that the representations of the rotation group are at most double-valued, we can conclude that the Poincaré group has two kinds of representations: single-valued and double-valued representations. In other words, an irreducible representation of the Poincaré group is either single-valued or double-valued corresponding to the representation of the rotation group which is a subgroup of the Poincaré group, and other representations are impossible.*)

At this stage, we shall give a brief discussion about the reason why the representations of the rotation group are at most double-valued.

As has been mentioned above, a spatial rotation is represented by a 3×3 matrix

$$R = \begin{pmatrix} R_{11} & R_{12} & R_{13} \\ R_{21} & R_{22} & R_{23} \\ R_{31} & R_{32} & R_{33} \end{pmatrix} , \tag{2.1.10}$$

where

$$\sum_i R_{ij} R_{ik} = \delta_{jk} , \tag{2.1.11}$$

$$\det (R) = 1 , \tag{2.1.12}$$

and R_{ij} are obviously real numbers. Since R is unitary, its diagonalization will give the eigenvalues ρ_1, ρ_2, ρ_3 whose absolute values are all 1. Since the secular equation which determines the eigenvalues is a cubic equation with real coefficients, it necessarily has one real solution and two solutions which are both real or complex conjugate of each other. Considering $\det(R) = \rho_1\rho_2\rho_3 = 1$ we find that at least one real solution must be 1.

If we denote the corresponding eigenvector as $\mathbf{n}^{(1)} = (n_1^{(1)}, n_2^{(1)}, n_3^{(1)})$, we have

$$\sum_j R_{ij} n_j^{(1)} = n_i^{(1)} , \tag{2.1.13}$$

where we can choose $n_i^{(1)}$ real because R_{ij} are real. That is, $\mathbf{n}^{(1)}$ is a real vector in a 3-dimensional space. As is seen in (2.1.13), it is invariant under the rotation R, and a line which is given by $c\mathbf{n}^{(1)} (-\infty < c < \infty)$ is called the rotation axis of R. Now, set the length of $\mathbf{n}^{(1)}$ to unity, and let $\mathbf{n}^{(2)} = (n_1^{(2)}, n_2^{(2)}, n_3^{(2)})$, $\mathbf{n}^{(3)} = (n_1^{(3)}, n_2^{(3)}, n_3^{(3)})$ be two unit vectors perpendicular to $\mathbf{n}^{(1)}$:

$$\mathbf{n}^{(l)} \mathbf{n}^{(m)} = \delta_{lm} \quad (l, m = 1, 2, 3) , \tag{2.1.14}$$

---

*)As is evident from the above discussion, this result has nothing to do with whether the representation of the Poincaré group is unitary or not.

where

$$\mathbf{n}^{(3)} = \mathbf{n}^{(1)} \times \mathbf{n}^{(2)} \,. \tag{2.1.15}$$

If we define W by

$$W = \begin{pmatrix} n_1^{(1)} & n_2^{(1)} & n_3^{(1)} \\ n_1^{(2)} & n_2^{(2)} & n_3^{(2)} \\ n_1^{(3)} & n_2^{(3)} & n_3^{(3)} \end{pmatrix} , \tag{2.1.16}$$

W^{-1} is obtained by interchanging the rows and the columns of W. Since they represent spatial rotations, WRW^{-1} is also a spatial rotation and is written as

$$WRW^{-1} = \begin{pmatrix} 1 & 0 & 0 \\ 0 & c_{22} & c_{23} \\ 0 & c_{32} & c_{33} \end{pmatrix} , \tag{2.1.17}$$

where

$$\left.\begin{aligned} c_{22} &= \sum_{i,j} n_i^{(2)} R_{ij} n_j^{(2)} , \quad c_{23} = \sum_{i,j} n_i^{(2)} R_{ij} n_j^{(3)} , \\ c_{32} &= \sum_{i,j} n_i^{(3)} R_{ij} n_j^{(2)} , \quad c_{33} = \sum_{i,j} n_i^{(3)} R_{ij} n_j^{(3)} . \end{aligned}\right\} \tag{2.1.18}$$

We have used (2.1.13), (2.1.14) and $\sum_i R_{ij} n_i^{(1)} = n_j^{(1)}$ to derive (2.1.17). The last relation follows from (2.1.11) by multiplying $n_k^{(1)}$, summing up with respect to k and applying (2.1.13). Here we can suppose $c_{23} \geq 0$. If c_{23} is negative, inverting the direction $\mathbf{n}^{(1)}$ we can invert $\mathbf{n}^{(3)}$ according to (2.1.15) and consequently we have a positive c_{23} according to (2.1.18). Hence, considering $(c_{22})^2 + (c_{23})^2 = (c_{32})^2 + (c_{33})^2 = 1$ and $c_{22}c_{32} + c_{23}c_{33} = 0$, we can write

$$WRW^{-1} = \begin{pmatrix} 1 & 0 & 0 \\ 0 & \cos w & \sin w \\ 0 & -\sin w & \cos w \end{pmatrix} , \tag{2.1.19}$$

$$0 \leq w \leq \pi \,. \tag{2.1.20}$$

This represents a rotation about $\mathbf{n}^{(1)}$ through rotation angle w. Since $\mathbf{n}^{(2)}$ is a mere unit vector perpendicular to $\mathbf{n}^{(1)}$, its choice is not unique. But it must be noticed that the value of w is independent of such an ununiqueness. In fact, if $\mathbf{n}^{(2)}$ and $\mathbf{n}^{(3)}$ were replaced by $\mathbf{n}^{(2)\prime} = \cos\delta \cdot \mathbf{n}^{(2)} + \sin\delta \cdot \mathbf{n}^{(3)}$ and $\mathbf{n}^{(3)\prime} = -\sin\delta \cdot \mathbf{n}^{(2)} + \cos\delta \cdot \mathbf{n}^{(3)}$, the invariance of c_{lm} $(l, m = 2, 3)$

can be easily confirmed. Therefore, for a given $R, \mathbf{n}^{(1)}$ and w are uniquely determined, and if $\mathbf{n}^{(1)}$ and w are given, the unique R can be determined by tracing the above argument conversely. As a result of some calculations, we can express the matrix element of R in terms of $\mathbf{n}^{(1)}$ and w as

$$R_{ij} = n_i^{(1)} n_j^{(1)} + (\delta_{ij} - n_i^{(1)} n_j^{(1)}) \cos w + \sum_k \varepsilon_{ijk} n_k^{(1)} \sin w \,, \qquad (2.1.21)$$

where ε_{ijk} is an anti-symmetric tensor given in §1.2. In short, any R is determined by the three parameters: w and two angles which describe the direction of $\mathbf{n}^{(1)}$. Therefore, in a 3-dimensional space, a point inside a sphere of radius π whose center is at the origin is uniquely specified by a vector $w\mathbf{n}^{(1)}$ $(0 \le w < \pi)$, and the corresponding spatial rotation always exists. A point P on the sphere must be identified with the symmetric point P' with respect to the center (Fig. 2.1(c)), because, as is seen in (2.1.21), both $\mathbf{n}^{(1)}$ and $-\mathbf{n}^{(1)}$ give the same R when $w = \pi$.

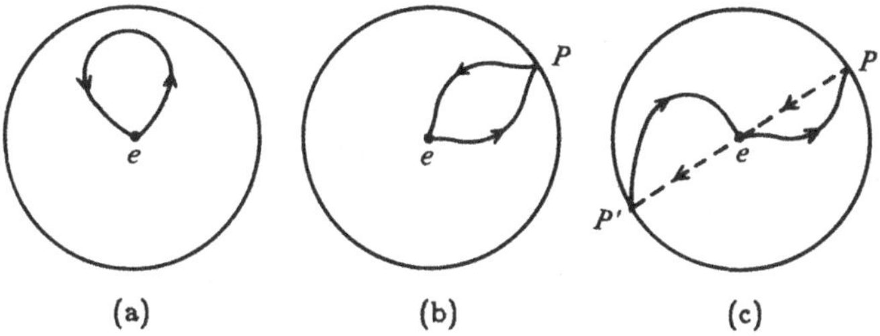

Fig. 2.1

This is a structure of the parameter space of the rotation group. The center e of the sphere represents the identity transformation. Closed curves which start from e and return to e again after tracing paths like (a) or (b) of Fig. 2.1, can be shrunk to e by continuous deformations. But if we consider a closed curve which starts from e, reaches to the point P on the sphere, jumps to the symmetric point P' and returns to e as shown in (c) of Fig. 2.1, we cannot shrink it to e continuously. A curve which jumps twice is, of course, shrinkable continuously (Fig. 2.2).

In general, closed curves with even number of jumps can be reduced to Fig. 2.1(a) by continuous deformations, and closed curves with odd number of jumps can be reduced to the type of Fig. 2.1(c).*) Among these two classes of closed curves in the parameter space, as has been mentioned already, the

*) This implies that the parameter space is not simply connected.

representation $D(e)$ of the starting point e is not always the same with the representation $D(e')$ of the end point e' (which is written as e' in order to distinguish it from e). In the case of the second class, however, if a curve restarts after arriving at e', makes a jump and returns to the center of the sphere again, then it finally reaches e. Hence we can write $D(e') \times D(e') = D(e)$. Since $D(e) = 1$, we obtain $D(e') = \pm 1$. The case $D(e') = 1$ gives a single-valued representation and the case $D(e') = -1$ gives a double-valued representation. Consequently we can conclude that the Poincaré group has only two classes of representation which are single-valued and double-valued.

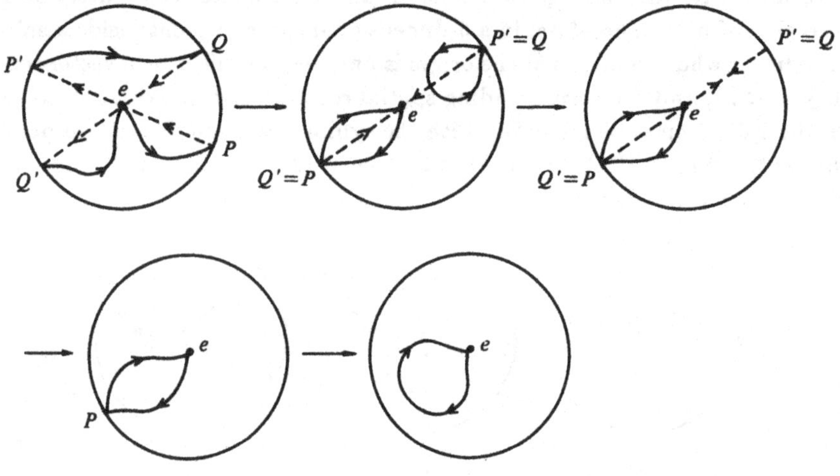

Fig. 2.2

§2.2 Spinor Representations

Spinor representations are typical examples of double-valued representations of the Lorentz group. They are not, however, unitary representations. So they are not our direct targets, but we shall discuss them in this section for the later usage.

Define 2×2 matrices σ_μ ($\mu = 1, 2, 3, 4$) by

$$\left.\begin{aligned} \sigma_1 &= \begin{pmatrix} 0 & 1 \\ 1 & 0 \end{pmatrix}, \quad & \sigma_2 &= \begin{pmatrix} 0 & -i \\ i & 0 \end{pmatrix}, \\ \sigma_3 &= \begin{pmatrix} 1 & 0 \\ 0 & -1 \end{pmatrix}, \quad & \sigma_4 &= \begin{pmatrix} i & 0 \\ 0 & i \end{pmatrix}. \end{aligned}\right\} \tag{2.2.1}$$

The matrices σ_i ($i = 1, 2, 3$) are usually called the Pauli matrices. If $\mathrm{Tr}(M)$ denotes the sum of all the diagonal elements of a matrix M, i.e., the trace of

M, the formula

$$\mathrm{Tr}\,(\sigma_\mu^\dagger \sigma_\nu) = 2\delta_{\mu\nu} \tag{2.2.2}$$

can be easily verified. The four σ's are linearly independent, and any 2×2 matrix M can be expressed as a linear combination of them:

$$\left.\begin{aligned} M &= c_\mu \sigma_\mu \ , \\ c_\mu &= \frac{1}{2}\mathrm{Tr}\,(\sigma_\mu^\dagger M) \ . \end{aligned}\right\} \tag{2.2.3}$$

If the transposed matrix of M is denoted as M^{T}, the relation

$$M\sigma_2 M^{\mathrm{T}}\sigma_2 = \det\,(M) \tag{2.2.4}$$

is verified by a direct calculation.

Let g be a 2×2 matrix with determinant equal to unity, and expand $g\sigma_\mu g^\dagger$ in terms of σ_ν as

$$g\sigma_\mu g^\dagger = \Lambda_{\nu\mu}\sigma_\nu \ , \tag{2.2.5}$$

$$\det\,(g) = 1 \ , \tag{2.2.6}$$

where $\Lambda_{\nu\mu}$ is a function of g, and is expressed as

$$\Lambda_{\nu\mu} = \frac{1}{2}\mathrm{Tr}\,(\sigma_\nu^\dagger g\sigma_\mu g^\dagger) \tag{2.2.7}$$

with the help of (2.2.3). It will be shown in the following way that the 4×4 matrix Λ defined in this manner gives a Lorentz transformation.

First of all, considering that the complex conjugate $[\mathrm{Tr}(M)]^*$ of $\mathrm{Tr}(M)$ is $\mathrm{Tr}(M^\dagger)$ and that $\sigma_i^\dagger = \sigma_i$ $(i = 1,2,3)$ and $\sigma_4^\dagger = -\sigma_4$, we immediately find that Λ_{ij}, Λ_{44} are real and Λ_{i4}, Λ_{4i} are purely imaginary.

Taking the transposed matrices of both sides of (2.2.5), sandwiching them by two σ_2's and applying (2.2.4), (2.2.6), we obtain

$$g^{-1\dagger}\sigma_\mu^\dagger g^{-1} = \Lambda_{\nu\mu}\sigma_\nu^\dagger \ , \tag{2.2.8}$$

where we have used relations $\det(\sigma_\mu) = -1$ and $\sigma_2\sigma_\mu^{\mathrm{T}}\sigma_2 = -\sigma_\mu^{-1} = -\sigma_\mu^\dagger$. Since we get, from (2.2.5) and (2.2.8),

$$\begin{aligned} g\sigma_\mu g^\dagger g^{-1\dagger}\sigma_\nu^\dagger g^{-1} &= g\sigma_\mu\sigma_\nu^\dagger g^{-1} \\ &= \Lambda_{\lambda\mu}\Lambda_{\rho\nu}\sigma_\lambda\sigma_\rho^\dagger \ , \end{aligned}$$

applying (2.2.2) to the trace of each side, we obtain

$$\delta_{\mu\nu} = \Lambda_{\lambda\mu}\Lambda_{\lambda\nu} \ . \tag{2.2.9}$$

Further from (2.2.7) we get $\Lambda_{44} = \mathrm{Tr}\,(g^\dagger g)/2 > 0$. This relation and (2.2.9) lead to

$$\Lambda_{44} \geq 1 \ . \tag{2.2.10}$$

Let us next calculate the determinant of Λ. If we make a product of four expressions

$$\begin{aligned} g\sigma_1 g^\dagger &= \Lambda_{\mu 1}\sigma_\mu \ , & g^{-1\dagger}\sigma_2^\dagger g^{-1} &= \Lambda_{\nu 2}\sigma_\nu^\dagger \ , \\ g\sigma_3 g^\dagger &= \Lambda_{\lambda 3}\sigma_\lambda \ , & g^{-1\dagger}\sigma_4^\dagger g^{-1} &= \Lambda_{\rho 4}\sigma_\rho^\dagger \ , \end{aligned}$$

which are derived from (2.2.5) and (2.2.8), we get

$$g\sigma_1\sigma_2^\dagger\sigma_3\sigma_4^\dagger g^{-1} = \Lambda_{\mu 1}\Lambda_{\nu 2}\Lambda_{\lambda 3}\Lambda_{\rho 4}\sigma_\mu\sigma_\nu^\dagger\sigma_\lambda\sigma_\rho^\dagger \ . \tag{2.2.11}$$

Now let $\varepsilon_{\mu\nu\lambda\rho}$ be completely anti-symmetric with respect to the indices μ, ν, λ, ρ, that is, its sign is changed by an exchange of any two indices, and $\varepsilon_{1234} = 1$, then the relation

$$\frac{1}{2}\mathrm{Tr}\,(\sigma_\mu\sigma_\nu^\dagger\sigma_\lambda\sigma_\rho^\dagger) = \delta_{\mu\nu}\delta_{\lambda\rho} - \delta_{\mu\lambda}\delta_{\nu\rho} + \delta_{\mu\rho}\delta_{\nu\lambda} + \varepsilon_{\mu\nu\lambda\rho} \tag{2.2.12}$$

is verified by a direct calculation. Therefore applications of (2.2.12) and (2.2.9) to the trace of each side of (2.2.11) lead to

$$1 = \varepsilon_{\mu\nu\lambda\rho}\Lambda_{\mu 1}\Lambda_{\nu 2}\lambda_{\lambda 3}\Lambda_{\rho 4} = \det\,(\Lambda) \ . \tag{2.2.13}$$

As a consequence of the above discussion, we find that $\Lambda_{\mu\nu}$ satisfies (1.2.7)–(1.2.10) and that an equation $x'_\mu = \Lambda_{\mu\nu}x_\nu$ gives a Lorentz transformation. In other words if (2.2.5) multiplied by x_μ is written as

$$\begin{pmatrix} ix'_4 + x'_3 & x'_1 - ix'_2 \\ x'_1 + ix'_2 & ix'_4 - x'_3 \end{pmatrix} = g \begin{pmatrix} ix_4 + x_3 & x_1 - ix_2 \\ x_1 + ix_2 & ix_4 - x_3 \end{pmatrix} g^\dagger \ , \tag{2.2.14}$$

this relation between x'_μ and x_μ represents a Lorentz transformation.

Conversely, does there always exist a g which corresponds to the Lorentz transformation (1.2.7)–(1.2.10)? This problem can be solved in the following way. Since any Lorentz transformation is written in the form of (2.1.4), it is sufficient to show the existence of $g_B(\tau)$ corresponding to $B(\tau)$ given by

(2.1.5) and g_R corresponding to R given by (2.1.21). By the use of (2.2.5) we find that $g_B(\tau)$ corresponding to $B(\tau)$ has a form

$$g_B(\tau) = \begin{pmatrix} \cosh\frac{\tau}{2} & \sinh\frac{\tau}{2} \\ \sinh\frac{\tau}{2} & \cosh\frac{\tau}{2} \end{pmatrix} \tag{2.2.15}$$

or its sign reversed expression, and by somewhat tedious calculation we have a solution

$$g_R = \begin{pmatrix} \cos\frac{w}{2} + in_3^{(1)}\sin\frac{w}{2} & (in_1^{(1)} + n_2^{(1)})\sin\frac{w}{2} \\ (in_1^{(1)} - n_2^{(1)})\sin\frac{w}{2} & \cos\frac{w}{2} - in_3^{(1)}\sin\frac{w}{2} \end{pmatrix} \tag{2.2.16}$$

or its sign reversed expression for g_R corresponding to R_{ij} of (2.1.21). That is, for g corresponding to the Lorentz transformation Λ, we have two solutions with the opposite signs. The fact that there is no other solution can be confirmed as follows. Using two 2×2 matrices g and g' whose determinants are both unity and which satisfy (2.2.14), we can write $x'_\mu\sigma_\mu = x_\mu g\sigma_\mu g^\dagger = x_\mu g'\sigma_\mu g'^\dagger$. Since this relation holds for any x_μ, we get $g'^{-1}g\sigma_\mu = \sigma_\mu g'^\dagger g^{\dagger-1} (\mu = 1,2,3,4)$. On the other hand, we have $g'^\dagger g^{\dagger-1} = (g'^{-1}g)^{\dagger-1} = \sigma_2(g'^{-1}g)^{\dagger\mathrm{T}}\sigma_2$ with the help of (2.2.4). Then if we expand $g'^{-1}g$ in terms of σ_ν to write $g'^{-1}g = c_\nu\sigma_\nu$, we get finally $c_\nu\sigma_\nu\sigma_\mu = -\sigma_\mu c_\nu^*\sigma_\nu$. Solving this equation, we easily find that $c_i = 0$ $(i = 1,2,3)$ and $c_4 = -c_4^*$. Therefore, writing $c_4 = ic$ (c: real) we get $g = -cg'$, and taking determinant of each side we obtain $c^2 = 1$, which leads to $g' = \pm g$.

Although such an ununiqueness of the sign can be included in g_R since g is expressed as a product of (2.2.15) and (2.2.16), it is not allowed to adopt only one of the signs and abandon the other. If we perform continuous variations of w and the direction of $\mathbf{n}^{(1)}$ from $w = w_0$ (corresponding to g_R^0) to $w = 2\pi - w_0$ in which case $\mathbf{n}^{(1)}$ turns opposite to the original direction, g_R changes to $-g_R^0$ as seen from (2.2.16). Then the exclusion of $-g_R^0$ contradicts the continuity of the representation (2.2.16). That is, analogously to the case of R_{ij} in the previous section, it is impossible to restrict the range of w to $0 \le w \le \pi$, and it must be set as

$$0 \le w \le 2\pi \tag{2.2.17}$$

in order to preserve the continuity of the representation. The parameter space is then represented by two spheres with radius π and with centers at e and e' respectively. The representation g_R is specified by a point $w\mathbf{n}^{(1)}$ in the sphere e when $0 \le w \le \pi$, and by a point $(2\pi - w)\mathbf{n}^{(1)}$ in the sphere e' when

$\pi \leq w \leq 2\pi$, where the vector $\mathbf{n}^{(1)}$ has its starting point at e or e' in each case, and a point $\pi\mathbf{n}^{(1)}$ on the sphere e is identical to a point $\pi\mathbf{n}^{(1)}$ on the sphere e'. As a result, the parameter space becomes simply connected (Fig. 2.3). When $0 \leq w' \leq \pi$, both a point $w'\mathbf{n}^{(1)}$ in the sphere e (where $w = w'$) and a point $-w'\mathbf{n}^{(1)}$ in the sphere e' (where $w = 2\pi - w'$) give the same R_{ij}, and two points which have no such relation give different R_{ij}'s. Hence g_R is a double-valued representation of the rotation group, and g in (2.2.14) is a double-valued representation of the Lorentz group. As far as only R_{ij} is concerned, the above two points $w'\mathbf{n}^{(1)}$ and $-w'\mathbf{n}^{(1)}$ in the spheres e and e' respectively are identical, then we can express them as a point in a sphere as was shown in Fig. 2.1.

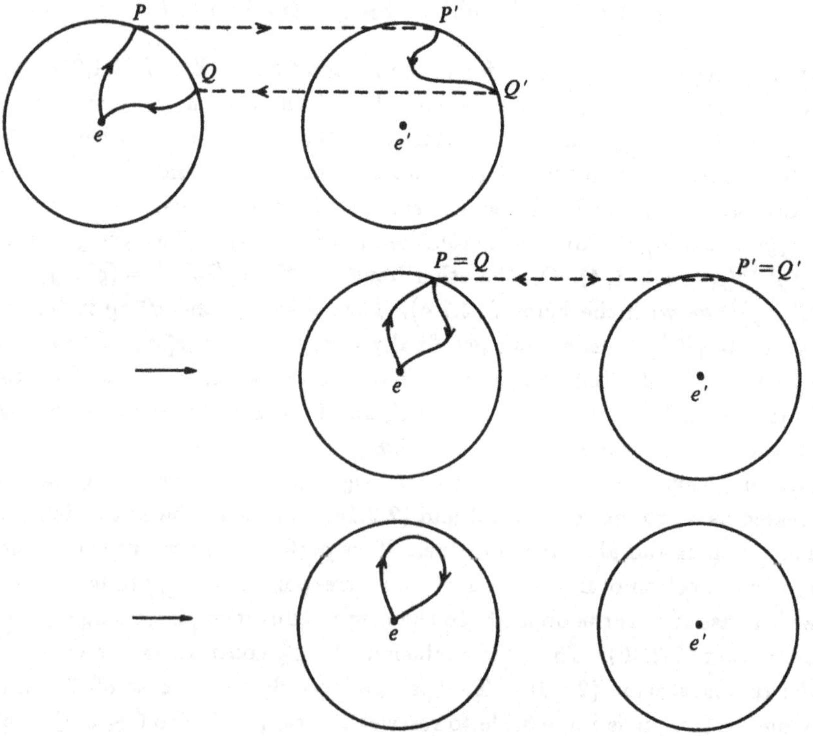

Fig. 2.3

According to (2.2.7) any g gives a Lorentz transformation which is written in the form of (2.1.4), then with the help of (2.2.15) and (2.2.16) we have

$$g = g_R g_B(\tau) g_{R'} \,. \tag{2.2.18}$$

A set of g, namely a set of 2×2 matrices whose determinants are unity, forms a group which is called SL(2, C).*) The above discussion implies that SL(2, C) is a double-valued representation of the Lorentz group, and that it is a spinor representation. A quantity which is transformed according to a spinor representation is called a spinor. This representation is not unitary, but is irreducible. Because we can easily show that 2×2 matrices which commute with all g are only constant multiples of the unit matrix, and hence by Schur's lemma**) we can conclude that the representation is irreducible.

Since $\Lambda_{\mu\nu}$ is uniquely determined corresponding to g, any representation of SL(2, C) is a representation of the Lorentz group. As is seen in (2.2.18), the parameter space of SL(2, C) has the same continuous property as that of g_R which is simply connected as shown in Fig. 2.3. Thus all representations of SL(2, C) are single-valued.

In this sense, it is convenient to consider $L(\Lambda)$ in (2.1.1) or (2.1.3) as a unitary representation $L(g)$ of SL(2, C). In a strict sense $L(\Lambda)$ must be written as

$$L(\Lambda)L(\Lambda^{-1}) = \pm 1 \tag{2.2.19}$$

in the case of double-valued representation, but we need not pay any attention to the signs because we can write, by the use of $L(g)$,

$$L(g)L(g') = L(gg') \tag{2.2.20}$$

and

$$L(g)L(g^{-1}) = 1\ . \tag{2.2.21}$$

Further, (2.1.3) is rewritten as

$$L(g)T(a) = T(\Lambda(g)a)L(g)\ , \tag{2.2.22}$$

where $\Lambda(g)$ is determined corresponding to g. It is, however, troublesome to write every expression like (2.2.20)–(2.2.22), thus we shall recognize all $L(\Lambda)$ to be $L(g)$ hereafter and shall omit the signs $\pm$.

Before closing this section, we shall give some additional discussion about spinor representations.

*)The S is the initial letter of a word "special" and means that the determinant is equal to 1. The L indicates a linear operator, the 2 in the parentheses indicates a 2×2 matrix and the C means that the matrix elements are complex numbers.

**)Schur's lemma is very useful in connection to the irreducibility of a representation. It implies that a representation $D(g)$ of a group $G(g, G)$ is irreducible, if and only if operators which commute with $D(g)$ for all g are only constant multiples of the unit matrix.

For $g \in \mathrm{SL}(2, C)$ define

$$\tilde{g} = g^{-1\dagger} \,. \tag{2.2.23}$$

If $gg' = g''$, we have $\tilde{g}'' = (gg')^{-1\dagger} = g^{-1\dagger}g'^{-1\dagger} = \tilde{g}\tilde{g}'$. Then a set of $\tilde{g}, \{\tilde{g}\}$, forms a representation of SL(2, C). The relation between $\tilde{g}$ and $\Lambda_{\mu\nu}$ is given by (2.2.8), and $\tilde{g}$ is also a double-valued representation of the Lorentz group. We call $\tilde{g}$ the conjugate representation of g. If we define g^* by replacing each matrix element of g by its complex conjugate, since $gg' = g''$ leads to $g^*g'^* = g''^*$, $\{g^*\}$ is also a representation of SL(2, C). The representation g^* is called the complex conjugate representation of g. Since we have $\sigma_2 g^{\mathrm{T}} \sigma_2 = g^{-1}$ from (2.2.4), taking the Hermitian conjugate of this equation we obtain

$$\sigma_2 g^* \sigma_2 = \tilde{g} \,, \tag{2.2.24}$$

which implies that g^* and $\tilde{g}$ are equivalent in a spinor representation. The irreducibility of g^* or $\tilde{g}$ can be proved in an analogous manner to the case of g. However, g and $\tilde{g}$ are not equivalent. This is because if they were equivalent, from (2.2.24) there must exist a 2×2 matrix A which satisfies $AgA^{-1} = g^*$ for all g and $\det(A) \neq 0$. Let g_{ij} denote the (i,j)-element of g, and take the trace of each side of $AgA^{-1} = g^*$, then we have $g_{11} + g_{22} = g_{11}^* + g_{22}^*$. But this equation does not hold for all g clearly because g has no restriction except for the condition that its determinant must be unity. Thus g and $\tilde{g}$ are not equivalent.

The whole set of unitary g's forms a subgroup of SL(2, C). This group is called a 2-dimensional special unitary group and is denoted as SU(2). From (2.2.7), $\Lambda_{\mu\nu}$ corresponding to $g_R \in \mathrm{SU}(2)$ is a spatial rotation. Conversely g_R corresponding to a spatial rotation is expressed by (2.2.16) or the sign changed expression, and is a unitary matrix. As is seen in the above discussion, SU(2) forms a double-valued representation of the rotation group, and its representation is irreducible according to Schur's lemma. It is clear that all representations of SU(2) are single-valued, and that they give all the representations of the rotation group. The conjugate representation of g_R, $\tilde{g}_R$, is equivalent to g_R because $\tilde{g}_R = g_R^{-1\dagger} = g_R$ is derived from the unitarity of g_R.

§2.3 Infinitesimal Transformations

Since any element of a continuous group is constructed by successive applications of infinitesimal transformations on the identity transformation, the representation of the continuous group can be obtained from the representations of the infinitesimal transformations if they are known. This method is frequently used in physics and especially its application to the rotation group is well known. Let us first review it briefly. A rotation through an angle θ_1 about the first axis is expressed as

$$\mathbf{x}' = \begin{pmatrix} 1 & 0 & 0 \\ 0 & \cos\theta_1 & \sin\theta_1 \\ 0 & -\sin\theta_1 & \cos\theta_1 \end{pmatrix} \mathbf{x} \,. \tag{2.3.1}$$

Supposing θ_1 to be an infinitesimal quantity, we expand it to the first order with respect to θ_1:

$$\begin{aligned} \mathbf{x}' &= (1 + i\theta_1 D_1)\mathbf{x} \,, \\ D_1 &= \begin{pmatrix} 0 & 0 & 0 \\ 0 & 0 & -i \\ 0 & i & 0 \end{pmatrix} \,. \end{aligned} \tag{2.3.2}$$

Similarly consider rotations through infinitesimal angles θ_2 and θ_3 about the second and the third axes respectively (Fig. 2.4). Then any infinitesimal rotation can be written as

$$\mathbf{x}' = (1 + i\boldsymbol{\theta}\mathbf{D})\mathbf{x} \tag{2.3.3}$$

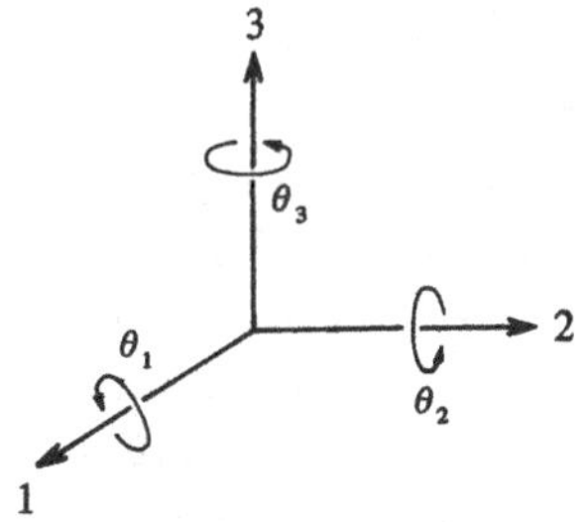

Fig. 2.4

in terms of three infinitesimal parameters, where we have introduced a vector $\boldsymbol{\theta}$ whose components are θ_1, θ_2 and θ_3, and have abbreviated $\sum_{i=1}^{3} \theta_i D_i$ as $\boldsymbol{\theta}\mathbf{D}$. The matrices D_2 and D_3 are defined by

$$D_2 = \begin{pmatrix} 0 & 0 & i \\ 0 & 0 & 0 \\ -i & 0 & 0 \end{pmatrix} \,, \quad D_3 = \begin{pmatrix} 0 & -i & 0 \\ i & 0 & 0 \\ 0 & 0 & 0 \end{pmatrix} \,, \tag{2.3.4}$$

and these D's satisfy a commutation relation

$$[D_i, D_j] = i \sum_{k=1}^{3} \varepsilon_{ijk} D_k \,. \tag{2.3.5}$$

This is an important relation which characterizes $\mathbf{D}$. In fact, it can be proved that every 3×3 irreducible matrix satisfying (2.3.5) is equivalent to $\mathbf{D}$ given by (2.3.2) and (2.3.4), and that it has a form ADA^{-1} ($\det(A) \neq 0$). Here the irreducibility of $\mathbf{D}$ implies that matrices commutable with three D's are only constant multiples of the unit matrix. Any rotation is given by successive applications of such infinitesimal rotations.

In order to get a general representation, expand it by the infinitesimal parameters and write

$$1 + i\boldsymbol{\theta}\mathbf{J} \,, \tag{2.3.6}$$

where $\mathbf{J}$ must satisfy the same commutation relation as that of $\mathbf{D}$ so that the above expression is a representation of an infinitesimal rotation. That is,

$$[J_i, J_j] = i \sum_{k=1}^{3} \varepsilon_{ijk} J_k \,, \tag{2.3.7}$$

and such $\mathbf{J}$ is a general angular momentum vector. If $\mathbf{J}$ is given, any representation of the rotation group is generated from the representation (2.3.6) of an infinitesimal rotation. So $\mathbf{J}$ is called the generator of the rotation group.

The operator $\mathbf{J}$ which gives a unitary representation of the rotation group is, of course, Hermitian. It is well known that the irreducible matrices of $\mathbf{J}$, and the corresponding irreducible representation of the rotation group, is uniquely specified by an eigenvalue of $\mathbf{J}^2$.

Let us investigate infinitesimal transformation of the Lorentz group in an analogous way. For this purpose, in addition to the above mentioned infinitesimal transformation, we have to consider infinitesimal Lorentz transformations in the directions of the first, the second and the third axes. For example, the Lorentz transformation into a reference frame which is moving with an infinitesimal velocity τ_1 in the direction of the first axis is given by

$$\left.\begin{aligned} \begin{pmatrix} x_1' \\ x_2' \\ x_3' \\ x_4' \end{pmatrix} &= (1 - i\tau_1 C_1) \begin{pmatrix} x_1 \\ x_2 \\ x_3 \\ x_4 \end{pmatrix} , \\ C_1 &= \begin{pmatrix} 0 & 0 & 0 & -1 \\ 0 & 0 & 0 & 0 \\ 0 & 0 & 0 & 0 \\ 1 & 0 & 0 & 0 \end{pmatrix} . \end{aligned}\right\} \tag{2.3.8}$$

Further, if τ_2 and τ_3 denote infinitesimal velocities of the reference frame in the direction of the second and the third axes respectively, the infinitesimal

Lorentz transformation is expressed with the help of Fig. 2.5 and Eq. (2.3.3) as

$$\begin{pmatrix} x_1' \\ x_2' \\ x_3' \\ x_4' \end{pmatrix} = (1 + i\theta \mathbf{D} - i\tau \mathbf{C}) \begin{pmatrix} x_1 \\ x_2 \\ x_3 \\ x_4 \end{pmatrix} , \tag{2.3.9}$$

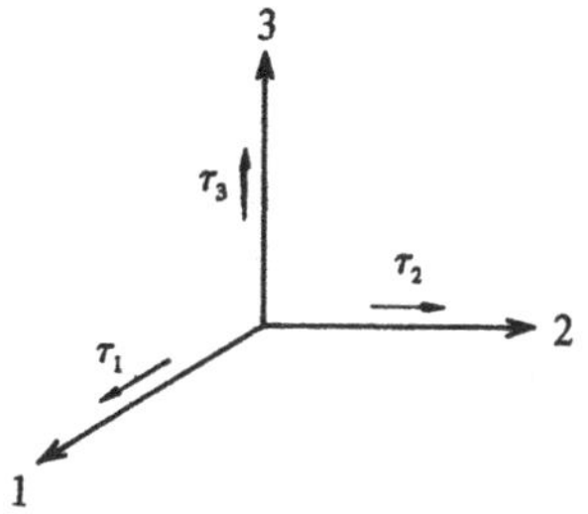

Fig. 2.5

where $\mathbf{D}$ in the above expression denotes a 4×4 matrix whose (i, j)-element $\mathbf{D}_{ij}$ $(i, j = 1, 2, 3)$ is the same as that of (2.3.2) or (2.3.4), and $\mathbf{D}_{i4} = \mathbf{D}_{4i} = \mathbf{D}_{44} = 0$. Furthermore, C_2 and C_3 are defined by

$$C_2 = \begin{pmatrix} 0 & 0 & 0 & 0 \\ 0 & 0 & 0 & -1 \\ 0 & 0 & 0 & 0 \\ 0 & 1 & 0 & 0 \end{pmatrix} , \quad C_3 = \begin{pmatrix} 0 & 0 & 0 & 0 \\ 0 & 0 & 0 & 0 \\ 0 & 0 & 0 & -1 \\ 0 & 0 & 1 & 0 \end{pmatrix} , \tag{2.3.10}$$

The commutation relations between these matrices are written as

$$\left. \begin{aligned} [D_i, D_j] &= i \sum_{k=1}^{3} \varepsilon_{ijk} D_k , \\ [D_i, C_j] &= i \sum_{k=1}^{3} \varepsilon_{ijk} C_k , \\ [C_i, C_j] &= -i \sum_{k=1}^{3} \varepsilon_{ijk} D_k . \end{aligned} \right\} \tag{2.3.11}$$

Hence if the infinitesimal transformation in a general representation is written as

$$1 + i\theta \mathbf{J} - i\tau \mathbf{K} , \tag{2.3.12}$$

the commutation relations which are satisfied by the generators **J** and **K** of the Lorentz group are given by

$$\left.\begin{aligned}
[J_i, J_j] &= i\sum_{k=1}^{3} \varepsilon_{ijk} J_k \ , \\
[J_i, K_j] &= i\sum_{k=1}^{3} \varepsilon_{ijk} K_k \ , \\
[K_i, K_j] &= -i\sum_{k=1}^{3} \varepsilon_{ijk} J_k \ .
\end{aligned}\right\} \tag{2.3.13}$$

In a unitary representation, **J** and **K** are, of course, Hermitian. If $J_{[\mu\nu]}$ are defined by

$$\left.\begin{aligned}
J_{[ij]} &= \sum_{k=1}^{3} \varepsilon_{ijk} J_k \ , \\
J_{[j4]} &= -J_{[4j]} = iK_j \ ,
\end{aligned}\right\} \tag{2.3.14}$$

they are antisymmetric with respect to the indices μ and ν, and then (2.3.13) can be written as

$$[J_{[\mu\nu]}, J_{[\lambda\rho]}] = i(\delta_{\mu\lambda} J_{[\nu\rho]} + \delta_{\mu\rho} J_{[\lambda\nu]} + \delta_{\nu\rho} J_{[\mu\lambda]} + \delta_{\nu\lambda} J_{[\rho\mu]}) \ . \tag{2.3.15}$$

If six infinitesimal quantities $\omega_{[\mu\nu]}$ are introduced by

$$\begin{aligned}
\omega_{[ij]} &= \sum_{k=1}^{3} \varepsilon_{ijk} \theta_k \ , \\
\omega_{[j4]} &= -\omega_{[4j]} = i\tau_j \ ,
\end{aligned} \tag{2.3.16}$$

the infinitesimal transformation (2.3.12) is expressed as

$$1 + \frac{i}{2} J_{[\mu\nu]} \omega_{[\mu\nu]} \ . \tag{2.3.17}$$

It is known that an irreducible representation of the Lorentz group is uniquely determined by eigenvalues of two operators

$$I_1 = \frac{1}{2} J_{[\mu\nu]} J_{[\mu\nu]} = \mathbf{J}^2 - \mathbf{K}^2 \ , \tag{2.3.18}$$

$$I_2 = \frac{1}{4i} \varepsilon_{\mu\nu\lambda\rho} J_{[\mu\nu]} J_{[\lambda\rho]} = \mathbf{JK} \ , \tag{2.3.19}$$

which are commutable with $J_{[\mu\nu]}$. These operators correspond to $\mathbf{J}^2$ in the rotation group. Such operators that commute with the generators and whose eigenvalues specify the irreducible representations are called Casimir operators.

From the argument of the previous section, we get

$$1 + \frac{i}{2}\boldsymbol{\theta\sigma} + \frac{1}{2}\boldsymbol{\tau\sigma} \tag{2.3.20}$$

for an infinitesimal transformation of SL(2, $\mathcal{C}$). The corresponding conjugate representation defined by (2.2.23) is

$$1 + \frac{i}{2}\boldsymbol{\theta\sigma} - \frac{1}{2}\boldsymbol{\tau\sigma}\,, \tag{2.3.21}$$

where $\boldsymbol{\sigma} = (\sigma_1, \sigma_2, \sigma_3)$, and σ_i $(i = 1, 2, 3)$ are the Pauli matrices defined by (2.2.1). Neither (2.3.20) nor (2.3.21) is a unitary representation. In fact, as will be discussed in §4.3, every unitary representation of the Lorentz group is infinite-dimensional representation except for the one-dimensional representation in which $\mathbf{J} = \mathbf{K} = 0$, i.e., $L(\Lambda) = 1$ for any Λ (cf. §4.3).

We shall apply the method of infinitesimal transformations, which has been mentioned in this section, to each specific problem later.

Chapter 3

Irreducible Representations of the Poincaré Group

§3.1 Translational Transformations

The starting point for finding unitary irreducible representation of the Poincaré group is based on Eqs. (2.1.1)–(2.1.3). That is, our present purpose is to determine all possible solutions for $L(\Lambda)$ and $T(a)$ which satisfy those equations, and are unitary and irreducible.

Let us first begin with the discussion of a subgroup of the Poincaré group, i.e., a group formed by the translations $T(a)$. The translations $T(a)$ satisfy (2.1.2) and are commutable with one another. Hence the transformation a_μ is accomplished by successive applications of transformations of a_1 in the direction of the first axis, a_2 in the second axis, . . . , a_0 in the fourth axis (the time axis). If $T^{(1)}(a_1)$ denotes a representation of the translation a_1 in the direction of the first axis, the parameter a_1 runs in the range $-\infty < a_1 < \infty$, and $T^{(1)}(a_1)$ satisfies $T^{(1)}(a_1)T^{(1)}(b_1) = T^{(1)}(a_1 + b_1)$. Since we can set $T^{(1)}(0) = 1$, we get a general solution $T^{(1)}(a_1) = \exp(ik_1a_1)$, where $T^{(1)}(a_1)$ is a unitary representation, and k_1 is a Hermitian operator. Applying a similar argument to the transformations in the direction of the second, the third and the fourth axes, we get finally

$$T(a) = e^{ik_\mu a_\mu} \ . \tag{3.1.1}$$

Of course, k_1, k_2, k_3 and $k_0(= k_4/i)$ are Hermitian operators and commutable with one another.

On the other hand, with the help of (2.1.3) we get

$$L(\Lambda)T(a)L(\Lambda)^{-1} = \exp[iL(\Lambda)k_\mu L(\Lambda)^{-1}a_\mu]$$
$$= \exp(ik_\mu \Lambda_{\mu\nu} a_\nu) = \exp[i(\Lambda^{-1})_{\mu\nu} k_\nu a_\mu] \ .$$

Since this equation holds for arbitrary a_μ, we have

$$L(\Lambda)k_\mu L(\Lambda)^{-1} = (\Lambda^{-1})_{\mu\nu} k_\nu \ , \tag{3.1.2}$$

and consequently we can conclude that k_μ^2 is commutable with $L(\Lambda)$ and $T(a)$:

$$[k_\mu^2, L(\Lambda)] = [k_\mu^2, T(a)] = 0 \ . \tag{3.1.3}$$

Therefore, according to Schur's lemma, in an irreducible representation of the Poincaré group we can set k_μ^2 as a real constant κ:

$$k^2 = \kappa \ . \tag{3.1.4}$$

That is, κ is one of the constants required for specifying an irreducible representation, and two irreducible representations with the different values of κ are not equivalent. In this sense, k_μ^2 is one of the Casimir operators of the Poincaré group, and the value of κ is determined corresponding to a given irreducible representation. We shall classify the irreducible representations by κ. In the following we shall use a symbol [ ] for describing the class.

$[M]$; $\kappa < 0$.

In this case, k_μ is called time-like, and the eigenvalue of k_0 does not take the value 0. Now let $\text{sgn}(k_0)$ be a symbol which has values 1 or -1 corresponding to the sign of the eigenvalue of k_0, then we easily find that $\text{sgn}(k_0)$ is commutable with any $L(\Lambda)$ and $T(a)$. Thus the following two different irreducible representations are possible in this case.

$$[M_+]; \quad \text{sgn}(k_0) = 1 \ ,$$
$$[M_-]; \quad \text{sgn}(k_0) = -1 \ .$$

$[0]$; $\kappa = 0$ and the eigenvalue of k_0 is not 0.

In this case, k_μ is called light-like or a null vector, and similarly to the case of $[M]$ there are two kinds of irreducible representations which are specified by

$$[0_+]; \quad \text{sgn}(k_0) = 1 \ ,$$
$$[0_-]; \quad \text{sgn}(k_0) = -1 \ .$$

Further there are cases where

$[L]; \kappa = 0$ and $k_\mu = 0$ $(\mu = 1, 2, 3, 4)$,

and

$[T]; \kappa > 0$.

In the case of $[T]$, k_μ is called space-like. Since the eigenvalue of k_0 changes its sign under an appropriate Lorentz transformation, the irreducible representations cannot be classified by $\text{sgn}(k_0)$ in contrast to the case of $[M]$ or $[0]$.

Needless to say k_μ is the generator of the translation and represents a 4-vector of momentum-energy. In a one-particle problem, therefore, $[M]$ describes a particle with finite mass and $[0]$ describes a massless particle. In the case of $[L]$, both momentum and energy are 0, and there is no corresponding particle picture. The class $[T]$ corresponds to a particle with imaginary mass.

We have classified the irreducible representations by k_μ only. But we have not specified the irreducible representations completely yet, because we have not used (2.1.1) in the above discussion at all. Therefore, even if an irreducible representation was given, it is generally impossible to completely specify the state vectors, which form a complete orthogonal basis in the representation space, by the eigenvalue $\bar{k}_\mu$ of the operator k_μ which has been classified above. To make up for the incompleteness, let us introduce a variable ξ which is independent of k_μ. Let $|\bar{k}, \bar{\xi}\rangle$ denote the state vector which forms a complete orthogonal basis of the given irreducible representation space. Namely,

$$k_\mu|\bar{k}, \bar{\xi}\rangle = \bar{k}_\mu|\bar{k}, \bar{\xi}\rangle \ . \tag{3.1.5}$$

Applying (3.1.2) we get

$$k_\mu L(\Lambda)|\bar{k}, \bar{\xi}\rangle = \Lambda_{\mu\nu}\bar{k}_\nu L(\Lambda)|\bar{k}, \bar{\xi}\rangle \ , \tag{3.1.6}$$

and we find that $L(\Lambda)|\bar{k}, \bar{\xi}\rangle$ is also an eigenstate of k_μ, and its eigenvalue is $\Lambda_{\mu\nu}\bar{k}_\nu$. We call the freedom of ξ the spin freedom.

Now let us consider a Lorentz invariant inner product of $|\bar{k}, \bar{\xi}\rangle$.

In the case of $[M]$, if $-\kappa = m^2$ $(m > 0)$, m represents the mass of a particle in a one-particle problem. Further if

$$\omega_{k} = \sqrt{\mathbf{k}^2 + m^2} \tag{3.1.7}$$

is introduced, (3.1.4) leads to $\bar{k}_0 = \pm\omega_{\bar{k}}$ where $+\omega_{\bar{k}}$ corresponds to $[M_+]$, and $-\omega_{\bar{k}}$ corresponds to $[M_-]$. Here $|\bar{k}, \bar{\xi}\rangle$ is written as $|\bar{k}, \bar{\xi}, \pm\rangle$ corresponding to $[M_\pm]$, and the inner product is given by

$$\langle\, \bar{k}, \bar{\xi}, \pm \,|\, \bar{k}', \bar{\xi}', \pm \,\rangle = \delta_{\bar{\xi}\bar{\xi}'}\omega_{\bar{k}}\delta(\bar{\mathbf{k}} - \bar{\mathbf{k}}') \ , \tag{3.1.8}$$

where $\delta(\bar{\mathbf{k}}-\bar{\mathbf{k}}')$ denotes $\delta(\bar{k}_1-\bar{k}_1')\delta(\bar{k}_2-\bar{k}_2')\,\delta(\bar{k}_3-\bar{k}_3')$. The Lorentz invariance of $\omega_{\bar{k}}\delta(\bar{\mathbf{k}}-\bar{\mathbf{k}}')$ on the right-hand side of the above expression can be proved as follows. If we write a Lorentz transformation as

$$\bar{q}_\mu = \Lambda_{\mu\nu}\bar{k}_\nu \ , \qquad \bar{q}'_\mu = \Lambda_{\mu\nu}\bar{k}' \ , \tag{3.1.9}$$

and if we write $d\bar{\mathbf{k}} = d\bar{k}_1 d\bar{k}_2 d\bar{k}_3$, using a Jacobian we have

$$d\bar{\mathbf{q}} = \left|\frac{\partial(\bar{q}_1,\bar{q}_2,\bar{q}_3)}{\partial(\bar{k}_1,\bar{k}_2,\bar{k}_3)}\right| d\bar{\mathbf{k}} \ . \tag{3.1.10}$$

Then, for a smooth function of $\bar{\mathbf{k}}$, $f(\bar{\mathbf{k}})$, the following equation holds:

$$\int d\bar{\mathbf{q}}\,\delta(\bar{\mathbf{q}}-\bar{\mathbf{q}}')f(\bar{\mathbf{q}}') = \int d\bar{\mathbf{k}}\left|\frac{\partial(\bar{q}_1,\bar{q}_2,\bar{q}_3)}{\partial(\bar{k}_1,\bar{k}_2,\bar{k}_3)}\right| \delta(\Lambda(\bar{\mathbf{k}}-\bar{\mathbf{k}}')f(\Lambda\bar{\mathbf{k}}) \ , \tag{3.1.11}$$

where $\Lambda\bar{\mathbf{k}}$ describes the space component of $\Lambda_{\mu\nu}\bar{k}_\nu$, namely $\Lambda\bar{\mathbf{k}} = (\Lambda_{1\nu}\bar{k}_\nu, \Lambda_{2\nu}\bar{k}_\nu, \Lambda_{3\nu}\bar{k}_\nu)$. On the other hand, the result of integration of the left-hand side of (3.1.11) is $f(\bar{\mathbf{q}}')$, and consequently, $f(\Lambda\mathbf{k}')$, which can be written as

$$f(\Lambda\bar{\mathbf{k}}') = \int d\bar{\mathbf{k}}\delta(\bar{\mathbf{k}}-\bar{\mathbf{k}}')f(\Lambda\mathbf{k}) \ . \tag{3.1.12}$$

Comparing this expression with the right-hand side of (3.1.11), we have

$$\left|\frac{\partial(\bar{q}_1,\bar{q}_2,\bar{q}_3)}{\partial(\bar{k}_1,\bar{k}_2,\bar{k}_3)}\right| \delta(\bar{\mathbf{q}}-\bar{\mathbf{q}}') = \delta(\bar{\mathbf{k}}-\bar{\mathbf{k}}') \ . \tag{3.1.13}$$

Considering $\partial\bar{k}_4/\partial\bar{k}_i = -\bar{k}_i/\bar{k}_4$ $(i = 1,2,3)$ we can calculate the Jacobian $\partial(\bar{q}_1,\bar{q}_2,\bar{q}_3)/\partial(\bar{k}_1,\bar{k}_2,\bar{k}_3)$ as

$$\begin{aligned}\Delta = \frac{\partial(\bar{q}_1,\bar{q}_2,\bar{q}_3)}{\partial(\bar{k}_1,\bar{k}_2,\bar{k}_3)} &= \begin{vmatrix} \Lambda_{11}-\Lambda_{14}\bar{k}_1/\bar{k}_4 & \Lambda_{12}-\Lambda_{14}\bar{k}_2/\bar{k}_4 & \Lambda_{13}-\Lambda_{14}\bar{k}_3/\bar{k}_4 \\ \Lambda_{21}-\Lambda_{24}\bar{k}_1/\bar{k}_4 & \Lambda_{22}-\Lambda_{24}\bar{k}_2/\bar{k}_4 & \Lambda_{23}-\Lambda_{24}\bar{k}_3/\bar{k}_4 \\ \Lambda_{31}-\Lambda_{34}\bar{k}_1/\bar{k}_4 & \Lambda_{32}-\Lambda_{34}\bar{k}_2/\bar{k}_4 & \Lambda_{33}-\Lambda_{34}\bar{k}_3/\bar{k}_4 \end{vmatrix} \\ &= \frac{1}{\bar{k}_4}\begin{vmatrix} \Lambda_{11} & \Lambda_{12} & \Lambda_{13} & \Lambda_{14} \\ \Lambda_{21} & \Lambda_{22} & \Lambda_{23} & \Lambda_{24} \\ \Lambda_{31} & \Lambda_{32} & \Lambda_{33} & \Lambda_{34} \\ \bar{k}_1 & \bar{k}_2 & \bar{k}_3 & \bar{k}_4 \end{vmatrix} \ .\end{aligned}$$

Here, applying $\det(\Lambda^{\mathrm{T}}) = 1$ to a product of determinants, $\Delta\det(\Lambda^{\mathrm{T}})$, we get

$$\Delta = \frac{1}{\bar{k}_4}\begin{vmatrix} 1 & 0 & 0 & 0 \\ 0 & 1 & 0 & 0 \\ 0 & 0 & 1 & 0 \\ \bar{q}_1 & \bar{q}_2 & \bar{q}_3 & \bar{q}_4 \end{vmatrix} = \frac{\bar{q}_0}{\bar{k}_0} \tag{3.1.14}$$

with the help of (1.2.8) and the first equation of (3.1.9). Therefore, from (3.1.13) we have $\bar{q}_0\delta(\bar{\mathbf{q}}-\bar{\mathbf{q}}') = \bar{k}_0\delta(\bar{k}-\bar{k}')$, and we can conclude that $\omega_{\bar{k}}\delta(\bar{\mathbf{k}}-\bar{\mathbf{k}}')$ is Lorentz invariant. The condition of completeness is written, by the use of Dirac's notation*), as

$$\sum_{\bar{\xi}} \int \frac{d\bar{\mathbf{k}}}{\omega_{\bar{k}}} |\bar{k}, \bar{\xi}, \pm\rangle\langle\bar{k}, \bar{\xi}, \pm| = 1 \, . \tag{3.1.15}$$

Here the invariance of $d\bar{\mathbf{k}}/\omega_{\bar{k}}$ is readily derived from (3.1.10) and (3.1.14).

In the case of $[0_\pm]$, an analogous argument is applicable if we substitute 0 for m in the case of $[M_\pm]$, and we have

$$\langle \bar{k}, \bar{\xi}, \pm | \bar{k}', \bar{\xi}', \pm\rangle = \delta_{\bar{\xi}\bar{\xi}'} |\bar{\mathbf{k}}| \delta(\bar{\mathbf{k}} - \bar{\mathbf{k}}') \, , \tag{3.1.16}$$

$$\sum_{\bar{\xi}} \int \frac{d\bar{\mathbf{k}}}{|\bar{\mathbf{k}}|} |\bar{\mathbf{k}}, \bar{\xi}, \pm\rangle\langle\bar{\mathbf{k}}, \bar{\xi}, \pm| = 1 \, . \tag{3.1.17}$$

In the case of $[L]$, since $\bar{k}_\mu = 0$, writing $|0, \bar{\xi}\rangle = |\bar{\xi}\rangle$ we have simply

$$\langle\bar{\xi}|\bar{\xi}'\rangle = \delta_{\bar{\xi}\bar{\xi}'} \, , \tag{3.1.18}$$

$$\sum_{\bar{\xi}} |\bar{\xi}\rangle\langle\bar{\xi}| = 1 \, . \tag{3.1.19}$$

In the case of $[T]$, m^2 in $[M]$ must be replaced by $-m^2$ since $\kappa > 0$. But such classifications of state vectors by $\pm$ as in the previous cases are not applicable because the two values of k_0, $\pm\sqrt{\bar{\mathbf{k}}^2 - m^2}$, can be connected by some Lorentz transformation. Then we define a Lorentz invariant inner product by

$$\langle \bar{k}, \bar{\xi}|\bar{k}', \bar{\xi}'\rangle\delta(\bar{k}_\mu^2 - m^2) = \delta^4(\bar{k} - \bar{k}')\delta_{\bar{\xi}\bar{\xi}'} \, , \tag{3.1.20}$$

where $\delta^4(\bar{k} - \bar{k}')$ denotes $\delta(\bar{k}_1 - \bar{k}_1')\,\delta(\bar{k}_2 - \bar{k}_2')\,\delta(\bar{k}_3 - \bar{k}_3')\,\delta(\bar{k}_0 - \bar{k}_0')$. Equation (3.1.20) defines essentially the same inner product as the previous one, because (3.1.8) is obtained by integrating (3.1.20) with respect to k_0 in the neighborhood of $\omega_{\bar{k}}$ or $-\omega_{\bar{k}}$ after changing the sign of m^2 in the δ-function on the left-hand side. The condition of the completeness is given by

$$\sum_{\bar{\xi}} \int d^4\bar{k}\delta(\bar{k}_\mu^2 - m^2)|\bar{k}, \bar{\xi}\rangle\langle\bar{k}, \bar{\xi}| = 1 \, , \tag{3.1.21}$$

*) P.A.M. Dirac: *The Principles of Quantum Mechanics* (fourth ed.), Oxford Uni. Press (1958).

where $d^4\bar{k} = d\bar{k}_1 d\bar{k}_2 d\bar{k}_3 d\bar{k}_0$.

In the above discussion, ξ has been assumed to have discrete value, but if it has continuous value, $\delta_{\bar{\xi}\bar{\xi}'}$ and the sum with respect to $\bar{\xi}$ must be replaced by $\delta(\bar{\xi} - \bar{\xi}')$ and the corresponding integration respectively.

§3.2 Lorentz Transformations

Let us introduce an operator $P(\Lambda)$ defined by the following equation:

$$P(\Lambda)|\bar{k}, \bar{\xi}\rangle = |\Lambda\bar{k}, \bar{\xi}\rangle \,. \tag{3.2.1}$$

The operator $P(\Lambda)$, which is independent of $\bar{\xi}$ and acts on $\bar{k}$ only, leaves an inner product invariant, and as a result it is a unitary operator. Especially in the case of the class $[L]$ we have $P(\Lambda) = 1$. Clearly from (3.2.1), $P(\Lambda)$ satisfies

$$P(\Lambda)P(\Lambda') = P(\Lambda\Lambda') \,, \tag{3.2.2}$$

and hence is one of the unitary representations of the Lorentz group.

Now let $F(k)$ be an operator defined in the irreducible representation space and commutable with k_μ. Then the following equation holds:

$$\begin{aligned} P(\Lambda)F(k)|\bar{k}, \bar{\xi}\rangle &= F(\bar{k})P(\Lambda)|\bar{k}, \bar{\xi}\rangle = F(\bar{k})|\Lambda\bar{k}, \bar{\xi}\rangle \\ &= F(\Lambda^{-1}k)|\Lambda\bar{k}, \bar{\xi}\rangle = F(\Lambda^{-1}k)P(\Lambda)|\bar{k}, \bar{\xi}\rangle \,. \end{aligned} \tag{3.2.3}$$

Considering the completeness of $|\bar{k}, \bar{\xi}\rangle$, we can remove $|\bar{k}, \bar{\xi}\rangle$ from each side of (3.2.3) to obtain

$$P(\Lambda)F(k) = F(\Lambda^{-1}k)P(\Lambda) \,. \tag{3.2.4}$$

Here substitute $T(\Lambda^{-1}a)$ for $F(k)$, then from the above equation and (3.1.1) we get

$$T(\Lambda^{-1}a) = P(\Lambda)^{-1}T(a)P(\Lambda) \,. \tag{3.2.5}$$

On the other hand, if we substitute $\Lambda^{-1}a$ for a in (2.1.3) we have

$$T(\Lambda^{-1}a) = L(\Lambda)^{-1}T(a)L(\Lambda) \,. \tag{3.2.6}$$

From these two equations we immediately get

$$P(\Lambda)^{-1}T(a)P(\Lambda) = L(\Lambda)^{-1}T(a)L(\Lambda)$$

which leads to

$$[L(\Lambda)P(\Lambda)^{-1}, T(a)] = 0 \,. \tag{3.2.7}$$

This relation holds for an arbitrary a_μ, and consequently $L(\Lambda)P(\Lambda)^{-1}$ becomes commutable with k_μ, that is, diagonal with respect to k_μ. If it is denoted as $Q(\Lambda, k)$, then $L(\Lambda)$ is always expressed as

$$L(\Lambda) = Q(\Lambda, k)P(\Lambda) . \tag{3.2.8}$$

Since both $L(\Lambda)$ and $P(\Lambda)$ are unitary operators, $Q(\Lambda, k)$ is also a unitary operator.

Equation (3.2.8) expresses the Lorentz transformation operator $L(\Lambda)$ as a product of $P(\Lambda)$ and $Q(\Lambda, k)$ which are diagonal with respect to ξ and k_μ respectively, where $P(\Lambda)$ has been given by (3.2.1) already, and is independent of ξ. As a result, if the other factor $Q(\Lambda, k)$ is known, the behavior of ξ under a Lorentz transformation is determined completely. Now let us summarize the properties of $Q(\Lambda, k)$.

Using (3.1.1), (3.2.4) and (3.2.8), we have

$$\begin{aligned} L(\Lambda\Lambda') = L(\Lambda)L(\Lambda') &= Q(\Lambda, k)P(\Lambda)Q(\Lambda', k)P(\Lambda') \\ &= Q(\Lambda, k)Q(\Lambda', \Lambda^{-1}k)P(\Lambda\Lambda') . \end{aligned} \tag{3.2.9}$$

Since the left-hand side of the above equation is equal to $Q(\Lambda\Lambda', k)P(\Lambda\Lambda')$, the following equation for Q holds:

$$Q(\Lambda\Lambda', k) = Q(\Lambda, k)Q(\Lambda', \Lambda^{-1}k) . \tag{3.2.10}$$

Let us put $\Lambda' = \Lambda^{-1}$. Here, as has been mentioned in §2.2, Λ^{-1} should read g^{-1} which is given by $\Lambda = \Lambda(g)$ for $g \in \mathrm{SL}(2, C)$. Then, with the help of (2.2.21), we can write

$$Q(\Lambda, k)^{-1} = Q(\Lambda^{-1}, \Lambda^{-1}k) . \tag{3.2.11}$$

Further, the equivalence of representation can be analyzed as follows.

When irreducible representations $L(\Lambda)$ and $T(a)$ are given, they can be replaced by $UL(\Lambda)U^{-1}$ and $UT(a)U^{-1}$ respectively, where U is defined in the irreducible representation space and must be a unitary operator so that the unitarity of the representation is preserved. This corresponds to the well known fact that quantum mechanics leaves itself invariant under a unitary transformation. We assume that U is restricted to be diagonal with respect to k_μ and a single-valued function of k_μ for the sake of later convenience, and we denote it as $U(k)$. A state vector $|\bar{k}, \bar{\xi}\rangle$ is a single-valued function of k_μ,*) and its single-valuedness will be preserved under the transformation of $U(k)$ if $U(k)$ is assumed to be a single-valued function of k_μ. The translation $T(a)$ is not transformed by such a unitary operator, and only $L(\Lambda)$ is transformed into $U(k)L(\Lambda)U(k)^{-1}$. Here using (3.2.8) and (3.2.4) we have

$$U(k)L(\Lambda)U(k)^{-1} = U(k)Q(\Lambda, k)U(\Lambda^{-1}k)^{-1}P(\Lambda) . \tag{3.2.12}$$

Therefore, if we leave $T(a)$ and $P(\Lambda)$ as they are, and use

$$Q'(\Lambda, k) = U(k)Q(\Lambda, k)U(\Lambda^{-1}k)^{-1} \tag{3.2.13}$$

in place of $Q(\Lambda, k)$, we have an equivalent representation. Needless to say, $Q'(\Lambda, k)$ is a unitary operator.

§3.3 Little Groups

On the basis of the preceding preparations, we introduce here the concept of little group following Wigner in order to determine the $Q(\Lambda, k)$.

When an irreducible representation of the Poincaré group is given, k_μ and its eigenvalue $\bar{k}_\mu$ are classified into one of $[M_+]$, $[M_-]$, $[0_+]$, $[0_-]$, $[L]$ and $[T]$ which have been mentioned in §3.1, and $\bar{k}_\mu$'s in each class are connected to one another by Lorentz transformations. Select an element of a set $\{\bar{k}\}$ which consists of $\bar{k}_\mu$ belonging to one class, and denote it as l_μ. The choice of l_μ is entirely arbitrary. A set of Lorentz transformations $\{\lambda\}$ which leaves l_μ invariant is a subset of the Lorentz group, and it is called a little group. That is, $\lambda_{\mu\nu}l_\nu = l_\mu$, which is simply written as

$$\lambda l = l . \tag{3.3.1}$$

On the other hand, an arbitrary $\bar{k}_\mu$ in $\{\bar{k}\}$ is connected to l_μ by a Lorentz transformation, and with an appropriate Lorentz transformation $\alpha_{\bar{k}}$ we can write

$$\bar{k} = \alpha_{\bar{k}} l \tag{3.3.2}$$

or

$$l = \alpha_{\bar{k}}^{-1} \bar{k} . \tag{3.3.3}$$

*)When k_μ has a degenerate eigenvalue $\bar{k}_\mu$, the degeneracy is removed by $\bar{\xi}$, and for a given $\bar{\xi}$, $|\bar{k}, \bar{\xi}\rangle$ is uniquely determined by $\bar{k}_\mu$ (apart from an arbitrary phase factor). It is not apparent that such a treatment is consistent with the continuity of representation, but we proceed here assuming it possible. See the footnote **) of p. 37 and §5.4 for the justification of this assumption.

Even if l and $\bar{k}$ are given, $\alpha_{\bar{k}}$ is not uniquely determined, but we arbitrarily choose $\alpha_{\bar{k}}$ which satisfies (3.3.2).

Now let us define $\lambda_{\bar{k}}$ by

$$\lambda_{\bar{k}} = \alpha_{\bar{k}}^{-1} \Lambda \alpha_{\Lambda^{-1}\bar{k}} \tag{3.3.4}$$

for a Lorentz transformation Λ. From (3.3.2) and (3.3.3) we easily get

$$\lambda_{\bar{k}} l = l \, , \tag{3.3.5}$$

and hence $\lambda_{\bar{k}}$ is an element of the little group. Since (3.3.4) leads to $\Lambda = \alpha_{\bar{k}} \lambda_{\bar{k}} (\alpha_{\Lambda^{-1}\bar{k}})^{-1}$, then

$$Q(\Lambda, k) = Q(\alpha_{\bar{k}} \lambda_{\bar{k}} (\alpha_{\Lambda^{-1}\bar{k}})^{-1}, k) \, . \tag{3.3.6}$$

Here using (3.2.10) we can modify the right-hand side to obtain

$$Q(\Lambda, k) = Q(\alpha_{\bar{k}}, k) Q(\lambda_{\bar{k}}, \alpha_{\bar{k}}^{-1} k) Q((\alpha_{\Lambda^{-1}\bar{k}})^{-1}, \lambda_{\bar{k}}^{-1} \alpha_{\bar{k}}^{-1} k) \, . \tag{3.3.7}$$

As is seen in the left-hand side, $\bar{k}$ on the right-hand side is a dummy variable, and the whole right-hand side must be independent of $\bar{k}$. We can therefore select $\bar{k}$ arbitrary, and put $\bar{k} = k$. Because $\bar{k}$ is merely a number and k is an operator, it might appear too rough to set them equal. But the diagonality of Q with respect to k permits such a treatment because k can be regarded to be an ordinary number in the discussion of the relations of Q's.*) As a result, from (3.3.3) and (3.3.1) we obtain

$$Q(\Lambda, k) = Q(\alpha_k, k) Q(\lambda_k, l) Q((\alpha_{\Lambda^{-1}k})^{-1}, l) \, . \tag{3.3.8}$$

Since $Q(\alpha_k, k)$ is a unitary operator, we write it as

$$Q(\alpha_k, k) = U(k)^{-1} \, . \tag{3.3.9}$$

Then from (3.2.11) and (3.3.3) we get

$$U(k) = Q(\alpha_k, k)^{-1} = Q(\alpha_k^{-1}, \alpha_k^{-1} k) = Q(\alpha_k^{-1}, l) \, . \tag{3.3.10}$$

---

*)If you do not like such a treatment, apply (3.3.7) on $|\bar{k}, \xi\rangle$ and substitute $\bar{k}$ for k in Q. Arrange the right-hand side with the help of (3.3.3) and (3.3.1), replace $\bar{k}$ with k in Q, and remove $|\bar{k}, \xi\rangle$ from each side. You will then obtain (3.3.8).

Hence

$$U(\Lambda^{-1}k) = Q((\alpha_{\Lambda^{-1}k})^{-1}, l) , \tag{3.3.11}$$

and using (3.3.8) we obtain

$$Q(\Lambda, k) = U(k)^{-1}Q(\lambda_k, l)U(\Lambda^{-1}, k) . \tag{3.3.12}$$

As is seen from (3.2.13) this result implies that the representation of the Poincaré group constructed from $Q(\lambda_k, l)$ is unitary equivalent to that constructed from $Q(\Lambda, k)$. Therefore, without any loss of generality, we can write*)

$$L(\Lambda) = Q(\lambda_k, l)P(\Lambda) . \tag{3.3.13}$$

The operator $Q(\lambda_k, l)$ has the following property. According to (3.3.4) we put

$$\left.\begin{aligned} \lambda_p &= \alpha_p^{-1}\Lambda\alpha_{\Lambda^{-1}p} , \\ \lambda'_p &= \alpha_p^{-1}\Lambda'\alpha_{\Lambda'^{-1}p} , \end{aligned}\right\} \tag{3.3.14}$$

and using (3.2.10) with λ_p and λ'_p substituted for Λ and Λ' respectively, we have the relation

$$Q(\lambda_p\lambda'_p, k) = Q(\lambda_p, k)Q(\lambda'_p, \lambda_p^{-1}k) . \tag{3.3.15}$$

If $k = l$ here,

$$Q(\lambda_p\lambda'_p, l) = Q(\lambda_p, l)Q(\lambda'_p, l) \tag{3.3.16}$$

holds. Hence $Q(\lambda_k, l)$ is a unitary representation of the little group. By the use of this, we can obtain all the unitary irreducible representations of the Poincaré group. It will be mentioned in the next section. The operator $Q(\lambda_k, l)$ in the irreducible representation of the Poincaré group is usually called a Wigner rotation.

We have mentioned in the first part of this section that the choice of l and α_k has some arbitrariness. But now it has become clear that the arbitrariness does not essentially affect the theory. In fact, as is seen from the absence of l and α_k in the left-hand side of (3.3.12), we obtain only a unitary equivalent representation by changing l or α_k. Therefore, when we seek for the explicit

---

*)Here $U(k)^{-1}$, i.e., $Q(\alpha_k, k)$, has been assumed to be a single-valued function of k_μ. This assumption is based on the fact that both α_k and $Q(\Lambda, k)$ are single-valued functions of k. In particular, the property of the latter comes from the assumption that $|\bar{k}, \xi\rangle$ is a single-valued function of $\bar{k}_\mu$.

form of $Q(\lambda_k, l)$, we can use l and α_k which are convenient for it. This will be done in Chapters 4 and 5.

§3.4 Irreducible Representations

In a unitary irreducible representation of the Poincaré group, $Q(\lambda_k, l)$ is a unitary irreducible representation of a little group. This is readily recognized if its contraposition is considered, i.e., let us suppose that $Q(\lambda_k, l)$ was reducible. All properties of $L(\Lambda)$ concerning ξ are connected to $Q(\lambda_k, l)$, since $P(\Lambda)$ is an operator independent of ξ. Because $Q(\lambda_k, l)$ is diagonal with respect to k, its reducibility must come from the freedom of ξ. Therefore the supposition of the reducibility of $Q(\lambda_k, l)$ implies that $L(\Lambda)$ is reducible with respect to the freedom of ξ, while $T(a)$ is diagonal with respect to k and ξ. These results lead to the reducibility of the representation of the Poincaré group, which contradicts the basic assumption. That is, if the unitary representation of the Poincaré group is irreducible, $Q(\lambda_k, l)$ must be a unitary irreducible representation of the little group.

Does the converse hold? That is, if $Q(\lambda_k, l)$ is an arbitrary unitary irreducible representation of the little group, do we always have the corresponding unitary irreducible representation of the Poincaré group? When $|\bar{k}, \xi\rangle$ is a single-valued function of $\bar{k}_\mu$, $P(\Lambda)$ is unitary, and then $Q(\lambda_k, l)P(\Lambda)$ is also unitary. Let us first examine whether this operator is a representation of the Lorentz group or not (apart from the irreducibility at present). To do this, we consider a product of $Q(\lambda_k, l)P(\Lambda)$ and $Q(\lambda'_k, l)P(\Lambda')$ to have

$$\begin{aligned} Q(\lambda_k, l)P(\Lambda)Q(\lambda'_k, l)P(\Lambda') &= Q(\lambda_k, l)Q(\lambda'_{\Lambda^{-1}k}, l)P(\Lambda)P(\Lambda') \\ &= Q(\lambda_k\lambda'_{\Lambda^{-1}k}, l)P(\Lambda\Lambda') \,. \end{aligned} \tag{3.4.1}$$

While

$$\begin{aligned} \lambda_k\lambda'_{\Lambda^{-1}k} &= \alpha_k^{-1}\Lambda\alpha_{\Lambda^{-1}k}(\alpha_{\Lambda^{-1}k})^{-1}\Lambda'\alpha_{\Lambda'^{-1}\Lambda^{-1}k} \\ &= \alpha_k^{-1}\Lambda\Lambda'\alpha_{(\Lambda\Lambda')^{-1}k} \,, \end{aligned} \tag{3.4.2}$$

then $Q(\lambda_k, l)P(\Lambda)$ is a representation of the Lorentz group. But it should be noted that there are cases in which the irreducible representations of the little group are multi-valued of degree more than two. It will be shown in the following section that such cases arise actually in $[0_\pm]$ and $[T]$. As has been mentioned in §2.1, however, the representations of the Lorentz group is originally at most double-valued, therefore such $Q(\lambda_k, l)$ must not contribute to the representations of the Lorentz group.*) That is, $Q(\lambda_k, l)P(\Lambda)$ is a representation of the Lorentz group only if the representation of the little group, $Q(\lambda_k, l)$, is at most double-valued.**)

In the following, $Q(\lambda_k, l)$ will be restricted to such a case. If $Q(\lambda_k, l)$ is unitary and irreducible, we can show that $L(\Lambda)$ expressed as $Q(\lambda_k, l)P(\Lambda)$, together with $T(a)$, gives an irreducible representation of the Poincaré group.

For an element f of the Poincaré group, let $D(f)$ be a unitary representation given in the above manner. We now assume that the representation is reducible. Then there exist state vectors $|\mathrm{I}\rangle$ and $|\mathrm{II}\rangle$ which satisfy

$$\langle \mathrm{II}|D(f)|\mathrm{I}\rangle = 0 \tag{3.4.3}$$

for any f in the representation space. Since $D(f)T(a)$ is also a representation of the Poincaré group,

$$\langle \mathrm{II}|D(f)T(a)|\mathrm{I}\rangle = 0 \tag{3.4.4}$$

for any f. On the other hand, $|\mathrm{I}\rangle$ is a superposition of $|\bar{k}, \bar{\xi}\rangle$, and we can write it as

$$|\mathrm{I}\rangle = \sum_{\bar{\xi}} \int_{\Sigma} d^4\bar{k} C_{\mathrm{I}}(\bar{k}, \bar{\xi})|\bar{k}, \bar{\xi}\rangle \ , \tag{3.4.5}$$

where the integration is performed in the area Σ which is formed by all $\bar{k}_\mu$ in a class of §3.1. Substituting (3.4.5) into (3.4.4) we get

$$\sum_{\bar{\xi}} \int_{\Sigma} d^4\bar{k} C_{\mathrm{I}}(\bar{k}, \bar{\xi})\langle \mathrm{II}|D(f)|\bar{k}, \bar{\xi}\rangle e^{i\bar{k}_\mu a_\mu} = 0 \ . \tag{3.4.6}$$

Since a_μ is arbitrary, we find that

$$\langle \mathrm{II}|D(f)|\bar{k}'\rangle_{\mathrm{I}} = 0 \tag{3.4.7}$$

*) The multi-valuedness of representations of the Lorentz group can be caused only by $Q(\lambda_k, l)$ if $|\bar{k}, \bar{\xi}\rangle$ is a single-valued function of $\bar{k}_\mu$ as has been mentioned previously.

**) When the representation of the little group is neither single-valued nor double-valued, $Q(\lambda_k, l)P(\Lambda)|\bar{k}, \bar{\xi}\rangle$ is not a single-valued function of $\bar{k}_\mu$. That is, $|\bar{k}, \bar{\xi}\rangle$ is preserved in a single-valued function under a Lorentz transformation, only if its representation is at most double-valued. This is justified by the fact that the representation of the rotation group is at most double-valued as has been mentioned in §2.1. In other words, if $Q(\lambda'_k, l)P(\Lambda')$ is correctly a representation of the rotation group for a spatial rotation Λ', $|\bar{k}, \bar{\xi}\rangle$ is a single-valued function of $\bar{k}_\mu$. See §5.4 concerning this point. In such circumstances $Q(\lambda_k, l)P(\Lambda)$ becomes a representation of the Lorentz group according to (3.4.1) and (3.4.2).

is satisfied by any f and by a non-zero state vector

$$|\bar{k}'\rangle_{\rm I} = \sum_{\xi} C_{\rm I}(\bar{k}', \xi)|\bar{k}', \xi'\rangle . \tag{3.4.8}$$

An analogous argument is applicable to $|{\rm II}\rangle$ and then a non-zero state vector $|\bar{k}''\rangle_{\rm II}$ can be defined to satisfy

$${}_{\rm II}\langle \bar{k}'' |D(f)|\bar{k}'\rangle_{\rm I} = 0 . \tag{3.4.9}$$

Let λ be an arbitrary element of the little group, and let us substitute $L(\alpha_{\bar{k}''})L(\lambda)L(\alpha_{\bar{k}'}^{-1})$ for $D(f)$ in (3.4.9):

$$\begin{aligned} L(\alpha_{\bar{k}'}^{-1})|\bar{k}'\rangle_{\rm I} &= Q(\alpha_k^{-1}\alpha_{\bar{k}'}^{-1}\alpha_{\alpha_{\bar{k}'}k}, l)P(\alpha_{\bar{k}'}^{-1})|\bar{k}'\rangle_{\rm I} \\ &= Q(\alpha_l^{-1}, l)|l\rangle_{\rm I} . \end{aligned} \tag{3.4.10}$$

Similarly we have

$$L(\alpha_{\bar{k}''})^{\dagger}|\bar{k}''\rangle_{\rm II} = L(\alpha_{\bar{k}''}^{-1})|\bar{k}''\rangle_{\rm II} = Q(\alpha_l^{-1}, l)|l\rangle_{\rm II} , \tag{3.4.11}$$

where $|l\rangle_{\rm I}$ and $|l\rangle_{II}$ are $P(\alpha_{\bar{k}'}^{-1})|\bar{k}'\rangle_{\rm I}$ and $P(\alpha_{\bar{k}''}^{-1})|\bar{k}''\rangle_{\rm II}$ respectively, and are not zero since P is a unitary operator. Further,

$$k_\mu|l\rangle_{\rm I,II} = l_\mu|l\rangle_{\rm I,II} . \tag{3.4.12}$$

With the help of $L(\lambda) = Q(\alpha_k^{-1}\lambda\alpha_{\lambda^{-1}k}, l)P(\lambda)$ and (3.4.10)–(3.4.12) and considering that α_l is an element of the little group, i.e., $\alpha_l l = l$, we get, from (3.3.16),

$$\begin{aligned} {}_{\rm II}\langle \bar{k}''|L(\alpha_{\bar{k}''})L(\lambda)L(\alpha_{\bar{k}'}^{-1})|\bar{k}'\rangle_{\rm I} &= {}_{\rm II}\langle l|Q(\alpha_l, l)Q(\alpha_l^{-1}\lambda\alpha_l, l)Q(\alpha_l^{-1}, l)|l\rangle_{\rm I} \\ &= {}_{\rm II}\langle l|Q(\lambda, l)|l\rangle_{\rm I} . \end{aligned} \tag{3.4.13}$$

The right-hand side vanishes according to (3.4.9). Now λ is an arbitrary element of the little group. This result implies that $Q(k, l)$ is a reducible representation of the little group, and contradicts the assumption. Hence the representation of the Poincaré group in the discussion must be irreducible.

As a consequence of the above discussion, we have the following conclusion: "The unitary representation of the Poincaré group, $L(\Lambda)$, is irreducible, if and only if $L(\Lambda)$ is given by (3.3.13), and $Q(\lambda_k, l)$ is a single-valued or double-valued unitary irreducible representation of the little group."

Chapter 4

Unitary Representations of Little Groups

§4.1 Rotation Group

According to the previous discussion, unitary irreducible representations of the Poincaré group are completely determined if all single and double-valued unitary irreducible representations of a little group are given. In this chapter, we shall investigate the little group corresponding to each class of k_μ, which was mentioned in §3.1, and we shall construct the unitary representations in each case.

Let us first consider the case of $[M_\pm]$. If we put

$$\left.\begin{aligned} k_\mu^2 &= \kappa = -m^2 , \\ m &> 0 , \end{aligned}\right\} \tag{4.1.1}$$

m is the mass of a particle in a one-particle problem. In order to construct a little group we chose the rest frame and write

$$l_\mu = (0, 0, 0, \pm im) . \tag{4.1.2}$$

All spatial components of l_μ are 0, and the signs $\pm$ in the fourth component correspond to the cases of $[M_+]$, $[M_-]$ respectively. It is clear that a subgroup of the Lorentz group which leaves l_μ invariant, namely a little group, is a rotation group in either case of $[M_+]$ or $[M_-]$.

The irreducible representations of this group are well known in connection to angular momenta. We here briefly summarize the results only. Let S be a generator of the rotation group. The operator **S** is Hermitian, and its components satisfy the same commutation relation as (2.3.7):

$$[S_i, S_j] = i\sum_{k=1}^{3} \varepsilon_{ijk} S_k \ . \tag{4.1.3}$$

The irreducible representation is uniquely specified by an eigenvalue of $\mathbf{S}^2$, $s(s+1)$ say, where s is one of the following numbers:

$$s = 0, \frac{1}{2}, 1, \frac{3}{2}, \dots \ . \tag{4.1.4}$$

An integral s corresponds to a single-valued representation, and a half-integral s corresponds to a double-valued representation. The complete orthogonal basis in an irreducible representation space is specified by an eigenvalue of S_3 and $\bar{\xi}$, and is written as $|s, \bar{\xi}\rangle$:

$$\left.\begin{aligned} \mathbf{S}^2|\,s, \bar{\xi}\rangle &= s(s+1)|\,s, \bar{\xi}\rangle \ , \\ S_3|\,s, \bar{\xi}\rangle &= \bar{\xi}|\,s, \bar{\xi}\rangle \ , \\ \langle\, s, \bar{\xi}\,|\,s, \bar{\xi}'\,\rangle &= \delta_{\bar{\xi}\bar{\xi}'} \ , \\ \bar{\xi} = s, s-1, s-2&, \dots -(s-1), -s \ . \end{aligned}\right\} \tag{4.1.5}$$

Further, if $S^{(\pm)}$ are defined by

$$S^{(\pm)} = S_1 \pm iS_2 \ , \tag{4.1.6}$$

Equation (4.1.5) leads to

$$\left.\begin{aligned} S^{(+)}|\,s, \bar{\xi}\rangle &= \sqrt{(s-\bar{\xi})(s+\bar{\xi}+1)}\,|\,s, \bar{\xi}+1\rangle \ , \\ S^{(-)}|\,s, \bar{\xi}\rangle &= \sqrt{(s+\bar{\xi})(s-\bar{\xi}+1)}\,|s, \bar{\xi}-1\rangle \ , \end{aligned}\right\} \tag{4.1.7}$$ *)

In one-particle problem, s is called the spin of the particle, and **S** is a spin vector. For a particle with finite mass, the unitary irreducible representation is uniquely determined if the mass m, the spin s and $\mathrm{sgn}(k_0)$ are given.

In the cases of spin 0 and $\frac{1}{2}$, we have $\mathbf{S} = 0$ and $\mathbf{S} = \boldsymbol{\sigma}/2$ respectively. In other cases we can also calculate **S** by the use of (4.1.5) and (4.1.7).

---

*) A brief derivation of these results will be found at the beginning of §5.4.

§4.2 Two-Dimensional Euclidean Group

We now consider the case of [0], i.e., the case $k_\mu^2 = 0$. In this case the mass is 0, and there exists no rest frame in contrast to the previous section. Then we use

$$l_\mu = (0, 0, 1, \pm i) \tag{4.2.1}$$

for l_μ. The signs of the fourth component of l_μ, + and −, correspond to $[0_+]$ and $[0_-]$ respectively, which were classified in §3.1.

To construct a little group, let us use the spinor representation of the Lorentz group mentioned in §2.2. To leave l_μ invariant, we replace both x_μ and x'_μ in (2.2.14) by l_μ. In the case of $[0_+]$, we solve

$$g^{(+)} \begin{pmatrix} 0 & 0 \\ 0 & -2 \end{pmatrix} g^{(+)\dagger} = \begin{pmatrix} 0 & 0 \\ 0 & -2 \end{pmatrix} \tag{4.2.2}$$

under the condition $\det(g^{(+)}) = 1$, to find

$$g^{(+)} = \delta(\beta) t^{(+)}(a_1, a_2) \ , \tag{4.2.3}$$

where

$$\delta(\beta) = \begin{pmatrix} e^{i\beta/2} & 0 \\ 0 & e^{-i\beta/2} \end{pmatrix} \ , \tag{4.2.4}$$

$$t^{(+)}(a_1, a_2) = \begin{pmatrix} 1 & 0 \\ a_1 + i a_2 & 1 \end{pmatrix} \ , \tag{4.2.5}$$

and a_1, a_2 and β are real numbers.

Similarly, in the case of $[0_-]$, noticing a relation

$$\sigma_2 \begin{pmatrix} 0 & 0 \\ 0 & 2 \end{pmatrix} \sigma_2 = \begin{pmatrix} 2 & 0 \\ 0 & 0 \end{pmatrix} \tag{4.2.6}$$

we get from the complex conjugate of (4.2.2)

$$\sigma_2 g^{(+)*} \sigma_2 \begin{pmatrix} 2 & 0 \\ 0 & 0 \end{pmatrix} \sigma_2 g^{(+)T} \sigma_2 = \begin{pmatrix} 2 & 0 \\ 0 & 0 \end{pmatrix} \ . \tag{4.2.7}$$

Hence we obtain

$$g^{(-)} = \sigma_2 g^{(+)*} \sigma_2 = \delta(\beta)\, t^{(-)}(a_1, a_2) \ , \tag{4.2.8}$$

$$t^{(-)}(a_1, a_2) = \begin{pmatrix} 1 & -(a_1 - i a_2) \\ 0 & 1 \end{pmatrix} \ , \tag{4.2.9}$$

where $g^{(+)*}$ is given by replacing each matrix element of $g^{(+)}$ by its complex conjugate.

Using these equations, we have the following relations between $\delta(\beta)$ and $t^{(\pm)}(a_1, a_2)$:

$$\delta(\beta)\delta(\beta') = \delta(\beta + \beta') , \qquad (4.2.10)$$

$$t^{(\pm)}(a_1, a_2)t^{(\pm)}(b_1, b_2) = t^{(\pm)}(a_1 + b_1, a_2 + b_2) , \qquad (4.2.11)$$

$$\delta(\beta)t^{(\pm)}(a_1, a_2) = t^{(\pm)}(a_1 \cos\beta + a_2 \sin\beta, -a_1 \sin\beta + a_2 \cos\beta)\delta(\beta) . \quad (4.2.12)$$

These relations imply that $\delta(\beta)$ and $t^{(\pm)}(a_1, a_2)$ give a representation of the group formed by the rotation

$$\left.\begin{array}{l} x_1 \longrightarrow x_1 \cos\beta + x_2 \sin\beta , \\ x_2 \longrightarrow -x_1 \sin\beta + x_2 \cos\beta , \end{array}\right\} \qquad (4.2.13)$$

and the translation

$$\left.\begin{array}{l} x_1 \longrightarrow x_1 + a_1 , \\ x_2 \longrightarrow x_2 + a_2 , \end{array}\right\} \qquad (4.2.14)$$

in a 2-dimensional Euclidean space. This group is called a 2-dimensional Euclidean group, and is denoted as $E(2)$. To investigate the connectedness of the parameter space, it is sufficient to consider only the transformation (4.2.13) due to the same reason as in the case of the Poincaré group. The parameter space is a circle with radius 1, and the point of $\beta = 0$ corresponds to the identity transformation (Fig. 4.1).

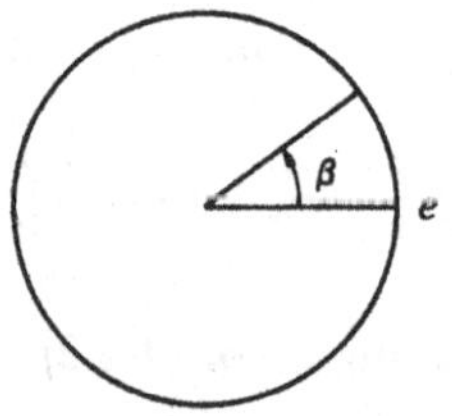

Fig. 4.1

In this space, there exists a closed curve which starts from e, turns back halfway and returns to e along the same path. There exists also a closed curve which starts from e to move round the circle several times and returns to e, where clockwise and anticlockwise windings are possible. The first kind of closed curve turning back along the same path is shrinkable to e by a continuous deformation, while other kinds of closed curve are not shrinkable.

There is no transition between the closed curves which have different winding numbers or different winding directions. That is, an infinite number of different kinds of closed curve is possible, which implies that the parameter space has a structure different from that of the rotation group. As a result, $E(2)$ has single-valued, double-valued,..., infinite number-valued representations. Among these representations, as has been mentioned in §3.4, only the single-valued and double-valued representations give the representations of the Poincaré group.

Let us construct the unitary representations of $E(2)$ by a method analogous to that of the unitary representations of the Poincaré group. The unitary representation of the translational transformation (4.2.14) is given by

$$t(a_1, a_2) = e^{i(\xi_1 a_1 + \xi_2 a_2)} , \tag{4.2.15}$$

where ξ_1 and ξ_2 are Hermitian operators commutable with each other, and correspond to k_μ in the Poincaré group. So they may be called "momenta" in the 2-dimensional space. Let $d(\beta)$ be a unitary representation of the 2-dimensional rotation (4.2.13). Since $\xi_1^2 + \xi_2^2$ commutes with both $t(a_1, a_2)$ and $d(\beta)$, it is simply a constant in an irreducible representation of $E(2)$. Then we have the following two kinds of irreducible representations:

$[0^0_\pm]$; $\xi_1^2 + \xi_2^2 = 0$,

$[\Xi^{\,0}_{\,\pm}]$; $\xi_1^2 + \xi_2^2 = \Xi > 0$,

where the signs $\pm$ correspond to $\mathrm{sgn}(k_0) = \pm$ respectively.

In the case of $[0^0_\pm]$, we have $\xi_1 = \xi_2 = 0$. Since

$$t(a_1, a_2) = 1 , \tag{4.2.16}$$

only $d(\beta)$ satisfying

$$d(\beta)d(\beta') = d(\beta + \beta') \tag{4.2.17}$$

is to be considered. It has only a unitary irreducible 1-dimensional representation which is written as (by Schur's lemma)

$$d(\beta) = e^{i\beta S} , \tag{4.2.18}$$

where S is an arbitrary real number, and different values of S give different irreducible representations. As has been mentioned, however, we are considering only single-valued and double-valued representations, then we can set

$$|S| = 0, \frac{1}{2}, 1, \frac{3}{2}, \cdots . \tag{4.2.19}$$

An integral value of S corresponds to a single-valued representation, and a half-integral S corresponds to a double-valued representation. If the value of S and $\mathrm{sgn}(k_0)$ are given, the corresponding unitary irreducible representation of the Poincaré group is fixed. A particle transforming according to one of these representations of the Poincaré group is called a massless particle with discrete spin. The spin used here is not directly connected to the conventional picture for the spin of a particle in the rest frame, which has been mentioned in the previous section. The physical content will be discussed in §5.4.

Next, we shall consider the case of $[\Xi^0_\pm]$. Since ξ will contribute in this case, the vector which forms the complete orthogonal basis of an irreducible representation space can be written as $|\,\bar{\xi},\bar{\sigma}\,\rangle$ analogously to the case of the Poincaré group. The variable $\bar{\xi}$ is the eigenvalue of ξ, namely $\xi_a|\,\bar{\xi},\bar{\sigma}\,\rangle = \bar{\xi}_a|\,\bar{\xi},\bar{\sigma}\,\rangle$ $(a = 1, 2)$, and the other variable $\bar{\sigma}$ was introduced to make $|\,\bar{\xi},\bar{\sigma}\,\rangle$ a complete basis. The variables $\bar{\xi}$ and $\bar{\sigma}$ correspond to $\bar{k}$ and $\bar{\xi}$ of the Poincaré group. Further, we assume that $|\,\bar{\xi},\bar{\sigma}\,\rangle$ is a single-valued continuous function of $\bar{\xi}$ and the multi-valuedness of the representation comes from $\bar{\sigma}$. If we introduce a polar coordinate to write $\xi_1 = \sqrt{\Xi}\cos\phi$, $\xi_2 = \sqrt{\Xi}\cos\phi$, we can use $|\,\bar{\phi},\bar{\sigma}\,\rangle$ in place of $|\,\bar{\xi},\bar{\sigma}\,\rangle$. By definition, it is a periodic function of period 2π with respect to ϕ:

$$|\,\bar{\phi}+2\pi,\bar{\sigma}\,\rangle = |\,\bar{\phi},\bar{\sigma}\,\rangle\,. \tag{4.2.20}$$

A 2-dimensional rotation operator $d(\beta)$ acting on the state is given by

$$d(\beta) = Q(\beta,\phi)P(\beta)\,, \tag{4.2.21}$$

$$P(\beta)|\,\bar{\phi},\bar{\sigma}\,\rangle = |\,\bar{\phi}-\beta,\bar{\sigma}\,\rangle \tag{4.2.22}$$

corresponding to (3.2.8). Noting that $Q(\beta',\phi)$ is a continuous function of ϕ, we have a similar equation to (3.2.4). Then the relation between $Q(\beta',\phi)$ and $P(\beta)$ can be written as

$$P(\beta)Q(\beta',\phi) = Q(\beta',\phi+\beta)P(\beta)\,. \tag{4.2.23}$$

Hence, from (4.2.17) we have

$$d(\beta+\beta') = d(\beta)d(\beta') = Q(\beta,\phi)Q(\beta',\phi+\beta)P(\beta+\beta')\,,$$

which leads to the equation for Q:

$$Q(\beta+\beta',\phi) = Q(\beta,\phi)Q(\beta',\phi+\beta)\,, \tag{4.2.24}$$

where $P(\beta)P(\beta') = P(\beta + \beta')$ has been used. Since $Q(\beta, \phi)$ commutes with ϕ, ϕ in (4.2.24) can be considered as a number like β. Thus, setting $\phi = 0$, we have

$$Q(\beta + \beta', 0) = Q(\beta, 0)Q(\beta', \beta) , \tag{4.2.25}$$

and performing a replacement $\beta' \to \beta, \beta \to \phi$, we obtain

$$Q(\beta, \phi) = Q(\phi, 0)^{-1}Q(\phi + \beta, 0) , \tag{4.2.26}$$

which is a general solution of (4.2.24). As a consequence, we can write

$$d(\beta) = Q(\phi, 0)^{-1}Q(\phi + \beta, 0)P(\beta) = Q(\phi, 0)^{-1}P(\beta)Q(\phi, 0) . \tag{4.2.27}$$

However, we should not consider this equation to imply unitary equivalence of $d(\beta)$ and $P(\beta)$, which may permit us to use the latter in place of the former. That is because, when $Q(\phi, 0)$ is not a periodic function of ϕ with period 2π, the new vector, which is unitary equivalent in the above sense, does not satisfy the condition (4.2.20).*)

We need only single and double-valued representations, and $P(\beta)$ is independent of the multi-valuedness of representation. Then $Q(\beta, \phi)$ is restricted to satisfy $Q(0, \phi) = Q(2\pi, \phi)$ in a single-valued representation, or $Q(0, \phi) = -Q(2\pi, \phi)$ in a double-valued representation. Therefore, putting $\beta = 2\pi$ in (4.2.26), we have

$$Q(\phi, 0) = Q(\phi + 2\pi, 0) \tag{4.2.28}$$

in a single-valued representation, which implies that $Q(\phi, 0)$ is a periodic function of period 2π, and this leads to the unitary equivalence of $d(\beta)$ and $P(\beta)$. Hence in a single-valued representation $P(\beta)$ can be used in place of $d(\beta)$:

$$d(\beta)|\bar{\phi}\rangle = |\bar{\phi} - \beta\rangle , \tag{4.2.29}$$

where $\bar{\sigma}$ has been omitted.

In a double-valued representation, according to (4.2.26), $Q(\phi, 0)$ satisfies

$$Q(\phi, 0) = -Q(\phi + 2\pi, 0) . \tag{4.2.30}$$

---

*)The operator $Q(\phi, 0)^{-1}$ corresponds to $Q(\alpha_k, k)$ in (3.3.8). But because of this reason, we do not use here the method of little group to construct the irreducible representations of $E(2)$.

If we introduce $\tilde{Q}(\phi)$ by

$$Q(\phi, 0) = e^{i\phi/2}\tilde{Q}(\phi) \,, \tag{4.2.31}$$

from (4.2.21), (4.2.26) and (4.2.30) we get

$$d(\beta) = \tilde{Q}(\phi)^{-1} e^{i\beta/2} P(\beta) \tilde{Q}(\phi) \,. \tag{4.2.32}$$

Since $\tilde{Q}(\phi)$ has a period 2π, we obtain finally

$$d(\beta) | \bar{\phi} \rangle = e^{i\beta/2} | \bar{\phi} - \beta \rangle \,, \tag{4.2.33}$$

where $\bar{\sigma}$ has been omitted because the value $\bar{\sigma}$ is unique. But it should be noticed that $| \bar{\phi} \rangle$ in the above equation is entirely different from that in (4.2.29). If we rewrite (4.2.29) and (4.2.33) in terms of $\bar{\xi}_1$ and $\bar{\xi}_2$, we obtain

$$d(\beta) | \bar{\xi}_1, \bar{\xi}_2 \rangle = | \bar{\xi}_1 \cos\beta + \bar{\xi}_2 \sin\beta, -\bar{\xi}_1 \sin\beta + \bar{\xi}_2 \cos\beta \rangle \tag{4.2.34}$$

for a single-valued representation, and

$$d(\beta) | \bar{\xi}_1, \bar{\xi}_2 \rangle_{\frac{1}{2}} = e^{i\beta/2} | \bar{\xi}_1 \cos\beta + \bar{\xi}_2 \sin\beta, -\bar{\xi}_1 \sin\beta + \bar{\xi}_2 \cos\beta \rangle_{\frac{1}{2}} \tag{4.2.35}$$

for a double-valued representation, where the index of the state vector indicates a double-valued representation explicitly. In place of (4.2.35), unitary equivalent

$$d(\beta) | \bar{\xi}_1, \bar{\xi}_2 \rangle_{-\frac{1}{2}} = e^{-i\beta/2} | \bar{\xi}_1 \cos\beta + \bar{\xi}_2 \sin\beta, -\bar{\xi}_1 \sin\beta + \bar{\xi}_2 \cos\beta \rangle_{-\frac{1}{2}} \tag{4.2.36}$$

is also possible.

The two kinds of unitary representation (4.2.34) and (4.2.35) (or (4.2.36)) are the irreducible representations of $E(2)$. If they were reducible, the contradiction is readily derived by the analogous argument as in §3.4.

Thus, when $\Xi > 0$, the Poincaré group has only two kinds of irreducible representations in each case of $\mathrm{sgn}(k_0) = \pm 1$. The different values of Ξ give different representations. The one-particle system transforming according to these representations is called a massless particle with continuous spin because the spin variable ξ has a continuous eigenvalue.*)

*) The continuous spin does not imply that the total angular momentum of the particle has a continuous eigenvalue. The value of the total angular momentum is an integer (> 0) in a single-valued representation, and half-integer in a double-valued representation (cf. §5.2 and §5.4).

§4.3 Lorentz Group

In the case of $[L]$, we have $k_\mu = 0$ ($\mu = 1, 2, 3, 4$) and $l_\mu = (0, 0, 0, 0)$. The little group is, therefore, nothing but the Lorentz group.

The unitary representation of the Lorentz group requires a somewhat complicated mathematical discussion, which is not the aim of this book.*) We investigate here only, what kind of unitary irreducible representation is possible.

The generators of the Lorentz group are **J** and **K** mentioned in §2.3, which satisfy the commutation relations (2.3.13), and which are Hermitian operators in a unitary representation. The operator **J** is, of course, the generator of the rotation group which is a subgroup of the Lorentz group. If a unitary irreducible representation of the Lorentz group is given, its representation space must have various subspaces which are irreducible representation spaces of the rotation group. Then choosing a subspace in which $\mathbf{J}^2$ has an eigenvalue $j(j+1)$, we denote a normalized state vector as $|\, j, \mu\,\rangle$ in this space, where μ is an eigenvalue of J_3 and takes a vlaue of $\mu = j, j-1, j-2, \ldots -j+1, -j$. If operators $J^{(\pm)} = J_1 \pm i J_2$ are introduced, we have the following equations which are familiar in the theory of angular momentum in quantum mechanics (cf. 4.1.7):

$$\left.\begin{aligned} J^{(+)}|\, j, \mu\,\rangle &= \sqrt{(j-\mu)(j+\mu+1)}\,|\, j, \mu+1\,\rangle \;, \\ J_3|\, j, \mu\,\rangle &= \mu\,|\, j, \mu\,\rangle \;, \\ J^{(-)}|\, j, \mu\,\rangle &= \sqrt{(j+\mu)(j-\mu+1)}\,|\, j, \mu-1\,\rangle \;. \end{aligned}\right\} \tag{4.3.1}$$

Although the matrix elements of **J** can be calculated from these equations, we must further calculate the matrix elements of **K**. Introducing $K^{(\pm)}$ defined by

$$K^{(\pm)} = K_1 \pm i K_2 \;, \tag{4.3.2}$$

we use $K^{(\pm)}$ and K_3, instead of K, which satisfy

$$\left.\begin{aligned} [K^{(+)}, K_3] &= J^{(+)} \;, \\ [K^{(-)}, K_3] &= -J^{(-)} \;, \\ [K^{(+)}, K^{(-)}] &= -2J_3 \;. \end{aligned}\right\} \tag{4.3.3}$$

To simplify our discussion, we assume that an irreducible representation space of the Lorentz group has at most one irreducible representation space of the rotation group which is specified by j.**)

---

*)For details, see reference [6] at the end of the volume.

**)This is not an assumption, but has to be proved. The proof however is omitted here. See the references [6] or [7] at the end of the volume.

The operator **K** is transformed by the spatial rotation as a 3-vector, that is, transformed in a similar manner to $\mathbf{x}/|\mathbf{x}|$. The normalized eigenfunctions of angular momentum 1 are expressed by $Y_1(\theta,\phi)$ $(m=1,0,-1)$ in quantum mechanics:

$$\left.\begin{aligned} Y_1^1(\theta,\phi) &= -\sqrt{\frac{3}{8\pi}}\,\frac{x_1+ix_2}{|\mathbf{x}|}\,, \\ Y_1^0(\theta,\phi) &= \sqrt{\frac{3}{4\pi}}\,\frac{x_3}{|\mathbf{x}|}\,, \\ Y_1^{-1}(\theta,\phi) &= \sqrt{\frac{3}{8\pi}}\,\frac{x_1-ix_2}{|\mathbf{x}|}\,. \end{aligned}\right\} \tag{4.3.4}$$

Then $-K^{(+)}/\sqrt{2}$, K_3 and $K^{(-)}/\sqrt{2}$ are transformed by the spatial rotation as $Y_1^1(\theta,\phi)$, $Y_1^0(\theta,\phi)$ and $Y_1^{-1}(\theta,\phi)$ respectively. Therefore the composition law of angular momentum gives

$$\begin{aligned} K^{(+)}|\,j,\mu\,\rangle = &-a_j\sqrt{(j+\mu+1)(j+\mu+2)}\,|\,j+1,\mu+1\,\rangle \\ &+b_j\sqrt{(j+\mu+1)(j-\mu)}\,|\,j,\mu+1\,\rangle \\ &-c_j\sqrt{(j-\mu-1)(j-\mu)}\,|\,j-1,\mu+1\,\rangle\,, \end{aligned} \tag{4.3.5}$$

$$\begin{aligned} K_3|\,j,\mu\,\rangle = &\,a_j\sqrt{(j-\mu+1)(j+\mu+1)}\,|\,j+1,\mu\,\rangle \\ &+b_j\mu|\,j,\mu\,\rangle - c_j\sqrt{j^2-\mu^2}\,|\,j-1,\mu\,\rangle\,, \end{aligned} \tag{4.3.6}$$

$$\begin{aligned} K^{(-)}|\,j,\mu\,\rangle = &\,a_j\sqrt{(j-\mu+1)(j-\mu+2)}\,|\,j+1,\mu-1\,\rangle \\ &+b_j\sqrt{(j-\mu+1)(j+\mu)}\,|\,j,\mu-1\,\rangle \\ &+c_j\sqrt{(j+\mu-1)(j+\mu)}\,|\,j-1,\mu-1\,\rangle\,, \end{aligned} \tag{4.3.7}$$

where the coefficients with square root on the right-hand sides are known as Clebsch-Gordan coefficients*) which appear in the theory for addition of angular momenta. In the above expressions the factors that depend on j only have been included in a_j, b_j and c_j. The unfixed factors a_j, b_j and c_j appear because $K^{(\pm)}$ and K_3 are not yet normalized like $Y_1^m(\theta,\phi)$, and they must be fixed by the condition of unitarity of the representations.

*)See p. 117 of the reference [5] at the end of the volume.

The phase factor $e^{i\delta_j}$ of $|\,j,\mu\rangle$, which depends on j only, cannot be fixed by (4.3.1) and can be taken as arbitrary. This freedom permits us to take c_j to be purely imaginary:

$$c_j^* = -c_j \qquad (-ic_j \geq 0)\,. \tag{4.3.8}$$

Since $K^{(+)}$ is the Hermitian conjugate of $K^{(-)}$, the following relations hold:

$$\left.\begin{aligned} \langle\, j+1,\mu+1\,|\,K^{(+)}\,|\,j,\mu\,\rangle^* &= \langle\, j,\mu\,|\,K^{(-)}\,|\,j+1,\mu+1\,\rangle\,, \\ \langle\, j,\mu+1\,|\,K^{(+)}\,|\,j,\mu\,\rangle^* &= \langle\, j,\mu\,|\,K^{(-)}\,|\,j,\mu+1\,\rangle\,, \\ \langle\, j-1,\mu+1\,|\,K^{(+)}\,|\,j,\mu\,\rangle^* &= \langle\, j,\mu\,|\,K^{(-)}\,|\,j-1,\mu+1\,\rangle\,. \end{aligned}\right\} \tag{4.3.9}$$

Substituting (4.3.5) and (4.3.7) into (4.3.9), and using (4.3.8), we obtain

$$b_j = b_j^* \qquad (j > 0)\,, \tag{4.3.10}$$

$$a_j = c_{j+1} \qquad (j \geq 0)\,. \tag{4.3.11}$$

The same relations are also derived from Hermiticity of K_3. These relations confirm the unitarity of the representation. The above relations tell us nothing about b_0, c_0 and c_{12}. But, we do not need such an information because the Clebsch-Gordan coefficients of the terms involving b_0, c_0 or $c_{1/2}$ in (4.3.5)–(4.3.7) all vanish. Calculating $[K^{(+)}, K^{(-)}]|\,j,\mu\,\rangle$ with the help of (4.3.5) and (4.3.7) we obtain

$$\begin{aligned} [K^{(+)}, K^{(-)}]|\,j,\mu\,\rangle =& 2\sqrt{(j+1)^2-\mu^2}\,c_{j+1}\{(j+2)b_{j+1} - jb_j\}|\,j+1,\mu\,\rangle \\ &+ 2\mu\{(2j+3)c_{k+1}^2 + b_j^2 - (2j-1)c_j^2\}|\,j,\mu\,\rangle \\ &- 2\sqrt{j^2-\mu^2}\,c_j\{(j+1)b_j - (j-1)b_{j-1}\}|\,j-1,\mu\,\rangle\,. \end{aligned} \tag{4.3.12}$$

Hence, applying the third equation of (4.3.3) to the above expression, we get

$$c_{j+1}\{(j+2)b_{j+1} - jb_j\} = 0 \qquad (j \geq 0)\,, \tag{4.3.13}$$

$$-(2j+3)c_{j+1}^2 - b_j^2 + (2j-1)c_j^2 = 1 \qquad (j > 0)\,. \tag{4.3.14}$$

The same relations are also derived from the first and the second equations of (4.3.3). Now we have used all properties of **J** and **K**. By the use of these results, let us determine b_j and c_j.

Let j_0 be the minimum value of j which appears in an irreducible representation of the Lorentz group.

First, we consider the case $j_0 > 0$. The definition of j_0 and (4.3.5)–(4.3.7) lead to $c_{j_0} = 0$. This relation holds for $j_0 > 1/2$ only, and we need no information about $c_{1/2}$ as has been noted. In fact, the term involving $c_{1/2}$ vanishes in (4.3.14). Here let $c_{j_0+1}, c_{j_0+2}, \ldots c_j \neq 0$, then (4.3.13) gives

$$b_j = \frac{j-1}{j+1} b_{j-1} = \frac{j_0(j_0+1)}{j(j+1)} b_{j_0} \,. \tag{4.3.15}$$

If

$$(j_0+1)b_{j_0} = i\nu \,, \tag{4.3.16}$$

we have

$$b_j = \frac{i\nu j_0}{j(j+1)} \,. \tag{4.3.17}$$

Since b_j is real due to (4.3.10), ν becomes purely imaginary or zero. Substituting (4.3.17) into (4.3.14), and multiplying both sides by $(2j+1)$ we get

$$-\{4(j+1)^2-1\}c_{j+1}^2+(4j^2-1)c_j^2 = (2j+1)-\nu^2 j_0^2\left\{\frac{1}{j^2}-\frac{1}{(j+1)^2}\right\} \,. \tag{4.3.18}$$

Summing up each side with respect to j from j_0 to j and rearranging, we obtain

$$c_{j+1}^2 = -\frac{\{(j+1)^2 - j_0^2\}\{(j+1)^2 - \nu^2\}}{\{4(j+1)^2-1\}(j+1)^2} \tag{4.3.19}$$

which is not zero. On the other hand, since (4.3.14) gives $c_{j_0+1}^2 = -(1 + b_{j_0}^2)/(2j_0+3) \neq 0$, the mathematical induction for j, satisfying $j \geq j_0+1 > 1$, shows that c_j does not vanish and is a purely imaginary number given by (4.3.19).

Next, let us consider the case $j_0 = 0$. In this case, we have $c_1 = 0$ or $b_1 = 0$, if we set $j = j_0 = 0$ in (4.3.13).

When $c_1 = 0$, from (4.3.11) we have $a_0 = 0$. Therefore, the right-hand sides of (4.3.1), (4.3.5)–(4.3.7) all vanish if $j = \mu = 0$. This implies that $J_i = K_i = 0$ $(i = 1, 2, 3)$, namely the 1-dimensional representation in which every element of the Lorentz group is equal to unity.

When $c_1 \neq 0$ and $b_1 = 0$, assuming again $c_1, c_2, \ldots c_j \neq 0$, we get $b_1 = b_2 = \ldots = b_j = 0$ from (4.3.13). Substituting these into (4.3.14), and multiplying each side by $(2j+1)$ and summing up with respect to j from 1 to j, we obtain, in a similar manner to (4.3.19),

$$-c_{j+1}^2 = \frac{j(j+2) - 3c_1^2}{4(j+1)^2 - 1} \,. \tag{4.3.20}$$

Here we set

$$-c_1^2 = \frac{1-\nu^2}{3} \,. \tag{4.3.21}$$

By definition c_1 is nonzero, and is chosen to be purely imaginary as has been mentioned. Then

$$1-\nu^2 > 0 \,. \tag{4.3.22}$$

Substituting (4.3.21) into (4.3.20) we get

$$-c_{j+1}^2 = \frac{(j+1)^2-\nu^2}{4(j+1)^2-1} \,. \tag{4.3.23}$$

That is, if $c_1 = c_2 = \ldots c_j \neq 0$ then $b_1 = b_2 = \ldots b_j = 0$, which gives $c_{j+1} \neq 0$. Therefore, from (4.3.13) we get $b_{j+1} = 0$, and hence according to the mathematical induction we obtain

$$b_j = 0 \tag{4.3.24}$$

for all $j > 0$. Equations (4.3.23) and (4.3.24) are nothing but (4.3.17) and (4.3.19) with $j_0 = 0$, which were derived for $j_0 > 0$. That is, we can conclude that (4.3.17) and (4.3.19) hold for any j_0. Therefore, in general,

$$a_{j-1} = c_j = \frac{i}{j}\sqrt{\frac{(j^2-j_0^2)(j^2-\nu^2)}{4j^2-1}} \quad (j \geq 1) \,, \tag{4.3.25}$$

$$b_j = \frac{i\nu j_0}{j(j+1)} \qquad (j > 0) \,. \tag{4.3.26}$$

The information about b_0, c_0 and $c_{1/2}$ is not needed as has been noted. Thus the matrix elements of $\mathbf{K}$ have been completely determined.*)

In the case $j_0 = 0$, from (4.3.22), ν is either a purely imaginary number or a real number satisfying $0 < \nu^2 < 1$. Then, as is seen from (4.3.25) and (4.3.26), both ν and $-\nu$ give the same representation. When ν is on the imaginary axis in the complex plane, we say that the unitary irreducible representation belongs to the principal series, and j_0 takes values of $j_0 = 0, 1/2, 1, 3/2, \ldots$. When ν is on the real axis and $0 < \nu^2 < 1$, we say that the unitary irreducible representation belongs to the supplementary series, and in this case, $j_0 = 0$ (Fig. 4.2). Further, when $j_0 = 0$ and $\nu^2 = 1$, we get $c_1 = 0$ which represents

*)To see the irreducibility of the given unitary representation, we can use Schur's lemma. That is, we can show that an operator commutable with both J_i and K_i $(i = 1, 2, 3)$ is only a constant.

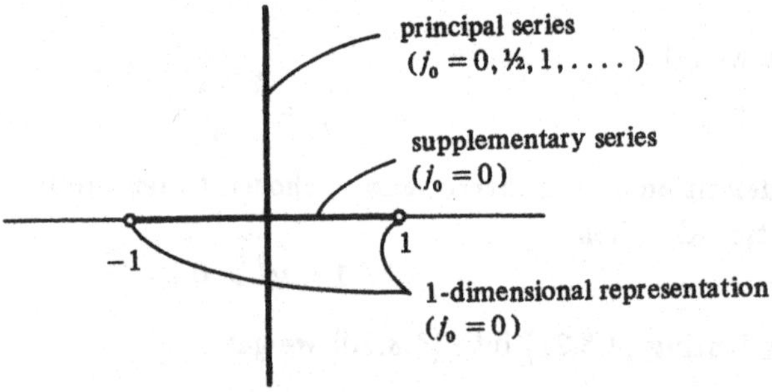

Fig. 4.2

the 1-dimensional representation mentioned previously. As is clear from the above discussion, except for the simplest 1-dimensional representation, all the unitary representations are infinite-dimensional.

Using the a_j, b_j and c_j determined above, and calculating $I_a \mid j_0, j_0 \rangle$ ($a = 1, 2$) we can show that the eigenvalues of I_1 and I_2 in (2.3.18) and (2.3.19) are given by

$$I_1 = j_0^2 + \nu^2 - 1 \, , \tag{4.3.27}$$

$$I_2 = i\nu j_0 \, . \tag{4.3.28}$$

§4.4 Three-Dimensional Lorentz Group

In the class $[T]$, we have $k_\mu = m^2 > 0$ and we take

$$l_\mu = (0, 0, m, 0) \, . \tag{4.4.1}$$

The subgroup of the Lorentz group which leaves l_μ invariant is given by the transformation:

$$\left.\begin{aligned} x'_\alpha &= \sum_\beta V_{\alpha\beta} x_\beta \qquad (\alpha, \beta = 1, 2, 4) \, , \\ x_4 &= i x_0 \, . \end{aligned}\right\} \tag{4.4.2}$$

The 3×3 matrix V satisfies

$$\left.\begin{aligned} &\sum_\beta V_{\alpha\beta} V_{\gamma\beta} = \delta_{\alpha\gamma} \, , \\ &\det(V) = 1 \, , \\ &V_{44} \geq 0 \, , \end{aligned}\right\} \tag{4.4.3}$$

where V_{11}, V_{12}, V_{21} and V_{44} are real, and V_{14}, V_{24}, V_{41} and V_{42} are purely imaginary. The transformation (4.4.2) leaves $x_1^2 + x_2^2 - x_0^2$ invariant, and the group is called a 3-dimensional Lorentz group.

According to the argument of §2.3, we shall construct infinitesimal transformations to obtain three generators H_0, H_1 and H_2, where H_0 is the generator of rotation in the plane formed by the first and the second axes, and H_1 and H_2 are the generators of (3-dimensional) Lorentz transformations in the directions of the first and the second axes respectively. These generators satisfy the commutation relations

$$\left.\begin{aligned} [H_1, H_2] &= -iH_0 \ , \\ [H_1, H_0] &= -iH_2 \ , \\ [H_2, H_0] &= iH_1 \ , \end{aligned}\right\} \tag{4.4.4}$$

and they are all Hermitian operators in a unitary representation.

By a similar procedure used in §2.1 to derive (2.1.4), any V satisfying (4.4.2) and (4.4.3) is written as

$$V = \gamma(\beta) b(\tau) \gamma(\beta') \ , \tag{4.4.5}$$

where

$$\gamma(\beta) = \begin{pmatrix} \cos\beta & \sin\beta & 0 \\ -\sin\beta & \cos\beta & 0 \\ 0 & 0 & 1 \end{pmatrix} \ ,$$

$$b(\tau) = \begin{pmatrix} \cosh\tau & 0 & i\sinh\tau \\ 0 & 1 & 0 \\ -\sinh\tau & 0 & \cosh\tau \end{pmatrix} \ .$$

The matrix $\gamma(\beta)$ is a rotation in the 1-2 plane, and $b(\tau)$ represents a Lorentz transformation in the direction of the first axis. Therefore the multi-valuedness of the 3-dimensional Lorentz group comes from the 2-dimensional rotation group. Although this group, as has been noted in §4.2, has an infinite number of multi-valued representations, we need only single-valued and double-valued representations, i.e., the eigenvalue of H_0 takes integral and half-integral (including negative) values. We shall hereafter restrict ourselves to these cases.*)

The Casimir operator of this group is

$$I = H_0^2 - H_1^2 - H_2^2 \ , \tag{4.4.6}$$

*) This can be proved by a direct calculation (see the end of §5.3).

and its eigenvalue q is fixed if an irreducible representation is given. The complete set of basic vectors, by which the irreducible representation space is spanned, is characterized by an eigenvalue μ of H_0. Hence there cannot exist two independent vectors which have the same value of μ in an irreducible representation space.**)

Let $|q,\mu\rangle$ be a vector spanning the irreducible representation space:

$$I|q,\mu\rangle = q|q,\mu\rangle\,, \tag{4.4.7}$$
$$H_0|q,\mu\rangle = \mu|q,\mu\rangle\,, \tag{4.4.8}$$
$$\langle q,\mu|q,\mu'\rangle = \delta_{\mu\mu'}\,, \tag{4.4.9}$$

where we do not yet know what values of μ are possible in the irreducible representation space. This is a problem to be solved hereafter.

Define $H^{(\pm)}$ by

$$H^{(\pm)} = H_1 \pm iH_2\,, \tag{4.4.10}$$

then the commutation relations can be written as

$$[H_0, H^{(\pm)}] = \pm H^{(\pm)}\,, \tag{4.4.11}$$
$$[H^{(+)}, H^{(-)}] = -2H_0\,, \tag{4.4.12}$$

and we have

$$H^{(\mp)}H^{(\pm)} = -I + H_0(H_0 \pm 1)\,. \tag{4.4.13}$$

Using (4.4.11) we get

$$H_0H^{(\pm)}|q,\mu\rangle = (\mu \pm 1)H^{(\pm)}|q,\mu\rangle\,, \tag{4.4.14}$$

that is, $H^{(+)}$ is an operator raising the value of μ by 1, and $H^{(-)}$ is an operator lowering the value of μ by 1. Therefore the values of μ in an irreducible representation are either all integers or all half-integers. Setting

$$H^{(\pm)}|q,\mu\rangle = c_{\pm}(q,\mu)|q,\mu \pm 1\rangle\,, \tag{4.4.15}$$

and estimating its norm we obtain

$$\begin{aligned} |c_{\pm}(q,\mu)|^2 &= \langle q,\mu|H^{(\mp)}H^{(\pm)}|q,\mu\rangle \\ &= \left(\mu \pm \frac{1}{2}\right)^2 - \left(q + \frac{1}{4}\right) \geq 0\,, \end{aligned} \tag{4.4.16}$$

)This is not necessarily self-evident, but the proof is omitted. The theory of unitary representations of 3-dimensional Lorentz group was accomplished for the first time by V. Bargmann. For details, see V. Bargmann: *Ann. Math.* **48, 568 (1947).

where we have used the fact that $H^{(+)}$ is the Hermitian conjugate of $H^{(-)}$, and (4.4.13). First, assume $c_\pm(q,\mu) \neq 0$ for any μ. If μ is an integer, the minimum value of $(\mu \pm 1/2)^2$ is 1/4, then

$$q < 0 \,. \tag{4.4.17}$$

If μ is a half-integer, the minimum value of $(\mu \pm 1/2)^2$ is 0, then

$$q < -1/4 \,. \tag{4.4.18}$$

These two representations are called continuous representations since q has a continuous value, and the irreducible representations of (4.4.17) and (4.4.18) are represented by C_q^0 and $C_q^{\frac{1}{2}}$ respectively.

Next, let us consider the case where $c_+(q,\mu)$ vanishes at $\mu = \mu_0$. Namely let

$$c_+(q,\mu_0) = 0 \,. \tag{4.4.19}$$

Then, from (4.4.16) we get

$$q = \mu_0(\mu_0 + 1) \,. \tag{4.4.20}$$

Equations (4.4.15) and (4.4.19) show that μ has the maximum value μ_0 in the representation: $\mu \leq \mu_0$. Under this condition, substituting (4.4.20) into (4.4.16) we can derive

$$|c_\pm(q,\mu)|^2 = \mu^2 \pm \mu - \mu_0^2 - \mu_0 \geq 0 \qquad (\mu \leq \mu_0) \,. \tag{4.4.21}$$

Hence setting $\mu = \mu_0$ in the above expression, we get

$$|c_-(q,\mu)|^2 = -2\mu_0 \geq 0 \,, \tag{4.4.22}$$

which gives $\mu_0 \leq 0$.

When $\mu_0 = 0$, (4.4.20) gives $q = 0$, and since $c_\pm(0,0) = 0$, we get $H_1 = H_2 = H_0 = 0$, which represents 1-dimensional representation.

In the case $\mu_1 < 0$, from (4.4.21) we have

$$|c_-(q,\mu)|^2 = (\mu + \mu_0)(\mu - \mu_0 - 1) > 0 \quad (\mu \leq \mu_0 < 0) \,, \tag{4.4.23}$$

$$|c_+(q,\mu)|^2 = (\mu - \mu_0)(\mu + \mu_0 + 1) > 0 \quad (\mu < \mu_0 < 0) \,. \tag{4.4.24}$$

This representation is denoted as $D_{\mu_0}^{(-)}$.

A similar argument is applicable to the case where

$$c_-(q,\mu_0) = 0\ , \tag{4.4.25}$$

and μ has the minimum value. In this case, from (4.4.16) we get

$$q = \mu_0(\mu_0 - 1)\ , \tag{4.4.26}$$

which gives

$$|c_+(q,\mu_0)|^2 = 2\mu_0 \geq 0\ . \tag{4.4.27}$$

When $\mu_0 = 0$, we have $q = 0$ and $c_\pm(0,0) = 0$, which is nothing but the previously given 1-dimensional representation. Hence, if $\mu_0 > 0$, (4.4.16) leads to

$$|c_+(q,\mu)|^2 = (\mu + \mu_0)(\mu - \mu_0 + 1) > 0 \quad (\mu \geq \mu_0 > 0)\ , \tag{4.4.28}$$

$$|c_-(q,\mu)|^2 = (\mu - \mu_0)(\mu + \mu_0 - 1) > 0 \quad (\mu > \mu_0 > 0)\ . \tag{4.4.29}$$

This expression completely determines the representation, which is denoted as $D^{(+)}_{\mu_0}$. Both $D^{(+)}_{\mu_0}$ and $D^{(-)}_{\mu_0}$ have a discrete value of q, and are called discrete representations.

The above discussion has given the single-valued and double-valued unitary irreducible representations of the 3-dimensional Lorentz group.*) The representations are summarized in Table 4.1, except for the 1-dimensional representation ($q = \mu = \mu_0 = 0$).

Table 4.1

	Representation	q	μ_0	μ	$c_\pm(q,\mu)$
continuous	$C_q^{\,0}$	$q<0$		$\ldots,-2,-1,0,1,2,\ldots$	$\sqrt{\mu(\mu\pm1)-q}$
	$C_q^{\,1/2}$	$q<-\frac{1}{4}$		$\ldots,-\frac{3}{2},-\frac{1}{2},\frac{1}{2},\frac{3}{2},\ldots$	$\sqrt{\mu(\mu\pm1)-q}$
discrete	$D_{\mu_0}^{\,(+)}$	$\mu_0(\mu_0-1)$	$\frac{1}{2},1,\frac{3}{2},\ldots$	$\mu_0,\ \mu_0+1,\ \mu_0+2,\ \mu_0+3,\ldots$	$\sqrt{(\mu\pm\mu_0)(\mu\mp\mu_0\pm1)}$
	$D_{\mu_0}^{\,(-)}$	$\mu_0(\mu_0+1)$	$-\frac{1}{2},-1,-\frac{3}{2},\ldots$	$\mu_0,\ \mu_0-1,\ \mu_0-2,\ \mu_0-3,\ldots$	$\sqrt{(\mu\mp\mu_0)(\mu\pm\mu_0\pm1)}$

*) The irreducibility of these representations is easily proved by Schur's lemma.

As is seen from the table, all the unitary representations are infinite-dimensional representations except for the 1-dimensional representation.

§4.5 Classification of Free Particles

In the discussions up to the previous section, we investigated what is the unitary irreducible representation of the little group for each class of k_μ, and we obtained all possible classes of unitary irreducible representations of the Poincaré group. The results will be summarized here briefly and some supplementary explanation will be given. In table 4.2, as has been noted, the freedom of spin means the freedom of ξ which is required, in addition to k_μ, to specify the state vector in an irreducible representation space.

Table 4.2

class	energy-momentum vector	little group	spin freedom
$[M_\pm]$	$k_0^2 - \mathbf{k}^2 = m^2 > 0$, $\mathrm{sgn}\,(k_0) = \begin{cases} 1 \\ -1 \end{cases}$	3-dimensional rotation group	finite dimension
$[0_\pm^0]$	$k_0^2 = \mathbf{k}^2 > 0$, $\mathrm{sgn}\,(k_0) = \begin{cases} 1 \\ -1 \end{cases}$	2-dimensional rotation group	1-dimension (discrete spin)
$[\Xi_\pm^0]$	$k_0^2 = \mathbf{k}^2 > 0$, $\mathrm{sgn}\,(k_0) = \begin{cases} 1 \\ -1 \end{cases}$	2-dimensional Euclidean group ($\Xi > 0$)	infinite dimension (continuous spin)
$[L]$	$k_\mu = 0$	Lorentz group	infinite dimension*)
$[T]$	$\mathbf{k}^2 - k_0^2 = m^2 > 0$	3-dimensional Lorentz group	infinite dimension*)

*) 1-dimension only for the identity representation.

In the classes of $[M_\pm], [0_\pm^0]$ and $[\Xi_\pm^0]$, there exist irreducible representations with $\mathrm{sgn}(k_0) = \pm 1$. If $\mathrm{sgn}(k_0) = 1$, a particle has a positive energy, but if $\mathrm{sgn}(k_0) = -1$, a particle has a negative energy, which causes the difficulty of physical interpretation of the one-particle state. Even if we reinterpreted it by Dirac's hole theory, in which the vacuum is filled with negative energy particles, we cannot have any picture of one-particle state which is realized when particles are separated far away from one another as has been mentioned in §2.1, and then the system becomes essentially a multi-particle system. To persist on the hole theory, we must assume all particles to be Fermions, but it is impossible because there actually exist particles like photons which are not Fermions. Although the hole theory played an epoch-making important role heuristically, its direct application clearly has its limitation. The proper interpretation of the case $\mathrm{sgn}(k_0) = -1$ becomes possible in quantum field theory which will be mentioned in Chapter 8. Hereafter we avoid calling k_0 energy if not necessary. The quantity k_0 is definitely energy when $\mathrm{sgn}(k_0) = 1$, but it is not appropriate to call it energy when $\mathrm{sgn}(k_0) = -1$. Then we call abstractly the states of $\mathrm{sgn}(k_0) = 1, -1$ positive and negative frequency states respectively. The latter does not describe, at this stage, any real particle, however in quantum field theory it will play an important role together with the former.

Almost all the free particles found in nature, including mesons, electrons and protons etc., belong to the class $[M_+]$ if they are regarded as stable. Only photons and probably neutrinos are considered to belong to the class $[0^0_+]$. The spin of the former is 1, and that of the latter is 1/2. No particle has been found belonging to the class $[\Xi^0_+]$. However, if such particles were found, and if they could be confined in a heat bath, peculiar phenomena such as infinite specific heat are conceivable.

In the class $[L]$, we have $k_\mu = 0$, and we have no ordinary particle picture. Such an object with zero energy and zero momentum is called a spurion in particle physics, and is sometimes imagined to be some kind of "particle". But it is in imagination only. It is a future problem to investigate the meaning of the obtained states, especially in an infinite-dimensional representation, even if they are not particles.

There is no evidence of particles belonging to the class $[T]$. They have imaginary masses and are usually called tachyons. This naming is from $\tau\alpha\chi\iota\sigma$ (fast) in Greek. If we consider a matter wave $\exp\{i(\bar{\mathbf{k}}\mathbf{x} - \bar{k}_0 t)\}$ with $\bar{\mathbf{k}}^2 = \bar{k}_0^2 + m^2$, its group velocity certainly exceeds the light velocity: $|\partial_0 \bar{k}/\partial \bar{\mathbf{k}}| > 1$. The tachyons have no reason to be excluded in such a large framework like the theory of unitary irreducible representation of the Poincaré group. However, if we try to consider them in a similar manner to the ordinary particles in quantum field theory, we will encounter troubles in the definition of vacuum and in the quantization of fields. We do not have any consistent theory yet, even in the simplest case of 1-dimensional representation of the little group. We have no knowledge whether such particles have any reality or not. Clarifying the properties of these particles is the subject for a future study.

Chapter 5

Wigner Rotations

§5.1 Particles with Finite Mass

In the discussions up to the previous section, we investigated the problem of what kinds of free particles are possible. However, to see how the state vectors transform under the Lorentz transformation, we have to get the explicit form of the Wigner rotation $Q(\lambda_k, l)$. In the case of $[L]$, we have $k_\mu = 0$, and in §4.3 we have already obtained the form of Q for an infinitesimal Lorentz transformation. Therefore we shall discuss the other cases here.

We shall use a wave function in momentum representation instead of a state vector $|\ \rangle$ hereafter, because they are convenient for the later discussions in connection to quantum field theory.

Let $|\ \rangle$ be a state vector belonging to an irreducible representation space of the Poincaré group. The transformation under the Lorentz group is given by

$$|\ \rangle' = Q(\lambda_k, l)P(\Lambda)|\ \rangle \tag{5.1.1}$$

according to (3.3.13). On the other hand, the wave functions in momentum representation are usually defined by

$$\phi_\xi^{(\pm)}(\mathbf{k}) = \langle\, k, \xi, \pm | \pm \rangle \tag{5.1.2}$$

corresponding to $|\pm\rangle$ in $[M_\pm]$, where k and ξ should actually be written as $\bar{k}$ and $\bar{\xi}$. The argument of $\phi_\xi^{(\pm)}$ is written as $\mathbf{k}$ since the value of k_0 is automatically fixed if **k** is given. The function $\phi_\xi^{(\pm)}(\mathbf{k})$ transforms as

$$\begin{aligned}\phi_\xi^{(\pm)'}(\mathbf{k}) &= \langle \mathbf{k}, \xi, \pm | \pm \rangle' \\ &= \sum_{\xi'} \int \frac{d\mathbf{k}'}{\omega_{k'}} \langle \mathbf{k}, \xi, \pm | Q(\lambda_k, l) | \mathbf{k}', \xi', \pm \rangle \langle \mathbf{k}', \xi', \pm | P(\Lambda) | \pm \rangle ,\end{aligned} \tag{5.1.3}$$

which is derived from (5.1.1) and (3.1.15). Since $Q(\lambda_k, l)$ is diagonal with respect to k, using (3.1.8) we have

$$\langle \mathbf{k}, \xi, \pm | Q(\lambda_k, l) | \mathbf{k}', \xi', \pm \rangle = \omega_k Q(\lambda_k, l)_{\xi\xi'} \delta(\mathbf{k} - \mathbf{k}') . \tag{5.1.4}$$

The relation $P(\Lambda)^\dagger = P(\Lambda^{-1})$ leads to

$$\langle \mathbf{k}, \xi, \pm | P(\Lambda) | \pm \rangle = \phi_\xi^{(\pm)}(\Lambda^{-1}\mathbf{k}) . \tag{5.1.5}$$

Thus the transformation of the wave function is given by

$$\phi_\xi^{(\pm)'}(\mathbf{k}) = \sum_{\xi'} Q(\lambda_k, l)_{\xi\xi'} \phi_{\xi'}^{(\pm)}(\Lambda^{-1}\mathbf{k}) , \tag{5.1.6}$$

where the symbol $\Lambda^{-1}\mathbf{k}$ has been defined in §3.1. The inner product of $|1, \pm\rangle$ and $|2, \pm\rangle$ is written, in terms of the corresponding wave functions $\phi_\xi^{(\pm)}(\mathbf{k})_1$ and $\phi_\xi^{(\pm)}(\mathbf{k})_2$, in the form

$$\langle 1, \pm | 2, \pm \rangle = \sum_\xi \int \frac{d\mathbf{k}}{\omega_k} \phi_\xi^{(\pm)*}(\mathbf{k})_1 \phi_\xi^{(\pm)}(\mathbf{k})_2 . \tag{5.1.7}$$

A similar argument on the relations between the state vectors and the corresponding wave functions is also applicable in the cases of $[0_\pm], [\Xi_\pm]$ and $[T]$.

Now let us try to construct $Q(\lambda_k, l)$ in the case of $[M_\pm]$, that is, in the case of finite mass. Our consideration is restricted to the infinitesimal Lorentz transformations. The result will readily lead to the Wigner rotation with finite Λ, which, for the present, is not necessary in this book.

To simplify our argument, we shall divide the infinitesimal transformation given by (2.3.9) into two parts. The first part consists of the infinitesimal rotations around the first, second and third axes (Fig. 2.4), which transform a 4-vector x_μ, according to (2.3.2)–(2.3.4), into

$$\begin{aligned}\mathbf{x}' &= \mathbf{x} + \mathbf{x} \times \boldsymbol{\theta} , \\ x_0' &= x_0 .\end{aligned} \tag{5.1.8}$$

The second part consists of the infinitesimal Lorentz transformations in the directions of the first, second and third axes (Fig. 2.5), which are written, according to (2.3.8)–(2.3.10), in the form

$$\left.\begin{aligned} \mathbf{x}' &= \mathbf{x} - \boldsymbol{\tau} x_0 \ , \\ x_0' &= x_0 - \boldsymbol{\tau}\mathbf{x} \ . \end{aligned}\right\} \tag{5.1.9}$$

We shall call (5.1.8) and (5.1.9) θ-transformation and τ-transformation respectively. The matrices of the Lorentz transformations are denoted as $\Lambda(\theta)$ and $\Lambda(\tau)$, and corresponding λ_k's are expressed as $\lambda_k(\theta)$ and $\lambda_k(\tau)$.

To get the Wigner rotations corresponding to each transformation, we must calculate

$$\lambda_k(\theta) = \alpha_k^{-1}\Lambda(\theta)\alpha_{\Lambda(\theta)^{-1}k} \ , \tag{5.1.10}$$

$$\lambda_k(\tau) = \alpha_k^{-1}\Lambda(\tau)\alpha_{\Lambda(\tau)^{-1}k} \ , \tag{5.1.11}$$

according to (3.3.4). If we now adopt (4.1.2) for l_μ, α_k satisfying (3.3.2) is given by

$$\alpha_k = \begin{pmatrix} 1+\rho_k k_1^2 & \rho_k k_1 k_2 & \rho_k k_1 k_3 & \mp i\,\frac{k_1}{m} \\ \rho_k k_2 k_1 & 1+\rho_k k_2^2 & \rho_k k_2 k_3 & \mp i\,\frac{k_2}{m} \\ \rho_k k_3 k_1 & \rho_k k_3 k_2 & 1+\rho_k k_3^2 & \mp i\,\frac{k_3}{m} \\ \pm i\,\frac{k_1}{m} & \pm i\,\frac{k_2}{m} & \pm i\,\frac{k_3}{m} & \frac{\omega_k}{m} \end{pmatrix} , \tag{5.1.12}$$

where ρ_k is defined by

$$\rho_k = \frac{1}{\mathbf{k}^2}\left(\frac{\omega_k}{m} - 1\right) . \tag{5.1.13}$$

The upper and the lower signs in the fourth row and the fourth column of (5.1.12) correspond to $[M_+]$ and $[M_-]$ respectively. It is easily verified that this expression satisfies (3.3.2). Let us calculate (5.1.10) and (5.1.11) by the use of (5.1.13). Since $\boldsymbol{\theta}$ and $\boldsymbol{\tau}$ are infinitesimal quantities, it is sufficient to consider the terms up to the first order with respect to them. The calculation may be somewhat tedious, but is quite straightforward. As a result, $\lambda_k(\theta)$ and $\lambda_k(\tau)$ give the following transformations for any 4-vector x_μ:

$$\left.\begin{aligned} \mathbf{x} &\xrightarrow{\lambda_k(\theta)} \mathbf{x} + \mathbf{x}\times\boldsymbol{\theta} \ , \\ x_0 &\xrightarrow{\lambda_k(\theta)} x_0 \ , \end{aligned}\right\} \tag{5.1.14}$$

$$\left.\begin{array}{l} \mathbf{x} \xrightarrow{\lambda_k(\tau)} \mathbf{x} + \mathbf{x} \times \left(\pm \dfrac{\mathbf{k} \times \boldsymbol{\tau}}{m + \omega_k} \right) , \\ x_0 \xrightarrow{\lambda_k(\tau)} x_0 \ . \end{array} \right\} \tag{5.1.15}$$

As is seen in these expressions, $\lambda_k(\theta)$ and $\lambda_k(\tau)$ are elements of the little group (the rotation group in this case), and give infinitesimal rotations through angles θ_1, θ_2 and θ_3 about the first, second and third axes for the θ-transformation $\Lambda(\theta)$, and give infinitesimal rotations through angles $\pm(\mathbf{k}\times\boldsymbol{\tau})_1/(m+\omega_k)$, $\pm(\mathbf{k}\times\boldsymbol{\tau})_2/(m+\omega_k)$ and $\pm(\mathbf{k}\times\boldsymbol{\tau})_3/(m+\omega_k)$ for the τ-transformation $\Lambda(\tau)$, where the signs $\pm$ correspond to $[M_\pm]$ respectively.

Since $Q(\lambda_k, l)$ in (5.1.6) is an irreducible representation of the 3-dimensional rotation through such rotation angles, according to the argument of §2.3, the wave function transforms under the θ-transformation as

$$\begin{aligned} \phi_\xi^{(\pm)\prime}(\mathbf{k}) &= \sum_{\xi'} (1 + i\mathbf{S}\boldsymbol{\theta})_{\xi\xi'} \phi_{\xi'}^{(\pm)}(\mathbf{k} - \mathbf{k} \times \boldsymbol{\theta}) \\ &= \sum_{\xi'} \left\{ 1 + i\boldsymbol{\theta}\left(\frac{1}{i}\mathbf{k} \times \frac{\partial}{\partial \mathbf{k}} + \mathbf{S} \right) \right\}_{\xi\xi'} \phi_{\xi'}^{(\pm)}(\mathbf{k}) \ , \end{aligned} \tag{5.1.16}$$

and under the τ-transformation as

$$\begin{aligned} \phi_\xi^{(\pm)\prime}(\mathbf{k}) &= \sum_{\xi'} \left(1 \pm i\mathbf{S}\frac{\mathbf{k}\times\boldsymbol{\tau}}{\omega_k + m} \right)_{\xi\xi'} \phi_{\xi'}^{(\pm)}(\mathbf{k} \pm \omega_k \boldsymbol{\tau}) \\ &= \sum_{\xi'} \left\{ 1 \pm i\omega_k \boldsymbol{\tau} \left(\frac{1}{i}\frac{\partial}{\partial \mathbf{k}} - \frac{\mathbf{k}\times\mathbf{S}}{\omega_k(m+\omega_k)} \right) \right\}_{\xi\xi'} \phi_{\xi'}^{(\pm)}(\mathbf{k}) \end{aligned} \tag{5.1.17}$$

where $\mathbf{S}$ is the generator of the rotation group, and at the same time, is an irreducible spin vector given by (4.1.3)–(4.1.7).

The transformation of (5.1.17) is written in the form depending on the positive and negative frequencies. Let us rewrite it in the form without any explicit dependence on the frequency in order to connect it to the later discussions. To do this, by the definition of k_0, we put

$$k_0 \phi_\xi^{(\pm)}(\mathbf{k}) = \pm\omega_k \phi_\xi^{(\pm)}(\mathbf{k}) \ . \tag{5.1.18}$$

Here it must be noticed that when we consider both positive and negative frequencies simultaneously $\phi_\xi^{(\pm)}(\mathbf{k})$ should be replaced by

$$\phi_\xi^{(\pm)}(\mathbf{k}) \longrightarrow \theta(\pm k_0)\phi_\xi^{(\pm)}(\mathbf{k}) \ , \tag{5.1.19}$$

where $\theta(x)$ is defined by

$$\theta(x) = \begin{cases} 1 & (x > 0) , \\ 0 & (x < 0) . \end{cases} \tag{5.1.20}$$

So far such a rewriting was unnecessary since we considered the positive and negative frequency states separately, in which case $\theta(\pm k_0)$ was always unity. Hereafter, $\phi_\xi^{(\pm)}(\mathbf{k})$ should always read $\theta(\pm k_0)\phi_\xi^{(\pm)}(\mathbf{k})$, also in the massless case, because the expression of the right-hand side of (5.1.19) is not simple. Of course, this assumption on the description of wave functions does not alter our arguments given up till now. If we introduce

$$\mathbf{J} = \frac{1}{i}\left(\mathbf{k} \times \frac{\partial}{\partial \mathbf{k}}\right) + \mathbf{S} , \tag{5.1.21}$$

$$\mathbf{K} = -k_0\left(\frac{1}{i}\frac{\partial}{\partial \mathbf{k}} - \frac{\mathbf{k} \times \mathbf{S}}{\omega_k(m+\omega_k)}\right) , \tag{5.1.22}$$

(5.1.16) and (5.1.17) are expressed as

$$\phi_\xi^{(\pm)\prime}(\mathbf{k}) = \sum_{\xi'}(1 + i\mathbf{J}\boldsymbol{\theta})_{\xi\xi'}\phi_{\xi'}^{(\pm)}(\mathbf{k}) \quad (\theta\text{-transformation}) , \tag{5.1.23}$$

$$\phi_\xi^{(\pm)\prime}(\mathbf{k}) = \sum_{\xi'}(1 - i\,\mathbf{K}\boldsymbol{\tau})_{\xi\xi'}\phi_{\xi'}^{(\pm)}(\mathbf{k}) \quad (\tau\text{-transformation}) , \tag{5.1.24}$$

where we have used

$$\left[\frac{\partial}{\partial \mathbf{k}}, k_0\right] = \frac{\mathbf{k}}{k_0} . \tag{5.1.25}$$

In these expressions, the operators in the parentheses on the right-hand sides of (5.1.23) and (5.1.24) are just $L[\Lambda(\theta)]$ and $L[\Lambda(\tau)]$ respectively. Hence, in the obtained representation, $\mathbf{J}$ and $\mathbf{K}$ are the generators of the Poincaré group. In fact, using (5.1.25) we can perform a direct calculation to verify that $\mathbf{J}$ and $\mathbf{K}$ given by (5.1.21) and (5.1.22) satisfy the commutation relations for the generators (2.3.13). Needless to say, $\mathbf{J}$ is the total angular momentum of a massive particle with spin s. The first term of (5.1.21) expresses the orbital angular momentum, and the second term is the spin angular momentum. Thus the behavior of such a particle in the Lorentz space is completely determined by (5.1.18), (5.1.23), (5.1.24) and (5.1.7).

If (5.1.7) is generalized to

$$\langle 1, \mp \,|\, 2, \pm \rangle = \sum_\xi \int \frac{d\mathbf{k}}{\omega_k}\phi_\xi^{(\mp)*}(\mathbf{k})_1\phi_\xi^{(\pm)}(\mathbf{k})_2 , \tag{5.1.26}$$

we find from (5.1.19) that the right-hand side of this equation vanishes, that is, $\langle 1, \mp | 2, \pm \rangle = 0$ holds automatically. A similar argument is also applicable to a massless particle.

§5.2 Particles with Zero Mass

Using (4.2.1) for l_μ, we shall construct the spinor representation of α_k satisfying (3.3.2). Let $\alpha_k^{(\pm)}$, corresponding to the positive and negative frequencies, denote 2×2 matrices with determinant equal to unity. In the case of $[0_+]$ in §3.1, we have

$$\alpha_k^{(+)} \begin{pmatrix} 0 & 0 \\ 0 & -2 \end{pmatrix} \alpha_k^{(+)\dagger} = \begin{pmatrix} -|\mathbf{k}| + k_3 & k_1 - ik_2 \\ k_1 + ik_2 & -|\mathbf{k}| - k_3 \end{pmatrix}, \tag{5.2.1}$$

which has a solution

$$\alpha_k^{(+)} = \frac{1}{\sqrt{2(|\mathbf{k}| + k_3)}} \begin{pmatrix} 1 + k_3/|\mathbf{k}| & -k_1 + ik_2 \\ (k_1 + ik_2)/|\mathbf{k}| & |\mathbf{k}| + k_3 \end{pmatrix}. \tag{5.2.2}$$

Since the spinor representations of θ- and τ-transformations have been given by (2.3.20), the spinor representations of $\lambda_k(\theta)$ and $\lambda_k(\tau)$ can be calculated by the use of the above $\alpha_k^{(+)}$. Although the calculations are somewhat long they are straightforward, and their details are not given here. The result is

$$\alpha_k^{(+)-1}\left(1 + \frac{i}{2}\sigma\theta\right)\alpha^{(+)}_{\Lambda(\theta)^{-1}k} = 1 + \frac{i}{2}\frac{\theta_3|\mathbf{k}| + \boldsymbol{\theta}\mathbf{k}}{|\mathbf{k}| + k_3}\sigma_3\,, \tag{5.2.3}$$

$$\alpha_k^{(+)-1}\left(1 + \frac{\sigma\tau}{2}\right)\alpha^{(+)}_{\Lambda(\tau)^{-1}k} = \left(1 + \frac{i}{2}\frac{(\mathbf{k}\times\boldsymbol{\tau})_3}{|\mathbf{k}| + k_3}\sigma_3\right)\left\{1 + \begin{pmatrix} 0 & 0 \\ x_\tau + iy_\tau & 0 \end{pmatrix}\right\}, \tag{5.2.4}$$

where

$$x_\tau = \frac{\tau_1}{|\mathbf{k}| + k_3} - \frac{(\boldsymbol{\tau}\times\mathbf{k})_2}{|\mathbf{k}|(|\mathbf{k}| + k_3)} - \frac{(\boldsymbol{\tau}\mathbf{k})k_1}{\mathbf{k}^2(|\mathbf{k}| + k_3)}, \tag{5.2.5}$$

$$y_\tau = \frac{\tau_2}{|\mathbf{k}| + k_3} + \frac{(\boldsymbol{\tau}\times\mathbf{k})_1}{|\mathbf{k}|(|\mathbf{k}| + k_3)} - \frac{(\boldsymbol{\tau}\mathbf{k})k_2}{\mathbf{k}^2(|\mathbf{k}| + k_3)}. \tag{5.2.6}$$

In the derivations of (5.2.3) and (5.2.4), the terms up to the first order with respect to the infinitesimal quantities $\boldsymbol{\theta}$ and $\boldsymbol{\tau}$ have been taken, and other terms have been omitted.

We can apply the above result to the case of $[0_-]$, i.e., the case of negative frequency. From a relation

$$\sigma_2 \begin{pmatrix} -|\mathbf{k}| + k_3 & k_1 + ik_2 \\ k_1 - ik_2 & -|\mathbf{k}| - k_3 \end{pmatrix} \sigma_2 = -\begin{pmatrix} |\mathbf{k}| + k_3 & k_1 - ik_2 \\ k_1 + ik_2 & |\mathbf{k}| - k_3 \end{pmatrix} \tag{5.2.7}$$

and (4.2.6), multiplying the complex conjugate of (5.2.1) by σ_2's on the both sides, we have

$$\sigma_2\alpha_k^{(+)*}\sigma_2\begin{pmatrix}2&0\\0&0\end{pmatrix}(\sigma_2\alpha_k^{(+)*}\sigma_2)^\dagger=\begin{pmatrix}|\mathbf{k}|+k_3 & k_1-ik_2\\ k_1+ik_2 & |\mathbf{k}|-k_3\end{pmatrix}\ . \qquad (5.2.8)$$

Thus

$$\alpha_k^{(-)}=\sigma_2\alpha_k^{(+)*}\sigma_2\ , \qquad (5.2.9)$$

where $\alpha_k^{(+)*}$ is a matrix formed from $\alpha_k^{(+)}$ with each element replaced by its complex conjugate.

Multiply the complex conjugate of (5.2.3) by σ_2's on the both sides and apply (5.2.9). Then in the θ-transformation we easily obtain

$$\alpha_k^{(-)-1}\left(1+i\frac{\boldsymbol{\sigma\theta}}{2}\right)\alpha^{(-)}_{\Lambda^{-1}(\theta)k}=1+\frac{i}{2}\frac{\theta_3|\mathbf{k}|+\boldsymbol{\theta}\mathbf{k}}{|\mathbf{k}|+k_3}\sigma_3\ . \qquad (5.2.10)$$

On the other hand, in the τ-transformation, according to (5.1.9) we have for the positive frequency

$$\left.\begin{aligned}\mathbf{k}&\xrightarrow{\Lambda(\tau)^{-1}}\mathbf{k}+\boldsymbol{\tau}|\mathbf{k}|\ ,\\ |\mathbf{k}|&\xrightarrow{\Lambda(\tau)^{-1}}|\mathbf{k}|+\boldsymbol{\tau}\mathbf{k}\ ,\end{aligned}\right\} \qquad (5.2.11)$$

and for the negative frequency

$$\left.\begin{aligned}\mathbf{k}&\xrightarrow{\Lambda(\tau)^{-1}}\mathbf{k}-\boldsymbol{\tau}|\mathbf{k}|\ ,\\ |\mathbf{k}|&\xrightarrow{\Lambda(\tau)^{-1}}|\mathbf{k}|-\boldsymbol{\tau}\mathbf{k}\ ,\end{aligned}\right\} \qquad (5.2.12)$$

which is just what is given in (5.2.11) with $\boldsymbol{\tau}$ replaced by $-\boldsymbol{\tau}$. Hence

$$\alpha^{(-)}_{\Lambda(\tau)^{-1}k}=\sigma_2\alpha^{(+)*}_{\Lambda(\tau)^{-1}k}\sigma_2\ , \qquad (5.2.13)$$

which leads to

$$\begin{aligned}\alpha_k^{(-)-1}\left(1+\frac{\boldsymbol{\sigma\tau}}{2}\right)\alpha^{(-)}_{\Lambda(\tau)^{-1}k}&=\sigma_2\left\{\alpha_k^{(+)*}\sigma_2\left(1+\frac{\boldsymbol{\sigma\tau}}{2}\right)\sigma_2\alpha^{(+)*}_{\Lambda(-\tau)^{-1}k}\right\}\sigma_2\\ &=\sigma_2\left(\alpha_k^{(+)-1}\left(1-\frac{\boldsymbol{\sigma\tau}}{2}\right)\alpha^{(+)}_{\Lambda(-\tau)^{-1}k}\right)^*\sigma_2\\ &=\left(1-\frac{i}{2}\frac{(\mathbf{k}\times\boldsymbol{\tau})_3}{|\mathbf{k}|+k_3}\sigma_3\right)\left\{1+\begin{pmatrix}0 & x_\tau-iy_\tau\\ 0 & 0\end{pmatrix}\right\}\ ,\end{aligned} \qquad (5.2.14)$$

where we have used an equation given in (5.2.4) with $\boldsymbol{\tau}$ replaced by $-\boldsymbol{\tau}$.

In this way we have obtained the spinor representations of the little group corresponding to the θ- and τ-transformations. Let us compare the above result with the argument of §4.2. If β, a_1 and a_2 are infinitesimal quantities in (4.2.3) and (4.2.8), the general forms of infinitesimal transformations of the little group in the spinor representation are written as

$$g^{(+)} = \left(1 + \frac{i}{2}\beta\sigma_3\right)\left\{1 + \begin{pmatrix} 0 & 0 \\ a_1 + ia_2 & 0 \end{pmatrix}\right\} , \tag{5.2.15}$$

$$g^{(-)} = \left(1 + \frac{i}{2}\beta\sigma_3\right)\left\{1 - \begin{pmatrix} 0 & a_1 - ia_2 \\ 0 & 0 \end{pmatrix}\right\} . \tag{5.2.16}$$

Comparing these expressions with (5.2.3) and (5.2.10), we get the following relations for the θ-transformation:

$$\left.\begin{aligned} \beta &= \frac{\theta_3|\mathbf{k}| + \mathbf{k}\boldsymbol{\theta}}{|\mathbf{k}| + k_3} , \\ a_1 &= a_2 = 0 , \end{aligned}\right\} \tag{5.2.17}$$

which are valid in both cases of positive and negative frequencies. For the τ-transformation, using (5.2.4) and (5.2.14), we have

$$\left.\begin{aligned} \beta &= \pm\frac{(\mathbf{k}\times\boldsymbol{\tau})_3}{|\mathbf{k}| + k_3} , \\ a_1 &= \pm x_\tau , \qquad a_2 = \pm y_\tau . \end{aligned}\right\} \tag{5.2.18}$$

The signs $\pm$ in (5.2.18) correspond to the cases of positive and negative frequencies respectively. By the use of these results, we shall investigate the behavior of wave functions under the Lorentz transformations.

First, let us consider the case of discrete spin, i.e., the case of $[0^0_\pm]$ in §4.2. Define the wave functions in momentum representation, similarly to (5.1.2), by

$$\phi_S^{(\pm)}(\mathbf{k}) = \langle \mathbf{k}, S, \pm | S, \pm \rangle , \tag{5.2.19}$$

where S is a constant specifying the irreducible representation of the Poincaré group, which has been mentioned in §4.2. From the requirement that the representation be at most double-valued, S must be one of the following numbers:

$$S = 0, \pm 1/2, \pm 1, \pm 3/2, \ldots \tag{5.2.20}$$

By the use of (5.2.17), (5.2.18) and an equation derived from (4.2.18) by expanding with respect to infinitesimal β, the transformations of $\phi_S^{(\pm)}(\mathbf{k})$ are given by

$$\phi_S^{(\pm)\prime}(\mathbf{k}) = \left(1 + iS\frac{\theta_3|\mathbf{k}| + \mathbf{k}\boldsymbol{\theta}}{|\mathbf{k}| + k_3}\right)\phi_S^{(\pm)}(\mathbf{k} - \mathbf{k}\times\boldsymbol{\theta}) \quad (\theta\text{- transformation}) , \tag{5.2.21}$$

$$\phi_S^{(\pm)\prime}(\mathbf{k}) = \left(1 \pm iS\frac{(\mathbf{k}\times\boldsymbol{\tau})_3}{|\mathbf{k}| + k_3}\right)\phi_S^{(\pm)}(\mathbf{k} \pm |\mathbf{k}|\boldsymbol{\tau}) \quad (\tau\text{-transformation}) . \tag{5.2.22}$$

Expanding $\phi_S^{(\pm)}$'s in the right-hand sides with respect to $\boldsymbol{\theta}$ and $\boldsymbol{\tau}$, and taking the terms up to the first order, we can write

$$\phi_S^{(\pm)\prime}(\mathbf{k}) = (1 + i\mathbf{J}\boldsymbol{\theta})\phi_S^{(\pm)}(\mathbf{k}) , \tag{5.2.23}$$

$$\phi_S^{(\pm)\prime}(\mathbf{k}) = (1 - i\mathbf{K}\boldsymbol{\tau})\phi_S^{(\pm)}(\mathbf{k}) , \tag{5.2.24}$$

where $\mathbf{J}$ and $\mathbf{K}$ are given by

$$\left.\begin{aligned} J_1 &= \frac{1}{i}\left(\mathbf{k}\times\frac{\partial}{\partial\mathbf{k}}\right)_1 + \frac{k_1}{|\mathbf{k}| + k_3}S , \\ J_2 &= \frac{1}{i}\left(\mathbf{k}\times\frac{\partial}{\partial\mathbf{k}}\right)_2 + \frac{k_2}{|\mathbf{k}| + k_3}S , \\ J_3 &= \frac{1}{i}\left(\mathbf{k}\times\frac{\partial}{\partial\mathbf{k}}\right)_3 + S , \end{aligned}\right\} \tag{5.2.25}$$

and

$$\left.\begin{aligned} K_1 &= -k_0\left(\frac{1}{i}\frac{\partial}{\partial k_1} - \frac{k_2}{|\mathbf{k}|(|\mathbf{k}| + k_3)}S\right) , \\ K_2 &= -k_0\left(\frac{1}{i}\frac{\partial}{\partial k_2} + \frac{k_1}{|\mathbf{k}|(|\mathbf{k}| + k_3)}S\right) , \\ K_3 &= -\frac{1}{i}k_0\frac{\partial}{\partial k_3} , \end{aligned}\right\} \tag{5.2.26}$$

which are derived with the help of

$$k_0\phi_S^{(\pm)}(\mathbf{k}) = \pm|\mathbf{k}|\phi_S^{(\pm)}(\mathbf{k}) . \tag{5.2.27}$$

If $\phi_S^{(\pm)}(k)_h$ denote the wave functions corresponding to $|S;h,\pm\rangle$ $(h = 1,2)$, their inner products are, from (3.1.17),

$$\langle S;1,\pm|S;2,\pm\rangle = \int\frac{d\mathbf{k}}{|\mathbf{k}|}\phi_S^{(\pm)*}(\mathbf{k})_1\phi_S^{(\pm)}(\mathbf{k})_2 . \tag{5.2.28}$$

Equations (5.2.23)–(5.2.28) derived in this way determine the behavior of a free massless particle with discrete spin completely.

It can be verified by a direct calculation that **J** and **K** in (5.2.25) and (5.2.26) have property of the generators of the Lorentz group and that they satisfy the commutation relations (2.3.13). However, the commutation relations cannot immediately lead to the restriction (5.2.20). In fact, (2.3.13) can be satisfied for any real number of S. The restriction that S takes only integral or half-integral values originally came from the fact that the Poincaré group has only single-valued and double-valued representations. Further tracing its root shows that it is based on the fact that the rotation group (a subgroup of the Poincaré group) has at most double-valued representation (§2.1). Therefore, when S does not satisfy (5.2.20), even if **J** satisfies the first equation of the commutation relations of angular momenta (2.3.13), operators of infinitesimal rotation and their multiplications cannot be representations of the rotation group. In this case Hermiticity of **J** is broken. In other words, the restriction of S is also derived by a direct calculation from the condition that **J** is Hermitian. A further detailed discussion on this subject will be given in §5.4.

Next, let us consider the case of continuous spin, i.e., the case of $[\Xi\,{}^{0}_{\pm}]$ in §4.2.

In a single-valued representation, define the wave functions, similarly to (5.1.2), by

$$\phi^{(\pm)}(\mathbf{k}, \xi_1, \xi_2) = \langle \mathbf{k}, \xi_1, \xi_2, \pm \,|\, \pm \rangle \,. \tag{5.2.29}$$

With the help of (4.2.15), (4.2.34), (5.2.17), (5.2.18), (5.2.5) and (5.2.6), the transformations of the wave functions are given by the following equations:

$$\phi^{(\pm)\prime}(\mathbf{k}, \xi_1, \xi_1) = \left\{1 + \boldsymbol{\theta}\left(\mathbf{k} \times \frac{\partial}{\partial \mathbf{k}}\right) + i\,\frac{\theta_3 |\,\mathbf{k}\,| + \boldsymbol{\theta}\mathbf{k}}{|\,\mathbf{k}\,| + k_3}\hat{l}_3\right\} \phi^{(\pm)}(\mathbf{k}, \xi_1, \xi_2)$$
$$(\theta\text{-transformation})\,, \tag{5.2.30}$$

$$\phi^{(\pm)\prime}(\mathbf{k}, \xi_1, \xi_2) = \left\{1 + k_0\boldsymbol{\tau}\frac{\partial}{\partial \mathbf{k}} + i\,\frac{(\mathbf{k} \times \boldsymbol{\tau})_3}{|\,\mathbf{k}\,| + k_3}\frac{k_0}{|\,\mathbf{k}\,|}\hat{l}_3 + i\,\frac{k_0}{\mathbf{k}^2}\mathcal{T}_\tau(\xi_1, \xi_2)\right\} \phi^{(\pm)}(\mathbf{k}, \xi_1, \xi_2)$$
$$(\tau\text{-transformation})\,, \tag{5.2.31}$$

where

$$\hat{l}_3 = \frac{1}{i}\left(\xi_1 \frac{\partial}{\partial \xi_2} - \xi_2 \frac{\partial}{\partial \xi_1}\right)\,, \tag{5.2.32}$$

$$\mathcal{F}_\tau(\xi_1,\xi_2) = \left\{(\tau_1\xi_1+\tau_2\xi_2) - \frac{k_1\xi_1+k_2\xi_2}{|\mathbf{k}|(|\mathbf{k}|+k_3)}(\tau\mathbf{k}+|\mathbf{k}|\tau_3)\right\} . \qquad (5.2.33)$$

We have used

$$k_0\phi^{(\pm)}(\mathbf{k},\xi_1,\xi_2) = \pm|\mathbf{k}|\phi^{(\pm)}(\mathbf{k},\xi_1,\xi_2) . \qquad (5.2.34)$$

Concerning the inner products, they are given by (3.1.16) with $\delta_{\bar{\xi}\bar{\xi}'}$ replaced by $\delta(\bar{\phi}-\bar{\phi}')$ when the state vectors are expressed as $|\mathbf{k},\bar{\phi},\pm\rangle$ (See §4.2 about $\bar{\phi}$), but when $|\mathbf{k},\xi_1,\xi_2,\pm\rangle$, since ξ_1 and ξ_2 are not independent, the inner products are defined in the following manner. That is, since

$$(\xi_1^2+\xi_2^2-\Xi)\phi^{(\pm)}(\mathbf{k},\xi_1,\xi_2) = 0 \qquad (5.2.35)$$

leads to

$$\phi^{(\pm)}(\mathbf{k},\xi_1,\xi_2) = \delta(\xi_1^2+\xi_2^2-\Xi)\tilde{\phi}^{(\pm)}(\mathbf{k},\xi_1,\xi_2) , \qquad (5.2.36)$$

we can express the inner products in the form

$$\langle 1,\pm|2,\pm\rangle = \int \frac{d\mathbf{k}}{|\mathbf{k}|}d\xi_1 d\xi_2\delta(\xi_1^2+\xi_2^2-\Xi)\tilde{\phi}^{(\pm)*}(\mathbf{k},\xi_1,\xi_2)_1\tilde{\phi}^{(\pm)}(\mathbf{k},\xi_1,\xi_2)_2 . \qquad (5.2.37)$$

Equations (5.2.30)–(5.2.37) characterize completely a continuous spin particle in a single-valued representation.

In the case of double-valued representation, we can proceed in a similar manner to the above discussion, and only the results will be given here. Using the result on the double-valued representation of continuous spin in §4.2 the representations can be derived along the same way mentioned above. The wave functions $\phi^{(\pm)}_{1/2}(\mathbf{k},\xi_1,\xi_2)$ and $\phi^{(\pm)}_{-1/2}(\mathbf{k},\xi_1,\xi_2)$ are obtained from the state vectors $|\mathbf{k},\xi_1,\xi_2,\pm\rangle_{1/2}$ and $|\mathbf{k},\xi_1,\xi_2,\pm\rangle_{-1/2}$ corresponding to $|\bar{\xi}_1,\bar{\xi}_2\rangle_{1/2}$ $|\bar{\xi}_1,\bar{\xi}_2\rangle_{-1/2}$ of (4.2.35) and (4.2.36) respectively. They are unitary equivalent to each other and will be expressed in a combined form $\phi^{(\pm)}_S(\mathbf{k},\xi_1,\xi_2)$ for later convenience. Of course, S is either $1/2$ or $-1/2$.

$$\phi^{(\pm)'}_S(\mathbf{k},\xi_1,\xi_2) = \left\{1+\boldsymbol{\theta}\left(\mathbf{k}\times\frac{\partial}{\partial\mathbf{k}}\right)+i\frac{\theta_3|\mathbf{k}|+\boldsymbol{\theta}\mathbf{k}}{|\mathbf{k}|+k_3}(\hat{l}_3+S)\right\}\phi^{(\pm)}_S(\mathbf{k},\xi_1,\xi_2)$$
$$(\theta\text{-transformation}) , \qquad (5.2.38)$$

$$\phi^{(\pm)'}_S(\mathbf{k},\xi_1,\xi_2) = \left\{1+k_0\boldsymbol{\tau}\frac{\partial}{\partial\mathbf{k}}+i\frac{(\mathbf{k}\times\boldsymbol{\tau})_3}{|\mathbf{k}|+k_3}\frac{k_0}{|\mathbf{k}|}(\hat{l}_3+S)\right.$$
$$\left.+i\frac{k_0}{\mathbf{k}^2}\mathcal{F}_\tau(\xi_1,\xi_2)\right\}\phi^{(\pm)}_S(\mathbf{k},\xi_1,\xi_2)$$
$$(\tau\text{-transformation}) , \qquad (5.2.39)$$

$$k_0\phi_S^{(\pm)}(\mathbf{k},\xi_1,\xi_2) = \pm|\mathbf{k}|\phi_S^{(\pm)}(\mathbf{k},\xi_1,\xi_2)\ , \tag{5.2.40}$$

$$\phi_S^{(\pm)}(\mathbf{k},\xi_1,\xi_2) = \delta(\xi_1^2+\xi_2^2-\Xi)\widetilde{\phi}_S^{(\pm)}(\mathbf{k},\xi_1,\xi_2)\ , \tag{5.2.41}$$

$$\begin{aligned} {}_S\langle 1,\pm|2,\pm\rangle_S = \int \frac{d\mathbf{k}}{|\mathbf{k}|} d\xi_1 d\xi_2 \delta(\xi_1^2+\xi_2^2-\Xi)\widetilde{\phi}_S^{(\pm)*}(\mathbf{k},\xi_1,\xi_2)_1 \\ \times \widetilde{\phi}_S^{(\pm)}(\mathbf{k},\xi_1,\xi_2)_2\ , \end{aligned} \tag{5.2.42}$$

where $\widehat{l}_3$ and $\mathcal{J}_\tau$ have been defined by (5.2.32) and (5.2.33) respectively.

§5.3 Particles with Imaginary Mass

Using (4.4.1) for l_μ, let us first construct α_k in a spinor representation. In this case we cannot divide the positive and negative frequency states in a Lorentz invariant way, so we do not force ourselves to eliminate k_0, but we shall proceed under the condition that k_μ used here satisfies the relation

$$|\mathbf{k}|^2 = k_0^2 + m^2\ . \tag{5.3.1}$$

Solving

$$\alpha_k \begin{pmatrix} m & 0 \\ 0 & -m \end{pmatrix} \alpha_k^\dagger = \begin{pmatrix} -k_0+k_3 & k_1-ik_2 \\ k_1+ik_2 & -k_0-k_3 \end{pmatrix} \tag{5.3.2}$$

we obtain

$$\alpha_k = \frac{1}{2\sqrt{m|\mathbf{k}|(|\mathbf{k}|+k_3)(|\mathbf{k}|+m)}}(\sigma_3|\mathbf{k}|+\boldsymbol{\sigma}\mathbf{k})\{-k_0+\sigma_3(|\mathbf{k}|+m)\}\ , \tag{5.3.3}$$

$$\alpha_k^{-1} = \frac{1}{2\sqrt{m|\mathbf{k}|(|\mathbf{k}|+k_3)(|\mathbf{k}|+m)}}\{k_0+\sigma_3(|\mathbf{k}|+m)\}(\sigma_3|\mathbf{k}|+\boldsymbol{\sigma}\mathbf{k})\ . \tag{5.3.4}$$

It can be readily verified by a direct calculation that this solution satisfies $\det(\alpha_k)=1$ and (5.3.2).

By the use of (5.3.3) and (5.3.4), we can derive the spinor representations of infinitesimal Wigner rotation for θ- and τ-transformations, and the result is given by the following equations:

$$\alpha_k^{-1}\left(1+\frac{i}{2}\boldsymbol{\sigma\theta}\right)\alpha_{\Lambda(\theta)^{-1}k} = 1 + i\frac{\theta_3|\mathbf{k}|+\boldsymbol{\theta}\mathbf{k}}{|\mathbf{k}|+k_3}\frac{\sigma_3}{2}\ , \tag{5.3.5}$$

$$\begin{aligned} \alpha_k^{-1}\left(1+\frac{\boldsymbol{\sigma\tau}}{2}\right)\alpha_{\Lambda(\tau)^{-1}k} = 1 + i\frac{k_0(\mathbf{k}\times\boldsymbol{\tau})_3}{|\mathbf{k}|(|\mathbf{k}|+k_3)}\frac{\sigma_3}{2} \\ +i\frac{m}{|\mathbf{k}|}\left(\tau_1 - \frac{\boldsymbol{\tau}\mathbf{k}+\tau_3|\mathbf{k}|}{|\mathbf{k}|(|\mathbf{k}|+k_3)}k_1\right)\frac{\sigma_1}{2i} + i\frac{m}{|\mathbf{k}|}\left(\tau_2 - \frac{\boldsymbol{\tau}\mathbf{k}+\tau_3|\mathbf{k}|}{|\mathbf{k}|(|\mathbf{k}|+k_3)}k_2\right)\frac{\sigma_2}{2i}\ , \end{aligned} \tag{5.3.6}$$

where $\sigma_3/2$, $\sigma_1/2i$ and $\sigma_2/2i$ are the generators of the 3-dimensional Lorentz group and satisfy the commutation relations (4.4.4). Since (5.3.6) is not a unitary operator itself, replacements $\sigma_3/2 \to H_0$, $\sigma_1/2i \to H_1$ and $\sigma_2/2i \to H_2$ are required in the above equations to obtain unitary and irreducible Wigner rotations. As a result, the Wigner rotations are given by

$$Q(\lambda_k(\theta), l) = 1 + i\,\frac{\theta_3|\mathbf{k}| + \boldsymbol{\theta}\mathbf{k}}{|\mathbf{k}| + k_3} H_0 \quad (\theta\text{-transformation})\ , \tag{5.3.7}$$

$$Q(\lambda_k(\tau), l) = 1 + i\,\frac{k_0(\mathbf{k}\times\boldsymbol{\tau})_3}{|\mathbf{k}|(|\mathbf{k}| + k_3)} H_0 + i\,\frac{m}{|\mathbf{k}|}\mathcal{F}_\tau(H_1, H_2) \quad (\tau\text{-transformation})\ , \tag{5.3.8}$$

where $\mathcal{F}_\tau(H_1, H_2)$ is obtained from $\mathcal{F}_\tau(\xi_1, \xi_2)$ of (5.2.33) with ξ_1 and ξ_2 replaced by H_1 and H_2.

On the other hand, if

$$\phi_\xi(\mathbf{k}, k_0) = \langle\, k, \xi \,|\ \rangle \tag{5.3.9}$$

denotes the wave function corresponding to a state vector $|\ \rangle$ in momentum representation, its inner product is given, from (3.1.20), by

$$\langle 1 | 2 \rangle = \sum_\xi \int d^4k\,\delta(k_\mu^2 - m^2)\phi_\xi^*(\mathbf{k}, k_0)_1 \phi_\xi(\mathbf{k}, k_0)_2\ . \tag{5.3.10}$$

The infinitesimal Lorentz transformations of this wave function are written, by the use of (5.3.7), (5.3.8) and a similar argument given previously, as

$$\phi'_\xi(\mathbf{k}, k_0) = \sum_{\xi'} \left\{1 + \left(\mathbf{k}\times\frac{\partial}{\partial\mathbf{k}}\right)\boldsymbol{\theta} + i\,\frac{\theta_3|\mathbf{k}| + \boldsymbol{\theta}\mathbf{k}}{|\mathbf{k}| + k_3} H_0\right\}_{\xi\xi'} \phi_{\xi'}(\mathbf{k}, k_0) \quad (\theta\text{-transformation})\ , \tag{5.3.11}$$

$$\phi'_\xi(\mathbf{k}, k_0) = \sum_{\xi'} \left\{1 + \left(k_0\frac{\partial}{\partial\mathbf{k}} + \mathbf{k}\frac{\partial}{\partial k_0}\right)\boldsymbol{\tau} + i\,\frac{k_0(\mathbf{k}\times\boldsymbol{\tau})_3}{|\mathbf{k}|(|\mathbf{k}| + k_3)} H_0 + i\,\frac{m}{|\mathbf{k}|}\mathcal{F}_\tau(H_1, H_2)\right\}_{\xi\xi'} \phi_{\xi'}(\mathbf{k}, k_0) \quad (\tau\text{-transformation})\ , \tag{5.3.12}$$

where the second term in the parentheses on the right-hand side of (5.3.12) has been obtained from $\phi_\xi(\Lambda(\tau)^{-1}\mathbf{k}, \Lambda(\tau)^{-1}k_0) = \phi_\xi(\mathbf{k}+\boldsymbol{\tau}k_0, k_0+\boldsymbol{\tau}\mathbf{k})$ by expanding up to the first order with respect to $\boldsymbol{\tau}$.

According to (5.3.11) the angular momentum of a particle with imaginary mass is given by

$$\left.\begin{aligned} J_1 &= \frac{1}{i}\left(\mathbf{k}\times\frac{\partial}{\partial\mathbf{k}}\right)_1 + \frac{k_1}{|\mathbf{k}|+k_3}H_0\,, \\ J_2 &= \frac{1}{i}\left(\mathbf{k}\times\frac{\partial}{\partial\mathbf{k}}\right)_2 + \frac{k_2}{|\mathbf{k}|+k_3}H_0\,, \\ J_3 &= \frac{1}{i}\left(\mathbf{k}\times\frac{\partial}{\partial\mathbf{k}}\right)_3 + H_0\,, \end{aligned}\right\} \tag{5.3.13}$$

which are just Eqs. (5.2.25) with H_0 substituted for S. Therefore the eigenvalue problem of the angular momenta given by (5.3.13) can be considered in a same manner to that of (5.2.25). In the next section, it will be derived that S must satisfy (5.2.20) from the Hermiticity condition of $\mathbf{J}$ of (5.2.25). If we apply the same argument to $\mathbf{J}$ of (5.3.13) we can derive the condition that the eigenvalue of H_0 takes either integral or half-integral numbers. In §4.4 we restricted the eigenvalue of H_0 in the above manner for the reason that the Poincaré group has only single-valued and double-valued representations. We can also derive this restriction from the Hermiticity of an angular momentum operator. In the case where the little group has the discrete representation $D^{(\pm)}_{\mu_0}$, it will also be derived from the result of the next section that the minimum value of j is $|\mu_0|$, $j(j+1)$ being the eigenvalue of $\mathbf{J}^2$.

§5.4 Angular Momenta of Massless Particles

The operator $\mathbf{J}$ given by (5.2.25) is a physical quantity with a significant meaning that it is the total angular momentum of a massless particle with discrete spin. The operator $\mathbf{J}$ has a different form from that of an usual angular momentum (5.1.21), and $\mathbf{k}$ transforms as a 3-vector, while $\partial/\partial\mathbf{k}$ is not a 3-vector. That is, $[J_i, k_j] = i\sum_l \varepsilon_{ijl}k_l$ does hold but $[J_i, \partial/\partial k_j] = i\sum_l \varepsilon_{ijl}\partial/\partial k_l$ does not. Moreover, $\mathbf{J}$ satisfies the commutation relation of angular momentum (2.3.7) for any real number S. It may not be meaningless from the physical point of view to study the properties of such $\mathbf{J}$ in detail. We do not assume that S satisfies (4.2.19), but assume only Hermiticity of $\mathbf{J}$ of (5.2.25), from which we shall derive (4.2.19). We shall construct the explicit form of a wave function in an irreducible representation space.

Before going on its calculation, however, it is convenient to fix the framework of our discussion. We shall now review briefly, the result derived from the assumption that $\mathbf{J}$ is a Hermitian operator satisfying the commutation relation (2.3.7), which is well known in quantum mechanics.

Since $\mathbf{J}^2$ commutes with J_i $(i = 1, 2, 3)$, it is a constant in an irreducible representation space of the rotation group. Therefore there always exists a state vector belonging to the maximum eigenvalue of J_3 in the irreducible representation space, because $\mathbf{J}^2 = J_1^2 + J_2^2 + J_3^2$, and the expectation value of $J_1^2 + J_2^2$ is not negative. Let j be the maximum value and $|\,j, j\,\rangle$ be the corresponding eigenstate satisfying $\langle\, j, j\,|\,j, j\,\rangle = 1$, where the first j in $|\,j, j\,\rangle$ is the characteristic number of the representation in the sense that it is the maximum value of J_3, and the second j is merely an eigenvalue of J_3. In general, $|\,j, \mu\,\rangle$ represents a normalized state vector belonging to an eigenvalue μ of J_3 in the representation space, i.e.,

$$J_3\,|\,j, \mu\,\rangle = \mu\,|\,j, \mu\,\rangle\,. \tag{5.4.1}$$

Introducing $J^{(\pm)} = J_1 \pm iJ_2$ we have

$$[J_3, J^{(\pm)}] = \pm J^{(\pm)}\,, \tag{5.4.2}$$

$$[J^{(+)}, J^{(-)}] = 2J_3\,, \tag{5.4.3}$$

which are the equivalent equations to (2.3.7). Applying (5.4.2) to $|\,j, \mu\rangle$, from (5.4.1), we have

$$J_3 J^{(\pm)}|\,j, \mu\,\rangle = (\mu \pm 1) J^{(\pm)}|\,j, \mu\,\rangle\,.$$

Thus $J^{(+)}$ is an operator raising the eigenvalue of J_3 by 1, and $J^{(-)}$ is a lowering operator. Therefore,

$$J_3(J^{(-)})^m|\,j, j\,\rangle = (j - m)(J^{(-)})^m|\,j, j\,\rangle \quad (m = 0, 1, 2, \ldots)\,, \tag{5.4.4}$$

and by definition

$$J^{(+)}|\,j, j\,\rangle = 0\,. \tag{5.4.5}$$

Now, with the help of (5.4.3) we get

$$\begin{aligned} J^{(+)}(J^{(-)})^n &= (J^{(-)})^n J^{(+)} + [J^{(+)}, (J^{(-)})^n] \\ &= (J^{(-)})^n J^{(+)} + 2\sum_{m=0}^{n-1} (J^{(-)})^{n-m-1} J_3 (J^{(-)})^m \\ &\qquad (n = 1, 2, \ldots)\,. \end{aligned} \tag{5.4.6}$$

Applying this operator to $|\,j, j\,\rangle$, from (5.4.4) and (5.4.5) we get

$$J^{(+)}(J^{(-)})^n|\,j, j\,\rangle = n(2j + 1 - n)(J^{(-)})^{n-1}|\,j, j\,\rangle\,. \tag{5.4.7}$$

A similar argument with the successive application of $J^{(+)}$ to the above equation leads to

$$\begin{aligned}&(J^{(+)})^n(J^{(-)})^n|\,j,j\,\rangle\\&\quad= n!(2j+1-n)(2j+2-n)\ldots(2j-1)(2j)|\,j,j\,\rangle\,,\end{aligned}$$

and a further multiplication of $\langle\,j,j\,|$ gives

$$n!(2j+1-n)(2j+2-n)\ldots(2j-1)(2j)\geq 0 \quad (n=1,2,\ldots)\,, \tag{5.4.8}$$

where we have used $\langle\,j,j\,|(J^{(+)})^n(J^{(-)})^n|\,j,j\,\rangle \geq 0$, which is derived from the fact that $J^{(+)}$ is Hermitian conjugate to $J^{(-)}$ because $\mathbf{J}$ is Hermitian. In order that (5.4.8) is satisfied, j must be one of the non-negative integral or half-integral numbers:

$$j = 0, 1/2, 1, 3/2, 2, \ldots\ . \tag{5.4.9}$$

Otherwise there exists n which makes the left-hand side of (5.4.8) negative. Thus

$$(J^{(-)})^{2j+1}|\,j,j\,\rangle = 0\,, \tag{5.4.10}$$

or in general

$$(J^{(-)})^n|\,j,j\,\rangle = \sqrt{\frac{n!(2j)!}{(2j-n)!}}\,|\,j,j-n\,\rangle \quad (n=0,1,\ldots,2j)\,, \tag{5.4.11}$$

from which we have the possible values of μ for each j of (5.4.9):

$$\mu = j, j-1, j-2, \ldots -j+1, -j\,. \tag{5.4.12}$$

Noting that we can write $\mathbf{J}^2 = J_3(J_3+1)+J^{(-)}J^{(+)}$, and applying it to $|\,j,j\,\rangle$ we obtain $j(j+1)$ for the eigenvalue of $\mathbf{J}^2$. It is also the eigenvalue to which $|\,j,\mu\,\rangle$ belongs since $J^{(-)}$ commutes with $\mathbf{J}^2$. If j is fixed, the matrix elements of $J^{(\pm)}$ are easily obtained from (5.4.11). They are shown to be irreducible from the fact that matrices commutable with $J^{(\pm)}$ and J_3 are only the constant multiples of the unit matrix. That is, the irreducible representation is characterized by j of (5.4.9).

The above result is derived generally independent of the explicit form of $\mathbf{J}$, and of course, the argument on $\mathbf{J}$ of (5.2.25) cannot go beyond this framework. However, for instance, we cannot get any information concerning the value of j in (5.4.9) without the explicit form of $\mathbf{J}$.

To proceed with explicit calculations let us rewrite **J** of (5.2.25) in terms of the polar coordinates $\mathbf{k} = (|\mathbf{k}|\sin\theta\cos\phi,\ |\mathbf{k}|\sin\theta\sin\phi,\ |\mathbf{k}|\cos\theta)$. Then we have

$$J_3 = \frac{1}{i}\frac{\partial}{\partial\phi} + S\ , \tag{5.4.13}$$

$$J^{(+)} = e^{i\phi}\left\{\sqrt{1-z^2}\left(-\frac{\partial}{\partial z} + \frac{S}{1+z}\right) + i\frac{z}{\sqrt{1-z^2}}\frac{\partial}{\partial\phi}\right\}\ , \tag{5.4.14}$$

$$J^{(-)} = e^{-i\phi}\left\{\sqrt{1-z^2}\left(\frac{\partial}{\partial z} + \frac{S}{1+z}\right) + i\frac{z}{\sqrt{1-z^2}}\frac{\partial}{\partial\phi}\right\}\ , \tag{5.4.15}$$

$$J^2 = -\left\{(1-z^2)\frac{\partial^2}{\partial z^2} - 2z\frac{\partial}{\partial z} + \frac{1}{1-z^2}\frac{\partial^2}{\partial\phi^2}\right\} + \frac{2}{1+z}S\left(S + \frac{1}{i}\frac{\partial}{\partial\phi}\right)\ , \tag{5.4.16}$$

$$z = \cos\theta\ . \tag{5.4.17}$$

Now let $Y^{\mu}_{j,S}(\theta,\phi)$ be an eigenfunction of $\mathbf{J}^2$ and J_3. Of course, according to the previous general argument, j is a non-negative integral or half-integral number and μ takes one of the $2j+1$ values running from j to $-j$. As is easily seen, the eigenfunction can be decomposed as

$$Y^{\mu}_{j,S}(\theta,\phi) = e^{i(\mu-S)\phi}X^{\mu}_{j,S}(z)\ , \tag{5.4.18}$$

where, needless to say, $\exp\{i(\mu-S)\phi\}$ is an eigenfunction of J_3. If we suppose here that $Y^{\mu}_{j,S}(\theta,\phi)$ is a single-valued function of **k**, it must be invariant under a transformation $\phi \to \phi + 2\pi$, and then S takes an integral or half-integral value corresponding to the value of μ which is an integer or half-integer. But our assumption is only the Hermiticity of the observable **J**, and the above supposition must be derived from this assumption if it is correct. In this sense, S has not yet been determined at this stage. We also want to know the explicit form of the function $Y^{\mu}_{j,S}(\theta,\phi)$ and which value of (5.4.9) j takes. So let us construct the explicit form of $X^{\mu}_{j,S}(z)$. The equation for $X_{j,S}(z)$ is now obtained from $\mathbf{J}^2 Y^{\mu}_{j,S}(\theta,\phi) = j(j+1)Y^{\mu}_{j,S}(\theta,\phi)$ by the use of (5.4.16) and (5.4.18), and is written as

$$\left\{(1-z^2)\frac{d^2}{dz^2} - 2z\frac{d}{dz} - \frac{(\mu-S)^2}{1-z^2} - \frac{2S\mu}{1+z} + j(j+1)\right\}X^{\mu}_{j,S}(z) = 0\ . \tag{5.4.19}$$

To solve this equation the boundary conditions at $z = \pm 1$ must be set so that each component of **J** becomes Hermitian. However, since it is not easy, we shall instead make use of (5.4.5) and (5.4.10) derived directly from the Hermiticity. Now let $\mu = j$ in (5.4.19), then it becomes

$$\left\{(1-z^2)\frac{d^2}{dz^2} - 2z\frac{d}{dz} - \frac{(j-S)^2}{1-z^2} - \frac{2Sj}{1+z} + j(j+1)\right\} X^j_{j,S}(z) = 0 , \tag{5.4.20}$$

But this equation is still not convenient. So we shall rewrite it into a hyper-geometric differential equation. To do this, let

$$X^j_{j,S}(z) = (1-z)^{(j-S)/2}(1+z)^{(j+S)/2} f_S(z) , \tag{5.4.21}$$

then (5.4.20) becomes

$$(1-z^2)\frac{d^2 f_S}{dz^2} + 2\{S-(j+1)z\}\frac{df_S}{dz} = 0 . \tag{5.4.22}$$

The general solution for this equation is expressed by a linear combination of two independent solutions, one of which is merely a constant and the other is expressed by a hyper-geometric function. However, we do not need to know the latter because, if $J^{(+)}$ of (5.4.14) is applied on $Y^j_{j,S}(\theta,\phi)$ given by (5.4.18) and (5.4.21), we have

$$J^{(+)}Y^j_{j,S}(\theta,\phi) = -e^{i(j+1-S)\phi}(1-z)^{(j+1-S)/2}(1+z)^{(j+1+S)/2}\frac{df_S}{dz} , \tag{5.4.23}$$

which vanishes according to (5.4.5), and hence $f_S(z)$ becomes a constant. Therefore we obtain

$$Y^j_{j,S}(\theta,\phi) \propto e^{i(j-S)\phi}(1-z)^{(j-S)/2}(1+z)^{(j+S)/2} . \tag{5.4.24}$$

Let us apply $J^{(-)}$ on $Y^j_{j,S}(\theta,\phi)$ n times. Since $J^{(-)}$ is an operator lowering the eigenvalue of J^3 by 1, we can derive, with the help of (5.4.15) and (5.4.24),

$$\begin{aligned}(J^{(-)})^n Y^j_{j,S}(\theta,\phi) &\propto (J^{(-)})^n\{e^{i(j-S)\phi}(1-z)^{(j-S)/2}(1+z)^{(j+S)/2}\}\\ &= e^{i(j-n-S)\phi}(1-z)^{-(j-n-S)/2}(1+z)^{-(j-n+S)/2}\\ &\quad\times D^{(n)}\{(1-z)^{j-S}(1+z)^{j+S}\}\\ &\propto Y^{j-n}_{j,S}(\theta,\phi) ,\end{aligned} \tag{5.4.25}$$

where $D^{(n)}$ is a brief account of d^n/dz^n. If $n = 2j+1$ here, (5.4.10) leads to

$$D^{(2j+1)}\{(1-z)^{j-S}(1+z)^{j+S}\} = 0 . \tag{5.4.26}$$

This equation implies that $(1-z)^{j-S}(1+z)^{j+S}$ is a polynomial at most of $2j$-th order with respect to z. Thus $j \pm S$ are non-negative integers, i.e.,

$$j = |S|, |S|+1, |S|+2, \ldots \ . \tag{5.4.27}$$

Since j is either integral or half-integral as has been shown generally, the possible value of S must be one of the following numbers:

$$S = 0, \pm\frac{1}{2}, \pm 1, \pm\frac{3}{2}, \pm 2, \ldots \ . \tag{5.4.28}$$

Therefore (5.2.20) has been derived from Hermiticity of **J**.

Consequently $Y^{\mu}_{j,S}(\theta,\phi)$ becomes a single-valued function of **k**, and the inner product of any two state vectors $\Phi_S(\theta,\phi)_1$ and $\Phi_S(\theta,\phi)_2$ in the representation space is given by

$$\langle S,1 | S,2 \rangle = \int_0^{2\pi} d\phi \int_0^{\pi} d\theta \sin\theta \Phi^*_S(\theta,\phi)_1 \Phi_S(\theta,\phi)_2 \ . \tag{5.4.29}$$

Needless to say, $d\phi d\theta \sin\theta$ is invariant under a spatial rotation because **k** is a vector.

By the use of (5.4.29) we can normalize $Y^{\mu}_{j,S}(\theta,\phi)$ which has been determined apart from the normalization constant. First, concerning $Y^{j}_{j,S}(\theta,\phi)$, according to (5.2.24) we write

$$Y^{j}_{j,S}(\theta,\phi) = \frac{1}{N} e^{i(j-S)\phi}(1-z)^{(j-S)/2}(1+z)^{(j+S)/2} \ , \tag{5.4.30}$$

and then we get

$$\begin{aligned} |N|^2 &= 2\pi \int_{-1}^{1} dz (1-z)^{j-S}(1+z)^{j+S} \\ &= 2^{2(j+1)}\pi \int_0^1 dz\, z^{j+S}(1-z)^{j-S} \\ &= 2^{2(j+1)}\pi \frac{\Gamma(j-S+1)\Gamma(j+S+1)}{\Gamma(2j+2)} \ . \end{aligned} \tag{5.4.31}$$

The integral in the above equation is known as Euler's integral of the first kind. The normalization constant of $Y^{\mu}_{j,S}(\theta,\phi)$ is readily obtained by successive applications of $J^{(-)}$ on (5.4.39) and by the use of (5.4.25), (5.4.30), (5.4.31) and

$$Y^{\mu}_{j,S}(\theta,\phi) = \sqrt{\frac{(j+\mu)!}{(2j)!(j-\mu)!}}\ (J^{(-)})^{j-\mu} Y^{j}_{j,S}(\theta,\phi) \ , \tag{5.4.32}$$

which is obtained from (5.4.11) with $n = j - \mu$. Of course, without making such a detour we can normalize (5.4.25) directly. In either way, as a result we get

$$Y^{\mu}_{j,S}(\theta,\phi) = \left(\frac{-1}{2}\right)^{\mu}\sqrt{\frac{2j+1}{4\pi}}\sqrt{\frac{(j-\mu)!(j+\mu)!}{(j-S)!(j+S)!}}\,e^{i(\mu-S)\phi}$$
$$\times\,(1-\cos\theta)^{(\mu-S)/2}(1+\cos\theta)^{(\mu+S)/2}P^{(\mu-S,\mu+S)}_{j-\mu}(\cos\theta)\ , \tag{5.4.33}$$

where $P^{(\alpha,\beta)}_n(z)$ is the Jacobi polynomial defined by

$$P^{(\alpha,\beta)}_n(z) = \frac{(-1)^n}{2^n n!}(1-z)^{-\alpha}(1+z)^{-\beta}D^{(n)}\{(1-z)^{\alpha+n}(1+z)^{\beta+n}\}\ . \tag{5.4.34}$$
*)

In this way, we found that S must be subject to the restriction (5.4.28) if **J** of (5.2.25) is Hermitian. As has been mentioned, the irreducible representations of the Poincaré group are specified by the value of S and $\mathrm{sgn}(k_0)$. The minimum value of the total angular momentum of a particle is $|\,S\,|$ as is seen in (5.4.27). In other words, the magnitude of the spin of a massless particle with discrete spin is the minimum value of possible total angular momentum of the particle. In the case of photon (spin 1) for example, it is well known that the minimum of the total angular momentum is 1 and the corresponding wave is expressed by a dipole radiation.**) While the component of **J** to the direction of **k** is readily calculated from (5.2.25) and given by

$$\mathbf{J}\frac{\mathbf{k}}{|\,\mathbf{k}\,|} = S\ , \tag{5.4.35}$$

which is a constant. Thus, $(\mathbf{Jk})/|\,\mathbf{k}\,|$ is an invariant quantity under the Poincaré group. As is seen in the above equation, S can be regarded as an angular momentum along the axis in the direction of momentum, and

---

*) The fact that the eigenfunction of angular momentum is expressed with two polar coordinates θ and ϕ in such a manner even in the case of half-integral j, is essentially based on the existence of the second term of (5.2.25). As a matter of fact, the right-hand side of (5.4.33) may be multiplied by an arbitrary constant phase factor with unit absolute value. It has been chosen here so that an ordinary spherical harmonic function appears when $S = 0$.

**) We must not identify the probability wave with the electro-magnetic wave existing in a real space, but since both waves satisfy the same equation in a covariant description (§7.3), the nature of one is reflected in the other.

then we have a picture that the spin is always directed parallel ($S > 0$) or anti-parallel ($S < 0$) to $\mathbf{k}$. The quantity S is usually called a polarization or helicity analogously to the polarization of light.

In this connection we shall note that we get no more restriction than (5.4.27) and (5.4.28) even if we considered $\mathbf{K}$ besides $\mathbf{J}$. In this case, by introducing $\mathbf{K}$, different j's get connected to one another, and the whole system forms a unitary irreducible representation of the Lorentz group. As to what kind of irreducible representation of the Lorentz group is possible in this case, it can easily be calculated in the following way. Since the arguments are similar in both cases for positive and negative frequencies, we shall discuss only the case of positive frequency. If $|\mathbf{k}|$ is substituted for k_0 in (5.2.26), the Casimir operators $\mathbf{J}^2 - \mathbf{K}^2$ and $\mathbf{JK}$ become, by a direct calculation,

$$\mathbf{J}^2 - \mathbf{K}^2 = S^2 + Z^2 - 1 \,, \tag{5.4.36}$$

$$\mathbf{JK} = iZ|S| \,, \tag{5.4.37}$$

where

$$Z = \begin{cases} -\left(|\mathbf{k}|\frac{\partial}{\partial|\mathbf{k}|} + 1\right) & (S \geq 0) \,, \\ |\mathbf{k}|\frac{\partial}{\partial|\mathbf{k}|} + 1 & (S < 0) \,. \end{cases} \tag{5.4.38)*}$$

Comparing (5.4.36) and (5.4.37) with (4.3.27) and (4.3.28), it is easily seen that $j_0 = |S|$ and that ν is an eigenvalue of Z. Thus, if $g_{\nu,S}(|\mathbf{k}|)$ denotes the eigenfunction of Z,

$$g_{\nu,S}(|\mathbf{k}|) = \frac{1}{\sqrt{2\pi}}|\mathbf{k}|^{\mp\nu-1} \,, \tag{5.4.39}$$

where the signs $-$ and $+$ in front of ν correspond to the cases $S \geq 0$ and $S < 0$ respectively, and $1/\sqrt{2\pi}$ is a normalization constant as will be seen later. Therefore, the eigenstate of (5.4.36) and (5.4.37) is

$$\Phi^{(+)}_{S,\nu,j,\mu}(\mathbf{k}) = \frac{1}{\sqrt{2\pi}}|\mathbf{k}|^{\mp\nu-1}Y^{\mu}_{j,S}(\theta,\phi) \,. \tag{5.4.40}$$

Further the inner product of this function is written, by the use of (5.2.28), as

$$\int \frac{d\mathbf{k}}{|\mathbf{k}|}\Phi^{(+)*}_{S,\nu,j,\mu}(\mathbf{k})\Phi^{(+)}_{S,\nu',j',\mu'}(\mathbf{k}) \tag{5.4.41}$$

---

*) When $S = 0$, either expression on the right-hand sides of (5.4.38) is adaptable as Z, but the first has been used here.

for a given value of S. According to the argument of §4.3, ν is purely imaginary when $S \neq 0$, and when $S = 0$ the two cases are possible in general, one of which is the case where ν is purely imaginary or zero and belongs to the principal series, and the other is the case where ν is real and $0 < \nu^2 < 1$ and belongs to the supplementary series.*) In our model, however, if the representation with $S = 0$ belongs to the supplementary series, the behavior of $g(\mathbf{k})$ becomes irregular at $|\mathbf{k}| = 0$ or ∞, and hence the inner product (5.4.41) is no longer definable. Thus the supplementary series is impossible in this case. So we conclude that ν takes the value of purely imaginary or zero for all S of (5.4.28). Consequently the inner product (5.4.41) is reduced, by rewriting $\mathbf{k}$ into the polar coordinate and setting $|\mathbf{k}| = e^t$, to

$$
\begin{aligned}
(5.4.41) &= \delta_{jj'}\delta_{\mu\mu'}\frac{1}{2\pi}\int_0^\infty d|\mathbf{k}||\mathbf{k}|^{\mp(\nu'-\nu)-1} \\
&= \delta_{jj'}\delta_{\mu\mu'}\frac{1}{2\pi}\int_{-\infty}^\infty dte^{\mp(\nu'-\nu)t} \\
&= \delta_{jj'}\delta_{\mu\mu'}\delta[i(\nu-\nu')] \,, \qquad (5.4.42)
\end{aligned}
$$

which is an ordinary form of normalization for the case of continuous eigenvalue. In this way, when S is given, all irreducible representations specified by ν being $-\infty < i\nu < \infty$ are possible for the irreducible representation of the Lorentz group. These are connected to one another by the translation of the coordinates, i.e., the transformation $T(a)$, and as a whole forms an irreducible representation of the Poincaré group.

In the above we studied the angular momentum of a massless particle with discrete spin. A similar argument is also applicable to the case of continuous spin and to the case of imaginary mass, but it is not given here in order to avoid the overlapping. The interested reader is encouraged to try to study this case by himself.

---

*)The 1-dimensional representation where $S = 0$ and $\nu^2 = 1$ is omitted because we consider only the case where the variable $\mathbf{k}$ has a physical meaning.

Chapter 6

Covariant Formalism I — Massive Particles

In the discussions up to the previous chapter, we looked for all possible kinds of free particles and investigated their behavior under the transformations of the Poincaré group. As has been seen in the previous chapter, the transformation properties of the wave functions are not so simple and the transformation coefficients depend generally on the variable k of the wave function. This comes from the fact that the Wigner rotation $Q(\lambda_k, l)$ depends on k, and it is quite different from the transformations of ordinary tensors or spinors. Further, in the case $k_0^2 \geq \mathbf{k}^2$, the transformation coefficients are different for the positive and negative frequencies. Although this is a self-contained formalism and it completely determines the properties of the state vectors, it may not be convenient by itself for a direct application in some cases. In particular, when we consider the correspondence to the covariant theory of relativistic quantum fields, we must rewrite our formalism into a covariant form and must make clear the connection with the former. Here, the covariant formalism means a formalism in which the transformation coefficients depend neither on the variables in the transformed functions nor on the sign of frequency. In connection to the field theory it is more convenient to express the formalisms in the x-representation rather than in the momentum representation, because the field is a function of a space-time point. The general theory of covariant formalism for a particle with imaginary mass is, however, not yet completed, except for the simplest case where the little group has a 1-dimensional representation.*) Therefore we shall discuss particles with finite mass and zero mass in the present and the next chapters.

§6.1 Particles with Spin 0

In this chapter we shall discuss covariant formalisms for particles with finite mass. Let us begin with a consideration of the simplest case of spin 0 particle. In this case, the transformations of a wave function are obtained from (5.1.16) and (5.1.17) with $\mathbf{S} = 0$, and then the infinitesimal Lorentz transformations are given by

$$\phi^{(\pm)'}(\mathbf{k}) = \phi^{(\pm)}(\mathbf{k} - \mathbf{k} \times \boldsymbol{\theta}) \quad (\theta\text{-transformation}), \tag{6.1.1}$$
$$\phi^{(\pm)'}(\mathbf{k}) = \phi^{(\pm)}(\mathbf{k} \pm \omega_k \boldsymbol{\tau}) \quad (\tau\text{-transformation}). \tag{6.1.2}$$

Let us introduce here a function in the x-representation by

$$U^{(\pm)}(x) = \frac{1}{(2\pi)^{3/2}} \int \frac{d\mathbf{k}}{\omega_k} \phi^{(\pm)}(\mathbf{k}) e^{i(\mathbf{k}\mathbf{x} \mp \omega_k t)} . \tag{6.1.3}$$

Since $d\mathbf{k}/\omega_k$ is a Lorentz invariant quantity as has been noted in §3.1, we get the transformations of $U^{(\pm)}(x)$ from (6.1.1) and (6.1.2) in the form

$$U^{(\pm)'}(x) = \frac{1}{(2\pi)^{3/2}} \int \frac{d\mathbf{k}}{\omega_k} \phi^{(\pm)'}(\mathbf{k}) e^{i(\mathbf{k}\mathbf{x} \mp \omega_k t)} = U^{(\pm)}(\Lambda^{-1} x) \tag{6.1.4}$$

for both θ- and τ-transformations, where Λ reads either $\Lambda(\theta)$ or $\Lambda(\tau)$. The inner product (5.1.7) is written as

$$\begin{aligned} \langle 1, \pm \,|\, 2, \pm \rangle &= \int \omega_k d\mathbf{k} \frac{\phi^{(\pm)*}(\mathbf{k})_1}{\omega_k} \frac{\phi^{(\pm)}(\mathbf{k})_2}{\omega_k} \\ &= \pm \frac{1}{2i} \int d\mathbf{x} \left(\frac{\partial U^{(\pm)*}(x)_1}{\partial t} U^{(\pm)}(x)_2 - U^{(\pm)*}(x)_1 \frac{\partial U^{(\pm)}(x)_2}{\partial t} \right) \end{aligned} \tag{6.1.5}$$

in the x-representation. The definitions of positive and negative frequencies, corresponding to (5.1.18), are given by

$$i \frac{\partial U^{(\pm)}(x)}{\partial t} = \pm \sqrt{m^2 - \nabla^2}\, U^{(\pm)}(x) , \tag{6.1.6}$$

*) In this case we can replace m^2 by $-m^2$ in the formalism for a spin 0 particle with finite mass, for example, in the Klein-Gordon equation (6.1.8) which will be mentioned in next section.

which are nothing but the equations of motion for $U^{(\pm)}(x)$. However, (6.1.5) and (6.1.6) depend on the sign of frequency explicitly, so they are not desirable from a view point of covariant formalism. Let us then modify the definition of the inner product and use a modified inner product $\langle\langle\ \rangle\rangle$ given by

$$\langle\langle 1,\pm\,|\,2,\pm\rangle\rangle = \pm\langle 1,\pm\,|\,2,\pm\rangle\ . \tag{6.1.7}$$

This inner product with no explicit frequency dependence is generally called a covariant inner product. Although (6.1.7) has a form independent of the sign of frequency, we must abandon its probability interpretation because the norm of a negative frequency state vector becomes negative. On the other hand the expectation value of k_0 in terms of the covariant inner product, $\langle\langle\pm\,|\,k_0\,|\,\pm\rangle\rangle$, has a positive value independent of the sign of frequency. This is quite curious for the particle picture in quantum mechanics. But, as has been mentioned in §4.5, the particle in a negative frequency state itself is not acceptable physically. The correct interpretation can be given in the quantum field theory which will be discussed later. Here, leaving this problem for the present, we shall give priority to the covariant formalism.

Instead of (6.1.6) we can use a covariant equation of motion

$$(\partial_\mu^2 - m^2)U(x) = 0\ , \tag{6.1.8}$$

where ∂_μ stands for $\partial/\partial x_\mu$ and this notation will be frequently used hereafter. The positive frequency part of a solution for this equation is $U^{(+)}$ and the negative frequency part is $U^{(-)}$. Equation (6.1.8) is called a Klein-Gordon equation. In terms of $U(x)$ the inner product is obtained from (6.1.7) in a frequency independent form:

$$\langle\langle 1\,|\,2\rangle\rangle = \frac{1}{2i}\int d\mathbf{x}\left(\frac{\partial U^*(x)_1}{\partial t}U(x)_2 - U^*(x)_1\frac{\partial U(x)_2}{\partial t}\right)\ . \tag{6.1.9}$$

If $U_1^{(+)}$ and $U_2^{(-)}$ are substituted for U_1 and U_2 respectively, the above equation leads to $\langle\langle 1+\,|\,2-\rangle\rangle = 0$. Concerning the Lorentz transformation, (6.1.4) leads to

$$U'(x) = U(\Lambda^{-1}x)\ . \tag{6.1.10}$$

Equations (6.1.8)–(6.1.10) give a covariant formalism for a spin 0 particle with finite mass.

Other various forms of covariant formalism are also possible. Although they look different, their contents are of course the same, because they are based on the irreducible representation of the Poincaré group only. We shall give one example here. Let

$$U_\mu(x) = \frac{1}{m}\partial_\mu U(x) , \tag{6.1.11}$$

then, from (6.1.10) the Lorentz transformation is given by

$$U'_\mu(x) = \Lambda_{\mu\nu}U_\nu(\Lambda^{-1}x) , \tag{6.1.12}$$

and the inner product is

$$\langle\langle 1 \mid 2 \rangle\rangle = -\frac{m}{2}\int d\mathbf{x}\{(U_4(x)_1)^* U(x)_2 + U^*(x)_1 U_4(x)_2\} . \tag{6.1.13}$$

On the other hand, from (6.1.8) and (6.1.11) we have

$$U(x) = \frac{1}{m}\partial_\mu U_\mu(x) . \tag{6.1.14}$$

If we combine (6.1.11) and (6.1.14) we obtain the same equation as (6.1.8). In this formalism the theory is described with five variables $U(x)$ and $U_\mu(x)$ ($\mu = 1, 2, 3, 4$). That is, if we introduce $\psi(x) = (U_1(x), U_2(x), U_3(x), U_4(x), U(x))$, the above equation is rewritten as

$$(\alpha_\mu\partial_\mu + m)\psi(x) = 0 . \tag{6.1.15}$$

This is called a Kemmer type equation for a spin 0 particle. The quantities α_μ ($\mu = 1, 2, 3, 4$) are 5×5 matrices and their elements at the p-th row and the q-th column ($p, q = 1, 2, \ldots 5$) are given by

$$(\alpha_\mu)_{pq} = -(\delta_{p\mu}\delta_{q5} + \delta_{p5}\delta_{q\mu}) . \tag{6.1.16}$$

In terms of α_μ the inner product is written as

$$\langle\langle 1 \mid 2 \rangle\rangle = \frac{m}{2}\int d\mathbf{x}\psi^*(x)_1\alpha_4\psi(x)_2 . \tag{6.1.17}$$

Here it must be noticed that the amplitude in a covariant formalism is not a probability amplitude by itself. In fact, a probability amplitude is directly connected to a state vector in an irreducible representation space of the Poincaré group. For example, in the case of massive particle with positive frequency, if we define its x-representation by

$$\phi_\xi(x) = \frac{1}{(2\pi)^{3/2}}\int d\mathbf{k}\omega_k^{-1/2}\phi_\xi^{(+)}(\mathbf{k})e^{i(\mathbf{k}\mathbf{x}-\omega_k t)} , \tag{6.1.18}$$

we obtain the inner product from (5.1.7) as

$$\langle 1,+|2,+\rangle = \sum_{\xi}\int dx\phi_{\xi}^{*}(x)_1\phi_{\xi}(x)_2 \ . \tag{6.1.19}$$

In this case we may interpret the above $\phi_\xi(x)$ as a probability amplitude in the x-representation. Further, it is easily seen that such $\phi_\xi(x)$ tends to the ordinary non-relativistic probability amplitude in the limit where the light velocity goes to infinity. However, $\mathbf{x}$ and t in (6.1.18) have no longer the transformation properties of a 4-vector.*) The quantity $\mathbf{x}$ is sometimes called a position operator of the particle.

§6.2 Dirac Particles

A particle with spin $\frac{1}{2}$ is called a Dirac particle. In this case we can put $\mathbf{S} = \sigma/2$ in (5.1.21) and (5.1.22), then $\phi^{(\pm)}(\mathbf{k})$ are two-component wave functions. As has been mentioned in the previous section, we must consider these positive and negative frequency wave functions not separately but together in order to get a covariant formalism. Now introducing

$$\varphi^{(+)}(\mathbf{k}) = \begin{pmatrix} \phi^{(+)}(\mathbf{k}) \\ 0 \\ 0 \end{pmatrix}, \quad \varphi^{(-)}(\mathbf{k}) = \begin{pmatrix} 0 \\ 0 \\ \phi^{(-)}(\mathbf{k}) \end{pmatrix}, \tag{6.2.1}$$

and combining them we define

$$\varphi(\mathbf{k}) = \varphi^{(+)}(\mathbf{k}) + \varphi^{(-)}(\mathbf{k}) \ . \tag{6.2.2}$$

The function $\varphi(\mathbf{k})$ has four components, and according to (5.1.18), $\varphi^{(\pm)}(\mathbf{k})$ are given as positive and negative frequency solutions for

$$k_0\varphi(\mathbf{k}) = \beta\omega_k\varphi(\mathbf{k}) \ , \tag{6.2.3}$$

where

$$\beta = \begin{pmatrix} 1 & 0 & 0 & 0 \\ 0 & 1 & 0 & 0 \\ 0 & 0 & -1 & 0 \\ 0 & 0 & 0 & -1 \end{pmatrix} . \tag{6.2.4}$$

*)This may seem curious. But, if a position operator x forming a 4-vector with t is introduced, it will be mixed with t by a Lorentz transformation, then we have a paradox that t which is a mere parameter in one reference frame becomes an operator in another frame. There were various arguments on the definition of a position operator for a relativistic particle, but they will not be given here. See for example T. D. Newton and E. P. Wigner: *Revs. Modern Phys.*, **21**, (1949) 400.

On the other hand, the Lorentz transformations of $\varphi(\mathbf{k})$ are obtained, from (5.1.21)–(5.1.24), as

$$\varphi'(\mathbf{k}) = (1 + i\mathbf{J}\boldsymbol{\theta})\varphi(\mathbf{k}) \qquad (\theta\text{-transformation}), \tag{6.2.5}$$

$$\varphi'(\mathbf{k}) = (1 - i\mathbf{K}\boldsymbol{\tau})\varphi(\mathbf{k}) \qquad (\tau\text{-transformation}), \tag{6.2.6}$$

$$\mathbf{J} = \frac{1}{i}\left(\mathbf{k} \times \frac{\partial}{\partial \mathbf{k}}\right) + \frac{\boldsymbol{\sigma}}{2}, \tag{6.2.7}$$

$$\mathbf{K} = -k_0\left(\frac{1}{i}\frac{\partial}{\partial \mathbf{k}} - \frac{1}{2}\frac{\mathbf{k} \times \boldsymbol{\sigma}}{\omega_k(m + \omega_k)}\right). \tag{6.2.8}$$

By the use of (5.1.7) the inner product is written as

$$\langle 1 | 2 \rangle = \int \frac{d\mathbf{k}}{\omega_k}\varphi^*(\mathbf{k})_1\varphi(\mathbf{k})_2 , \tag{6.2.9}$$

where $\boldsymbol{\sigma}$ is a brief account of

$$\begin{pmatrix} \boldsymbol{\sigma} & \begin{matrix} 0 & 0 \\ 0 & 0 \end{matrix} \\ \begin{matrix} 0 & 0 \\ 0 & 0 \end{matrix} & \boldsymbol{\sigma} \end{pmatrix}$$

which is a 4×4 matrix constructed from the 2×2 Pauli matrix. Hereafter $\boldsymbol{\sigma}$ shall denote the 4×4 matrix when it acts on a 4-component quantity, and the 2×2 Pauli matrix when it acts on a 2-component quantity. In this way the explicit dependence on the sign of frequency has been eliminated, but we have not yet obtained any covariant formalism because the second term on the right-hand side of (6.2.8) involves $\mathbf{k}$. However, since $\omega_k\beta$ in (6.2.3) is nothing but the Hamiltonian of Dirac equation diagonalized by a unitary transformation, we may obtain a covariant formalism by tracing back the diagonalization process. Put

$$\chi(\mathbf{k}) = U_F(\mathbf{k})\varphi(\mathbf{k}) , \tag{6.2.10}$$

where $U_F(\mathbf{k})$ is a unitary operator defined by

$$\begin{aligned} U_F(\mathbf{k}) &= \exp\left\{-\frac{\beta(\boldsymbol{\alpha}\mathbf{k})}{2|\mathbf{k}|}\tan^{-1}\frac{|\mathbf{k}|}{m}\right\} \\ &= \frac{\omega_k + m - \beta(\boldsymbol{\alpha}\mathbf{k})}{\sqrt{2\omega_k(\omega_k + m)}} , \end{aligned} \tag{6.2.11}$$

and its inverse is usually called a Foldy transformation operator. Here $\boldsymbol{\alpha} = (\alpha_1, \alpha_2, \alpha_3)$ is constructed by 4×4 matrices different from (6.1.16):

$$\alpha_i = \begin{pmatrix} \begin{matrix} 0 & 0 \\ 0 & 0 \end{matrix} & \sigma_i \\ \sigma_i & \begin{matrix} 0 & 0 \\ 0 & 0 \end{matrix} \end{pmatrix} \qquad (i = 1, 2, 3) \ , \tag{6.2.12}$$

which satisfy

$$\left.\begin{aligned} \alpha_i \alpha_j &= \delta_{ij} + i \sum_{k=1}^{3} \varepsilon_{ijk} \sigma_k \ , \\ \{\alpha_i, \alpha_j\} &= 2\delta_{ij} \ , \quad \{\alpha_i, \beta\} = 0 \ , \end{aligned}\right\} \tag{6.2.13}$$

where

$$\{A, B\} = AB + BA \ . \tag{6.2.14}$$

Using (6.2.13) we can easily derive

$$\begin{aligned} U_F(\mathbf{k})\beta U_F(\mathbf{k})^{-1} &= \frac{\{\omega_k + m - \beta(\boldsymbol{\alpha}\mathbf{k})\}\beta\{\omega_k + m + \beta(\boldsymbol{\alpha}\mathbf{k})\}}{2\omega_k(\omega_k + m)} \\ &= \frac{\boldsymbol{\alpha}\mathbf{k} + \beta m}{\omega_k} \ , \end{aligned} \tag{6.2.15}$$

and from (6.2.3) and (6.2.10) we get

$$k_0 \chi(\mathbf{k}) = (\boldsymbol{\alpha}\mathbf{k} + \beta m)\chi(\mathbf{k}) \ . \tag{6.2.16}$$

The part in the parentheses on the right-hand side of the above equation is nothing but the Hamiltonian of Dirac equation.

To get the transformation property of $\chi(\mathbf{k})$ under a Lorentz transformation, we must have the knowledge of $U_F(\mathbf{k})\mathbf{J}U_F(\mathbf{k})^{-1}$ and $U_F(\mathbf{k})\mathbf{K}U_F(\mathbf{k})^{-1}$. Since $U_F(\mathbf{k})$ is commutable with $\mathbf{k}$, let us first calculate $U_F(\mathbf{k})\left(\frac{1}{i}\frac{\partial}{\partial \mathbf{k}}\right)U_F(\mathbf{k})^{-1}$ and $\frac{1}{2}U_F(\mathbf{k})\boldsymbol{\sigma}U_F(\mathbf{k})^{-1}$. After a somewhat tedious calculation we get the following relations:

$$\begin{aligned} U_F(\mathbf{k})\left(\frac{1}{i}\frac{\partial}{\partial \mathbf{k}}\right)U_F(\mathbf{k})^{-1} &= \frac{1}{i}\frac{\partial}{\partial \mathbf{k}} - iU_F(\mathbf{k})\frac{\partial U_F(\mathbf{k})^{-1}}{\partial \mathbf{k}} \\ &= \frac{1}{i}\frac{\partial}{\partial \mathbf{k}} - \frac{1}{2\omega_k^2(\omega_k + m)}\{i\mathbf{k}(\boldsymbol{\alpha}\mathbf{k})\beta + i\beta\omega_k(\omega_k + m)\boldsymbol{\alpha} \\ &\quad - \omega_k(\boldsymbol{\sigma} \times \mathbf{k})\} \ , \end{aligned} \tag{6.2.17}$$

$$\frac{1}{2}U_F(\mathbf{k})\boldsymbol{\sigma}U_F(\mathbf{k})^{-1} = \frac{1}{2\omega_k(\omega_k+m)}\{m(\omega_k+m)\boldsymbol{\sigma}+i(\omega_k+m)\beta(\mathbf{k}\times\boldsymbol{\alpha})+\mathbf{k}(\mathbf{k}\boldsymbol{\sigma})\} \tag{6.2.18}$$

where we have used

$$\alpha_i\sigma_j\alpha_k = \sigma_i\sigma_j\sigma_k = i\varepsilon_{ijk} + \delta_{ij}\sigma_k - \delta_{ik}\sigma_j + \delta_{jk}\sigma_i \ . \tag{6.2.19}$$

Hence we obtain

$$\begin{aligned} U_F(\mathbf{k})\mathbf{J}U_F(\mathbf{k})^{-1} &= \mathbf{k}\times U_F(\mathbf{k})\left(\frac{1}{i}\frac{\partial}{\partial\mathbf{k}}\right)U_F(\mathbf{k})^{-1} + \frac{1}{2}U_F(\mathbf{k})\boldsymbol{\sigma}U_F(\mathbf{k})^{-1} \\ &= \frac{1}{i}\mathbf{k}\times\frac{\partial}{\partial\mathbf{k}} + \frac{m(\omega_k+m)+\mathbf{k}^2}{2\omega_k(\omega_k+m)}\boldsymbol{\sigma} \\ &= \frac{1}{i}\mathbf{k}\times\frac{\partial}{\partial\mathbf{k}} + \frac{\boldsymbol{\sigma}}{2} , \end{aligned} \tag{6.2.20}$$

and similarly we have

$$\begin{aligned} U_F(\mathbf{k})\mathbf{K}U_F(\mathbf{k})^{-1} &= -k_0\left\{U_F(\mathbf{k})\left(\frac{1}{i}\frac{\partial}{\partial\mathbf{k}}\right)U_F(\mathbf{k})^{-1} - \frac{\mathbf{k}\times U_F(\mathbf{k})\boldsymbol{\sigma}U_F(\mathbf{k})^{-1}}{2\omega_k(\omega_k+m)}\right\} \\ &= ik_0\frac{\partial}{\partial\mathbf{k}} - \frac{i}{2}\frac{k_0}{\omega_k^2}\{i(\mathbf{k}\times\boldsymbol{\sigma})+\boldsymbol{\alpha}\beta m\} \ . \end{aligned} \tag{6.2.21}$$

Applying an identity

$$i(\mathbf{k}\times\boldsymbol{\sigma}) + \boldsymbol{\alpha}\beta m = \boldsymbol{\alpha}\{(\boldsymbol{\alpha}\mathbf{k})+\beta m\} - \mathbf{k} \tag{6.2.22}$$

and Eq. (6.2.16), the transformations of $\chi(\mathbf{k})$ are derived from (6.2.5) and (6.2.6):

$$\chi'(\mathbf{k}) = \left\{1 + i\left(\frac{1}{i}\mathbf{k}\times\frac{\partial}{\partial\mathbf{k}} + \frac{\boldsymbol{\sigma}}{2}\right)\boldsymbol{\theta}\right\}\chi(\mathbf{k}) \qquad (\theta\text{-transformation}) , \tag{6.2.23}$$

$$\chi'(\mathbf{k}) = \left\{1 - \frac{\boldsymbol{\sigma}\boldsymbol{\tau}}{2} + k_0\left(\frac{\partial}{\partial\mathbf{k}} + \frac{\mathbf{k}}{2\omega_k^2}\right)\boldsymbol{\tau}\right\}\chi(\mathbf{k}) \qquad (\tau\text{-transformation}) . \tag{6.2.24}$$

These expressions are much simpler, but the description is not yet covariant completely because (6.2.24) contains the term $(k_0\mathbf{k}\boldsymbol{\tau}/2\omega_k^2)\chi(\mathbf{k})$. This term is, however, easily eliminated by the help of a relation

$$\sqrt{\omega_k}\left(\frac{\partial}{\partial\mathbf{k}} + \frac{\mathbf{k}}{2\omega_k^2}\right)\chi(\mathbf{k}) = \frac{\partial}{\partial\mathbf{k}}(\sqrt{\omega_k}\,\chi(\mathbf{k})) \ . \tag{6.2.25}$$

That is, if we put

$$\psi(\mathbf{k}) = \sqrt{\omega_k}\chi(\mathbf{k}) , \tag{6.2.26}$$

since ω_k is commutable with $\mathbf{k} \times \frac{\partial}{\partial \mathbf{k}}$, we have

$$\psi'(\mathbf{k}) = \left\{1 + i\left(\frac{1}{i}\mathbf{k} \times \frac{\partial}{\partial \mathbf{k}} + \frac{\boldsymbol{\sigma}}{2}\right)\boldsymbol{\theta}\right\}\psi(\mathbf{k}) \qquad (\theta\text{-transformation}) , \tag{6.2.27}$$

$$\psi'(\mathbf{k}) = \left\{1 + i\left(k_0\frac{\partial}{\partial \mathbf{k}} - \frac{\boldsymbol{\alpha}}{2}\right)\boldsymbol{\tau}\right\}\psi(\mathbf{k}) \qquad (\tau\text{-transformation}) , \tag{6.2.28}$$

or

$$\psi'(\mathbf{k}) = \left(1 + \frac{i}{2}\boldsymbol{\sigma\theta}\right)\psi(\mathbf{k} - \mathbf{k} \times \boldsymbol{\theta}) \qquad (\theta\text{-transformation}) , \tag{6.2.29}$$

$$\psi'(\mathbf{k}) = \left(1 - \frac{1}{2}\boldsymbol{\alpha\tau}\right)\psi(\mathbf{k} + k_0\boldsymbol{\tau}) \qquad (\tau\text{-transformation}) . \tag{6.2.30}$$

In terms of $\psi(\mathbf{k})$, (6.2.16) can be written as

$$k_0\psi(\mathbf{k}) = (\boldsymbol{\alpha}\mathbf{k} + \beta m)\psi(\mathbf{k}) , \tag{6.2.31}$$

and the inner product is

$$\langle 1 | 2 \rangle = \int \frac{d\mathbf{k}}{\omega_k^2}\psi^*(\mathbf{k})_1\psi(\mathbf{k})_2 . \tag{6.2.32}$$

Equations (6.2.29)–(6.2.32) obtained in this way have completely covariant forms, and are now easily converted into the x-representation. Define $\psi(x)$ by

$$\psi(x) = \frac{1}{(2\pi)^{3/2}}\int \frac{d\mathbf{k}}{\omega_k}e^{i(\mathbf{kx} - k_0 t)}\psi(\mathbf{k}) . \tag{6.2.33}$$

Of course, k_0 takes values of ω_k and $-\omega_k$ corresponding to the positive and negative frequencies of $\psi(x)$ respectively. Then we get immediately

$$\psi'(x) = \left(1 + \frac{i}{2}\boldsymbol{\sigma\theta}\right)\psi(\Lambda(\theta)^{-1}x) \quad (\theta\text{-transformation}) , \tag{6.2.34}$$

$$\psi'(x) = \left(1 - \frac{1}{2}\boldsymbol{\alpha\tau}\right)\psi(\Lambda(\tau)^{-1}x) \quad (\tau\text{-transformation}) , \tag{6.2.35}$$

$$i\frac{\partial\psi(x)}{\partial t} = \left(\frac{1}{i}\boldsymbol{\alpha}\nabla + \beta m\right)\psi(x) , \tag{6.2.36}$$

$$\langle 1 | 2 \rangle = \int d\mathbf{x}\psi^*(x)_1\psi(x)_2 . \tag{6.2.37}$$

Although these equations are usually used to describe a spin $\frac{1}{2}$ particle, the above discussion reveals the relation between $\psi(x)$ and the basic vectors of an irreducible representation space of the Poincaré group. When $\psi^{(\pm)}(\mathbf{k})$ denote the positive and negative frequency parts of $\psi(\mathbf{k})$ in the momentum representation, i.e.,

$$\pm\omega_k\psi^{(\pm)}(\mathbf{k}) = (\boldsymbol{\alpha}\mathbf{k} + \beta m)\psi^{(\pm)}(\mathbf{k}) \ , \qquad (6.2.38)$$

the relation is given by

$$\psi^{(\pm)}(\mathbf{k}) = \sqrt{\omega_k}\, U_F(\mathbf{k})\varphi^{(\pm)}(\mathbf{k}) \ , \qquad (6.2.39)$$

where $\psi(\mathbf{k}) = \psi^{(+)}(\mathbf{k}) + \psi^{(-)}(\mathbf{k})$. The functions $\varphi^{(\pm)}(\mathbf{k})$ have the forms (6.2.1), and the upper two components of the first equation and the lower two components of the second equation are the positive and negative frequency vectors in the irreducible representation spaces respectively.

Since the inner product (6.2.37) is itself covariant, the covariant inner product is nothing but (6.2.37):

$$\langle\langle 1 \,|\, 2 \rangle\rangle = \langle 1 \,|\, 2 \rangle \ . \qquad (6.2.40)$$

This is quite different from the case of spin 0 particle. It is, however, a special case of more general covariant inner product which will be mentioned in the next section.

In the covariant formalism the τ-transformation matrix $(1 - \boldsymbol{\alpha\tau}/2)$ is no more unitary. Nevertheless, it leaves the inner product (6.2.37) invariant. This is due essentially to the existence of the Dirac equation (6.2.36). Actually the transformation matrix has become non-unitary because (6.2.24) has been derived from (6.2.6) with the help of (6.2.16) corresponding to the equation of motion. Thus the unitarity of the transformation (6.2.35) is guaranteed by (6.2.36). The Dirac equation is frequently used in the form

$$(\gamma_\mu \partial_\mu + m)\psi(x) = 0 \qquad (6.2.41)$$

which is obtained from (6.2.36) multiplied on the left by β. The matrices γ_μ $(\mu = 1, 2, 3, 4)$ are now defined by

$$\left.\begin{aligned} \gamma_j &= \frac{1}{i}\beta\alpha_j \quad (j = 1, 2, 3) \ , \\ \gamma_4 &= \beta \ . \end{aligned}\right\} \qquad (6.2.42)$$

As the covariant formalism for spin 0 particle was not unique, various covariant formalisms are also possible in the case of a spin $\frac{1}{2}$ particle. These will be mentioned in a more general discussion later.

§6.3 Particles with Higher Spin

In the case of spin $n/2$ ($n > 1$), a $(n+1)\times(n+1)$ spin matrix is used for **S** in (5.1.21) and (5.1.22). Such **S** has, however, a complicated property when n becomes large, and it is not easy to rewrite our theory into a covariant formalism directly. We shall therefore avoid this procedure, and shall rather generalize the argument of the previous section to get a covariant description of a spin $n/2$ particle. As a preparation let us begin with the following consideration.

A spinor $u = (u_1, u_2)$ transforms as

$$u'_{\xi'} = \sum_{\xi=1,2} \left(1 + \frac{i}{2}\sigma\theta\right)_{\xi'\xi} u_\xi \tag{6.3.1}$$

under an infinitesimal spatial rotation. Spinors $u^{(1)} = (u_1, 0)$ and $u^{(2)} = (0, u_2)$ are the eigenstates of σ_3 belonging to the eigenvalues 1 and -1 respectively. Consider a direct product of such spinors and let $u_{\xi_1\xi_2\ldots\xi_n}$ denote its component. By definition it transforms as

$$u'_{\xi'_1\xi'_2\ldots\xi'_n} = \sum_{\xi_1\ldots\xi_n} \left\{1 + \frac{i}{2}(\sigma^{(1)} + \sigma^{(2)} + \ldots + \sigma^{(n)})\theta\right\}_{\xi'_1\ldots\xi'_n\xi_1\ldots\xi_n} u_{\xi_1\xi_2\ldots\xi_n} \,, \tag{6.3.2}$$

where $\sigma^{(i)}$ $(i = 1, 2, \ldots, n)$ are the Pauli matrices acting on the indices ξ_i, and their matrix elements are given by

$$(\sigma^{(i)})_{\xi'_1\ldots\xi'_n\xi_1\ldots\xi_n} = \delta_{\xi'_1\xi_1}\delta_{\xi'_2\xi_2}\cdots\delta_{\xi'_{i-1}\xi_{i-1}}\delta_{\xi'_{i+1}\xi_{i+1}}\cdots\delta_{\xi'_n\xi_n}\sigma_{\xi'_i\xi_i} \,. \tag{6.3.3}$$

The direct product space has a dimension 2^n, and consists of various irreducible representation spaces of the rotation group. Let us consider a component $u_{(\xi_1\xi_2\ldots\xi_n)}$ with totally symmetrized indices to extract one of the irreducible representation spaces. That is, $u_{(\xi_1\xi_2\ldots\xi_n)}$ is invariant under an exchange of any two indices. It is denoted as $u_{(\ldots)_n}$ for simplicity. When each $\xi_1, \xi_2, \ldots, \xi_n$ has the value 1 or 2, the number of independent $u_{(\ldots)_n}$'s is $n+1$, which implies that $u_{(\ldots)_n}$'s are the components of a state vector in a $(n+1)$-dimensional space (denoted as $V^{(n+1)}$). From (6.3.2) the transformation property of $u_{(\ldots)_n}$ is given by

$$u'_{(\ldots)_n} = \left\{1 + \frac{i}{2}(\sigma^{(1)} + \sigma^{(2)} + \ldots + \sigma^{(n)})\theta\right\} u_{(\ldots)_n} \,. \tag{6.3.4}$$

Although the explicit description of the indices $\xi_1, \xi_2, \ldots, \xi_n$ was left out in the above expression, the meaning is clear. Here, if we consider a state vector in which $u_{(11\ldots1)} \neq 0$ and all the other components are 0, among the states in $V^{(n+1)}$ it belongs to the maximum eigenvalue of the angular momentum $(\sigma_3^{(1)} + \sigma_3^{(2)} + \ldots + \sigma_3^{(n)})/2$ around the third axis. Since the eigenvalue is $n/2$ and since any state vector in $V^{(n+1)}$ stays in $V^{(n+1)}$ after a 3-dimensional rotation according to (6.3.4), the irreducible representation space $D_{n/2}$ of the rotation group with angular momentum $n/2$ is included in $V^{(n+1)}$. On the other hand $D_{n/2}$ has a dimension $n+1$. Since this coincides with that of $V^{(n+1)}$, $V^{(n+1)}$ and $D_{n/2}$ must be the same space. Therefore, although $(\boldsymbol{\sigma}^{(1)} + \boldsymbol{\sigma}^{(2)} + \ldots + \boldsymbol{\sigma}^{(n)})/2$ in (6.3.4) is a reducible operator, the state vector transforms always as a vector in the irreducible representation space $D_{n/2}$ under the rotation group. Thus we can use (6.3.4) to describe the transformation of an irreducible representation with angular momentum $n/2$. Using $(\boldsymbol{\sigma}^{(1)} + \boldsymbol{\sigma}^{(2)} + \ldots + \boldsymbol{\sigma}^{(n)})/2$ for $\mathbf{S}$ in (5.1.21) and (5.1.22), and using $\phi^{(\pm)}_{(\ldots)_n}(\mathbf{k})$ symmetric with respect to n indices for $\phi^{(\pm)}_{\xi}(\mathbf{k})$, we obtain the infinitesimal Lorentz transformation of a spin $n/2$ particle.

To generalize the argument of the previous section, we assume here that each index of $\phi^{(+)}_{(\ldots)_n}$ has the value 1 or 2, and that of $\phi^{(-)}_{(\ldots)_n}$ has the value 3 or 4, and we introduce $\varphi^{(\pm)}_{(a_1a_2\ldots a_n)}(\mathbf{k})$ and $\varphi_{(a_1a_2\ldots a_n)}(\mathbf{k})$, which seem to have more numbers of components, by the following equations:

$$\varphi^{(+)}_{(a_1a_2\ldots a_n)}(\mathbf{k}) = \begin{cases} N\phi^{(+)}_{(a_1a_2\ldots a_n)}(\mathbf{k}) & (a_1, a_2, \ldots a_n = 1 \,\mathrm{or}\, 2)\,, \\ 0 & (\mathrm{other\ cases})\,, \end{cases} \tag{6.3.5}$$

$$\varphi^{(-)}_{(a_1a_2\ldots a_n)}(\mathbf{k}) = \begin{cases} N\phi^{(-)}_{(a_1a_2\ldots a_n)}(\mathbf{k}) & (a_1, a_2 \ldots a_n = 3 \,\mathrm{or}\, 4)\,, \\ 0 & (\mathrm{other\ cases})\,, \end{cases} \tag{6.3.6}$$

$$\varphi^{(\mathbf{k})}_{(a_1a_2\ldots a_n)} = \varphi^{(+)}_{(a_1a_2\ldots a_n)}(\mathbf{k}) + \varphi^{(-)}_{(a_1a_2\ldots a_n)}(\mathbf{k})\,, \tag{6.3.7}$$

where $N(\neq 0)$ has been introduced as a normalization constant.

The transformations of these functions are given by

$$\varphi'_{(\ldots)_n}(\mathbf{k}) = (1 + i\mathbf{J}\boldsymbol{\theta})\varphi_{(\ldots)_n}(\mathbf{k}) \quad (\theta\text{-transformation})\,, \tag{6.3.8}$$

$$\varphi'_{(\ldots)_n}(\mathbf{k}) = (1 - i\mathbf{K}\boldsymbol{\tau})\varphi_{(\ldots)_n}(\mathbf{k}) \quad (\tau\text{-transformation})\,, \tag{6.3.9}$$

$$\mathbf{J} = \frac{1}{i}\left(\mathbf{k} \times \frac{\partial}{\partial \mathbf{k}}\right) + \frac{1}{2}\sum_{i=1}^{n} \boldsymbol{\sigma}^{(i)} , \tag{6.3.10}$$

$$\mathbf{K} = -k_0 \left(\frac{1}{i}\frac{\partial}{\partial \mathbf{k}} - \frac{1}{2}\sum_{i=1}^{n} \frac{\mathbf{k} \times \boldsymbol{\sigma}^{(i)}}{\omega_k(\omega_k + m)} \right) , \tag{6.3.11}$$

where $\varphi_{(\ldots)_n}(\mathbf{k})$ has been used for $\varphi_{(a_1 a_2 \ldots a_n)}(\mathbf{k})$, and $\boldsymbol{\sigma}^{(i)}$ acts on a_i and is the same 4×4 matrix introduced in the case of spin $\frac{1}{2}$.

As we have introduced Eq. (6.2.3) which is a solution (6.2.1) in the case of spin $\frac{1}{2}$, we shall assume a set of equations

$$k_0 \varphi_{(\ldots)_n}(\mathbf{k}) = \omega_k \beta^{(i)} \varphi_{(\ldots)_n}(\mathbf{k}) \quad (i = 1, 2, \ldots n) , \tag{6.3.12}$$

for a spin $n/2$ particle, where $\beta^{(i)}$ are β matrices acting on a_i. It is easily verified in the following way that this set of equations has the solutions (6.3.5) and (6.3.6).

Consider the case $i \neq j$ in (6.3.12). When a_i is 1 or 2 and a_j is 3 or 4, we have

$$k_0 \varphi_{(\ldots)_n}(\mathbf{k}) = \omega_k \beta^{(i)} \varphi_{(\ldots)_n}(\mathbf{k}) = \omega_k \varphi_{(\ldots)_n}(\mathbf{k}) , \tag{6.3.13}$$

$$k_0 \varphi_{(\ldots)_n}(\mathbf{k}) = \omega_k \beta^{(j)} \varphi_{(\ldots)_n}(\mathbf{k}) = -\omega_k \varphi_{(\ldots)_n}(\mathbf{k}) . \tag{6.3.14}$$

Hence $\varphi_{(\ldots)_n}(\mathbf{k})$ vanishes when the index 1 or 2 coexists with 3 or 4. Further, from (6.3.12) we have

$$k_0 \varphi_{(\ldots)_n}(\mathbf{k}) = \omega_k \varphi_{(\ldots)_n}(\mathbf{k}) \quad (a_1, a_2 \ldots a_n = 1 \,\mathrm{or}\, 2) , \tag{6.3.15}$$

$$k_0 \varphi_{(\ldots)_n}(\mathbf{k}) = -\omega_k \varphi_{(\ldots)_n}(\mathbf{k}) \quad (a_1, a_2 \ldots a_n = 3 \,\mathrm{or}\, 4) . \tag{6.3.16}$$

On the other hand, multiplying (6.3.12) by k_0 we get $k_0^2 = \omega_k^2$, and hence $k_0 = \pm\omega_k$. Therefore, from the above argument we can conclude that all positive frequency $\varphi_{(\ldots)_n}(\mathbf{k})$ with $k_0 = \omega_k$ vanish when they have the indices of the value 3 or 4, and that all negative frequency $\varphi_{(\ldots)_n}(\mathbf{k})$ with $k_0 = -\omega_k$ vanish when they have the indices of the value 1 or 2. This result is nothing but (6.3.5) and (6.3.6).

We shall use (6.3.12) instead of (6.3.5) and (6.3.6). Since $\varphi_{(\ldots)_n}(\mathbf{k})$ is a symmetrized direct product of n $\varphi(\mathbf{k})$'s of spin $\frac{1}{2}$, the transformation (6.2.39) with respect to each index, converts (6.3.8), (6.3.9) and (6.3.12) into the covariant forms. That is, let

$$\psi_{(\dots)_n}(\mathbf{k}) = \omega_k^{n/2} \prod_{j=1}^{n} U_F^{(j)}(\mathbf{k})\varphi_{(\dots)_n}(\mathbf{k}) \,, \tag{6.3.17}$$

where $U^{(j)}(\mathbf{k})$ acts on a_j and is obtained from (6.2.11) with $\boldsymbol{\alpha}$ and β replaced by $\boldsymbol{\alpha}^{(j)}$ and $\beta^{(j)}$ respectively. Using this function in a calculation analogous to that of the previous section we get

$$\left\{\omega_k^{n/2} \prod_{j=1}^{n} U_F^{(j)}(\mathbf{k})\right\} \beta^{(i)} \left\{\omega_k^{n/2} \prod_{j=1}^{n} U_F^{(j)}(\mathbf{k})\right\}^{-1} = \frac{\boldsymbol{\alpha}^{(i)}\mathbf{k} + \beta^{(i)}m}{\omega_k} \,, \tag{6.3.18}$$

and for $\mathbf{K}$ and $\mathbf{J}$ in (6.3.10) and (6.3.11) we get

$$\left\{\omega_k^{n/2} \prod_{j=1}^{n} U_F^{(j)}(\mathbf{k})\right\} \mathbf{J} \left\{\omega_k^{n/2} \prod_{j=1}^{n} U_F^{(j)}(\mathbf{k})\right\}^{-1} = \frac{1}{i}\left(\mathbf{k} \times \frac{\partial}{\partial \mathbf{k}}\right) + \frac{1}{2}\sum_{i=1}^{n} \boldsymbol{\sigma}^{(i)} \,, \tag{6.3.19}$$

$$\left\{\omega_k^{n/2} \prod_{j=1}^{n} U_F^{(j)}(\mathbf{k})\right\} \mathbf{K} \left\{\omega_k^{n/2} \prod_{j=1}^{n} U_F^{(j)}(\mathbf{k})\right\}^{-1}$$
$$= i\left[k_0 \frac{\partial}{\partial \mathbf{k}} - \frac{k_0}{2\omega_k^2} \sum_{i=1}^{n} \boldsymbol{\alpha}^{(i)}\{(\boldsymbol{\alpha}^{(i)}\mathbf{k}) + \beta^{(i)}m\}\right] \,. \tag{6.3.20}$$

Consequently we obtain

$$k_0\psi_{(\dots)_n}(\mathbf{k}) = (\boldsymbol{\alpha}^{(i)}\mathbf{k} + \beta^{(i)}m)\psi_{(\dots)_n}(\mathbf{k}) \quad (i = 1, 2, \dots n) \,, \tag{6.3.21}$$

$$\psi'_{(\dots)_n}(\mathbf{k}) = \left(1 + \frac{i}{2}\sum_{i=1}^{n} \boldsymbol{\sigma}^{(i)}\boldsymbol{\theta}\right) \psi_{(\dots)_n}(\mathbf{k} - \mathbf{k} \times \boldsymbol{\theta}) \quad (\theta\text{-transformation}) \,, \tag{6.3.22}$$

$$\psi'_{(\dots)_n}(\mathbf{k}) = \left(1 - \frac{1}{2}\sum_{i=1}^{n} \boldsymbol{\alpha}^{(i)}\boldsymbol{\tau}\right) \psi_{(\dots)_n}(\mathbf{k} + k_0\boldsymbol{\tau}) \quad (\tau\text{-transformation}) \,. \tag{6.3.23}$$

If we introduce here a function in the x-representation

$$\psi_{(\dots)_n}(x) = \frac{1}{(2\pi)^{3/2}} \int \frac{d\mathbf{k}}{\omega_k} e^{i(\mathbf{k}\mathbf{x} - k_0 t)} \psi_{(\dots)_n}(\mathbf{k}) \,, \tag{6.3.24}$$

the above equations are rewritten as

$$(\gamma_\mu^{(i)}\partial_\mu + m)\psi_{(\dots)_n}(x) = 0 \quad (i = 1, 2, \dots n) , \tag{6.3.25}$$

$$\psi'_{(\dots)_n}(x) = \left(1 + \frac{i}{2}\sum_{i=1}^{n}\sigma^{(i)}\theta\right) \psi_{(\dots)_n}(\Lambda(\theta)^{-1}x) , \tag{6.3.26}$$

$$\psi'_{(\dots)_n}(x) = \left(1 - \frac{1}{2}\sum_{i=1}^{n}\alpha^{(i)}\tau\right) \psi_{(\dots)_n}(\Lambda(\tau)^{-1}x) . \tag{6.3.27}$$

Needless to say, $\gamma^{(i)}$ is a γ matrix acting on a_i. The equations of motion (6.3.25) are called Bargmann-Wigner equations, and are known as equations describing a spin $n/2$ particle. The function $\psi_{(\dots)_n}(x)$ is called a Bargmann-Wigner amplitude.

Next, let us consider the inner product. Using (6.3.5), (6.3.6) and (6.3.17) we have, from (5.1.7),

$$\langle 1, \pm \,|\, 2, \pm \rangle = \frac{1}{|N|^2}\int \frac{d\mathbf{k}}{\omega_k^{n+1}}\psi^{(\pm)*}_{(\dots)_n}(\mathbf{k})_1\psi^{(\pm)}_{(\dots)_n}(\mathbf{k})_2 , \tag{6.3.28}$$

where the integrand on the right-hand side is a brief account of the sum $\sum_{a_1a_2\dots a_n}\psi^{(\pm)*}_{(a_1a_2\dots a_n)}(\mathbf{k})_1\psi^{(\pm)}_{(a_1a_2\dots a_n)}(\mathbf{k})_2/\omega_k^{n+1}$. In order to convert this expression into the x-representation, let us first rewrite it into

$$\langle 1, \pm \,|\, 2, \pm \rangle = \frac{(\pm 1)^{n-1}}{|N|^2}\int \frac{d\mathbf{k}}{k_0^{n-1}}\frac{\psi^{(\pm)*}_{(\dots)_n}(\mathbf{k})_1}{\omega_k}\frac{\psi^{(\pm)}_{(\dots)_n}(\mathbf{k})_2}{\omega_k} . \tag{6.3.29}$$

This cannot easily be converted into the x-representation because of the factor $1/k_0^{n-1}$ in the integrand. We shall use the following relation to solve the problem.

For $\chi_{(\dots)_n}(\mathbf{k})_h$ $(h = 1, 2)$ satisfying (6.3.21), the relation

$$\begin{aligned} &k_0^l\chi^*_{(\dots)_n}(\mathbf{k})_1\beta^{(i_1)}\beta^{(i_2)}\dots\beta^{(i_l)}\chi_{(\dots)_n}(\mathbf{k})_2 \\ &= m^l\chi^*_{(\dots)_n}(\mathbf{k})_1\chi_{(\dots)_n}(\mathbf{k})_2 \qquad (0 \le l \le n) \end{aligned} \tag{6.3.30}$$

holds, where $i_1, i_2, \dots i_l$ are different from one another.

To prove this relation let us consider first $k_0\chi^*_{(\dots)_n}(\mathbf{k})_1\beta^{(i_1)}\beta^{(i_2)}\dots$ $\beta^{(i_l)}\chi_{(\dots)_n}(\mathbf{k})_2$. Since $\chi_{(\dots)_n}(\mathbf{k})_h$ $(h = 1, 2)$ satisfies (6.3.21), we have

$$k_0\beta^{(i_1)}\chi_{(\dots)_n}(\mathbf{k})_2 = (\beta^{(i_1)}\alpha^{(i_1)}\mathbf{k} + m)\chi_{(\dots)_n}(\mathbf{k})_2 , \tag{6.3.31}$$

$$\chi^*_{(\dots)_n}(\mathbf{k})_1(\beta^{(i_1)}\alpha^{(i_1)}\mathbf{k}+m) = \chi^*_{(\dots)_n}(\mathbf{k})_1(-\alpha^{(i_1)}\mathbf{k}\,\beta^{(i_1)}+m)$$
$$= -k_0\chi^*_{(\dots)_n}(\mathbf{k})_1\beta^{(i_1)} + 2m\chi^*_{(\dots)_n}(\mathbf{k})_1\,. \qquad (6.3.32)$$

Using (6.3.31) and (6.3.32) we get

$$\begin{aligned}&k_0\chi^*_{(\dots)_n}(\mathbf{k})_1\beta^{(i_1)}\beta^{(i_2)}\dots\beta^{(i_l)}\chi_{(\dots)_n}(\mathbf{k})_2\\ &= \chi^*_{(\dots)_n}(\mathbf{k})_1(\beta^{(i_1)}\alpha^{(i_1)}\mathbf{k}+m)\beta^{(i_2)}\beta^{(i_3)}\dots\beta^{(i_l)}\chi_{(\dots)_n}(\mathbf{k})_2\\ &= -k_0\chi^*_{(\dots)_n}(\mathbf{k})_1\beta^{(i_1)}\beta^{(i_2)}\dots\beta^{(i_l)}\chi_{(\dots)_n}(\mathbf{k})_2\\ &\quad + 2m\chi^*_{(\dots)_n}(\mathbf{k})_1\beta^{(i_2)}\beta^{(i_3)}\dots\beta^{(i_l)}\chi_{(\dots)_n}(\mathbf{k})_2\,. \end{aligned}\qquad (6.3.33)$$

Hence we obtain

$$\begin{aligned}&k_0\chi^*_{(\dots)_n}(\mathbf{k})_1\beta^{(i_1)}\beta^{(i_2)}\dots\beta^{(i_l)}\chi_{(\dots)_n}(\mathbf{k})_2\\ &= m\chi^*_{(\dots)_n}(\mathbf{k})_1\beta^{(i_2)}\beta^{(i_3)}\dots\beta^{(i_l)}\chi_{(\dots)_n}(\mathbf{k})_2\,. \end{aligned}\qquad (6.3.34)$$

Multiplying the above equation by k_0 successively and performing a similar calculation we obtain (6.3.30).

Putting $l = n-1$ in (6.3.30) and replacing $\chi_{(\dots)_n}(\mathbf{k})_h$ by $\psi^{(\pm)}_{(\dots)_n}(\mathbf{k})_h$ we have

$$\begin{aligned}&\frac{1}{k_0^{n-1}}\psi^{(\pm)*}_{(\dots)_n}(\mathbf{k})_1\psi^{(\pm)}_{(\dots)_n}(\mathbf{k})_2\\ &= \frac{1}{m^{n-1}}\psi^{(\pm)*}_{(\dots)_n}(\mathbf{k})_1\beta^{(1)}\beta^{(2)}\dots\beta^{(i-1)}\beta^{(i+1)}\dots\beta^{(n)}\psi^{(\pm)}_{(\dots)_n}(\mathbf{k})_2\\ &= \frac{1}{m^{n-1}}\overline{\psi^{(\pm)}_{(\dots)_n}}(\mathbf{k})_1\beta^{(i)}\psi^{(\pm)}_{(\dots)_n}(\mathbf{k})_2\,, \end{aligned}\qquad (6.3.35)$$

where we have defined

$$\overline{\psi^{(\pm)}_{(\dots)_n}}(\mathbf{k}) = \psi^{(\pm)*}_{(\dots)_n}(\mathbf{k})\prod_{j=1}^{n}\beta^{(j)}\,. \qquad (6.3.36)$$

Since (6.3.35) holds for any i, by the use of (6.3.24) the inner product is written as

$$\langle 1, \pm \,|\, 2, \pm \rangle = \frac{(\pm 1)^{n-1}}{n|N|^2 m^{n-1}} \sum_{i=1}^{n} \int d\mathbf{x} \overline{\psi^{(\pm)}_{(\dots)_n}}(x)_1 \beta^{(i)} \psi^{(\pm)}_{(\dots)_n}(x)_2 \, . \tag{6.3.37}$$ *)

This expression is not covariant when n is an even number, i.e., in the case of integral spin, because of the factor $(\pm)^{n-1}$. Then, as we have done previously, we define a covariant inner product

$$\langle\langle 1, \pm \,|\, 2, \pm \rangle\rangle = (\pm 1)^{n-1} \langle 1, \pm \,|\, 2, \pm \rangle \, . \tag{6.3.38}$$

Since the inner product has now a form independent of the sign of frequency, we can write generally

$$\langle\langle 1 \,|\, 2 \rangle\rangle = \frac{1}{n|N|^2 m^{n-1}} \sum_{i=1}^{n} \int d\mathbf{x} \, \overline{\psi_{(\dots)_n}}(x)_1 \beta^{(i)} \psi_{(\dots)_n}(x)_2 \, , \tag{6.3.39}$$

where $\overline{\psi_{(\dots)_n}}(x) = \overline{\psi^{(+)}_{(\dots)_n}}(x) + \overline{\psi^{(-)}_{(\dots)_n}}(x)$. Equations (6.3.25), (6.3.26), (6.3.27) and (6.3.29) give a covariant formalism for a spin $n/2$ particle written with the Bargmann-Wigner amplitude.

From the above argument we can conclude, for example, that for an integral spin particle in a negative frequency state the norm defined by the covariant inner product is negative and that the expectation value of k_0 is positive. This is nothing but a generalization of the theory for a spin 0 particle. We can summarize our results as in Table 6.1. Our results have an intimate connection to the statistical property of particles which will be mentioned in Chapter 8.

§6.4 Generalized Bargmann-Wigner Equations

In the previous section we constructed the covariant formalism for a spin $n/2$ particle by extracting an irreducible representation space belonging to the angular momentum $n/2$ from the direct product space of n spinors. This direct product space, however, as is seen from the composition law of angular momenta, consists of irreducible representation spaces belonging to angular momenta $\frac{n}{2} - r$ (r takes integers in the range $0 \le r \le n/2$). Thus, if we extract

*) In this expression we have taken the sum over all the indices and have divided it by n. But such a procedure is not necessary. We have written it merely in the symmetric form with respect to all indices.

Table 6.1 Norms by the covariant inner product and the expectation values of k_0. The symbols (+) and (–) in the left column represent the positive and negative frequencies.

	norm		expectation value of k_0	
	integral spin	half-integral spin	integral spin	half-integral spin
(+)	positive	positive	positive	positive
(–)	negative	positive	positive	negative

one of such irreducible representation spaces, by an analogous argument to that of the previous section we can obtain another covariant formalism which belongs to spin $\frac{n}{2} - r$ and which is different in its apparent form.

Let us consider the simplest case $n = 2$. The quantity $u_{\xi_1\xi_2}$ is decomposed into the symmetric part $u_{(\xi_1\xi_2)}$ and the anti-symmetric part $u_{[\xi_1\xi_2]}$ with respect to the exchange of the indices ξ_1 and ξ_2. These functions are not transformed to each other by any transformation of the rotation group. The symmetric $u_{(\xi_1\xi_2)}$ has been introduced in the argument of the previous section. While, in terms of $\varepsilon_{\xi_1\xi_2}$ satisfying $\varepsilon_{\xi_1\xi_2} = -\varepsilon_{\xi_2\xi_1}$ and $\varepsilon_{12} = 1$, $u_{[\xi_1\xi_2]}$ is expressed as

$$u_{[\xi_1\xi_2]} = \varepsilon_{\xi_1\xi_2} u \,, \tag{6.4.1}$$

where u can be written as

$$u = \frac{1}{2} \sum_{\xi_1,\xi_2} \varepsilon_{\xi_1\xi_2} u_{\xi_1\xi_2} \,.$$

Further, in terms of a 2×2 matrix g satisfying $\det(g) = 1$ we have

$$\sum_{\xi_1',\xi_2'} g_{\xi_1\xi_1'} g_{\xi_2\xi_2'} \varepsilon_{\xi_1'\xi_2'} = \det(g)\varepsilon_{\xi_1\xi_2} = \varepsilon_{\xi_1\xi_2} \,, \tag{6.4.2}$$

which implies that $u_{[\xi_1\xi_2]}$ has only one independent component and that it forms a 1-dimensional space of angular momentum 0. Needless to say, this fact shows that $u_{(\xi_1\xi_2)}$ and $u_{[\xi_1\xi_2]}$ correspond to the irreducible representations of angular momenta 1 and 0 respectively, which are obtained from the composition of two states with angular momentum 1/2. In the argument of the previous section, we used the symmetry with respect to the indices to select the spin of a particle, and we proceeded our calculation to derive the covariant formalism independently of the symmetry. Taking account of this fact, we get a covariant formalism for a spin 0 particle in terms of $u_{[\xi_1\xi_2]}$ (simply denoted as $u_{[..]}$ hereafter) in a similar manner to that of $u_{(\ldots)_n}$. The result is easily derived as

$$\psi'_{[..]}(x) = \left\{1 + \frac{i}{2}(\sigma^{(1)} + \sigma^{(2)})\theta\right\}\psi_{[..]}(x) \quad (\theta\text{-transformation}), \tag{6.4.3}$$

$$\psi'_{[..]}(x) = \left\{1 - \frac{1}{2}(\alpha^{(1)} + \alpha^{(2)})\tau\right\}\psi_{[..]}(x) \quad (\tau\text{-transformation}), \tag{6.4.4}$$

$$(\gamma_\mu^{(i)}\partial_\mu + m)\psi_{[..]}(x) = 0 \qquad (i = 1, 2), \tag{6.4.5}$$

$$\langle\langle 1 \,|\, 2 \rangle\rangle = \frac{1}{2|N|^2 m}\sum_{i=1,2}\int dx\,\bar{\psi}_{[..]}(x)_1\beta^{(i)}\psi_{[..]}(x)_2, \tag{6.4.6}$$

where $\psi_{[..]}(x)$ means $\psi_{[a_1a_2]}(x)$ $(a_1, a_2 = 1,2,3,4)$, and $\bar{\psi}_{[..]}(x)$ is $\psi^*_{[..]}(x)$ $\times\beta^{(1)}\beta^{(2)}$ similarly to (6.3.36). Equations (6.4.3)–(6.4.6) have an equivalent content to that of the covariant formalism for a spin 0 particle mentioned in §6.1.

In the above discussion we have extracted the irreducible representation of angular momentum 0 by anti-symmetrizing $u_{\xi_1\xi_2}$ with respect to its two indices. If an analogous method of symmetrization or anti-symmetrization with respect to appropriate indices is applicable in extracting an irreducible representation of angular momentum $\frac{n}{2} - r$ from $u_{\xi_1\xi_2\ldots\xi_n}$, we can generalize the above argument. For a systematic procedure of extracting an angular momentum, Young's symmetrizer is available. In the following we shall summarize it briefly. The proofs may be found in other appropriate text books.*)

When we consider a quantity with n indices, we shall first divide n into a set of positive integers $n_1, n_2, \ldots n_a$:

$$\begin{aligned} &n_1 + n_2 + \ldots + n_a = n, \\ &n_1 \geq n_2 \geq \ldots \geq n_a > 0. \end{aligned} \tag{6.4.7}$$

There are various ways of division and furthermore any number in the range from 1 to n is possible for the value of a. If a division (6.4.7) is given, we

---

*)For example, S. Iyanaga and M. Sugiura: *Algebra for Applied Mathematicians*, Iwanami (1960).

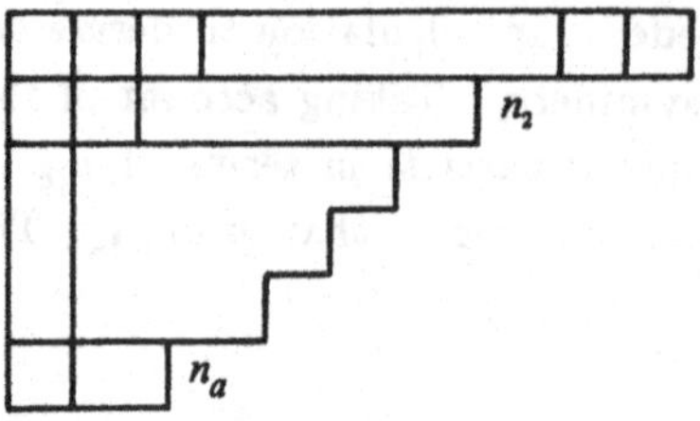

Fig. 6.1

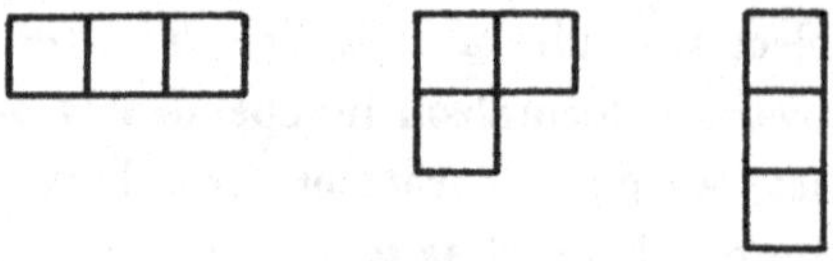

(a) (b) (c)

Fig. 6.2

can draw a diagram which consists of n_1 boxes in the first row, n_2 boxes in the second row, ... and n_a boxes in the a-th row (Fig. 6.1). This is called a Young diagram. For example, when $n = 3$, three kinds of the diagram can be drawn corresponding to the three possible divisions (Fig. 6.2). Next, integers from 1 to n are placed in the boxes such that the numbers increase in going from left to right in a row, and increase in going from top to bottom in a column. Such an arrangement of integers is not unique even if a Young diagram is fixed. For example, in the case $n = 3$, there are two arrangements for the diagram of Fig. 6.2(b) as shown in Fig. 6.3. The Young diagram in which integers are arranged by the above rule is called a standard Young tableau.

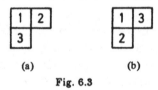

(a) (b)

Fig. 6.3

The number of different standard Young tableaus constructed from a Young diagram ($n = n_1 + n_2 + \ldots + n_a$) with n_i boxes in the i-th row ($i = 1, 2, \ldots a$) is known to be given by

$$n!\,\frac{\prod\limits_{i<j}(l_i - l_j)}{l_1!l_2!\ldots l_a!}\,, \tag{6.4.8}$$

where $l_i = n_i + a - i$.

When a standard Young tableau is given, let $p_i^{(s)}$ be a permutation operator for the integers in the i-th row, where s specifies a particular permutation. Summing up the operators over all s we introduce

$$P_i = \sum_s p_i^{(s)} \tag{6.4.9}$$

for each row, and define a product of all $\mathcal{P}_i$:

$$\mathcal{P} = \prod_{i=1}^{a} \mathcal{P}_i \,. \tag{6.4.10}$$

Similarly, if $q_j^{(s)}$ is a permutation operator for the integers in the j-th column, using an operator

$$\mathcal{Q}_j = \sum_s \varepsilon_s q_j^{(s)} \tag{6.4.11}$$

we define

$$\mathcal{Q} = \prod_j \mathcal{Q}_j \,, \tag{6.4.12}$$

where ε_s is 1 when $q_j^{(s)}$ is an even permutation and is -1 when $q_j^{(s)}$ is an odd permutation. An operator defined as a product of such $\mathcal{Q}$ and $\mathcal{P}$ is called Young's symmetrizer:

$$Y = \mathcal{Q}\mathcal{P} \,. \tag{6.4.13}$$ *)

There exists one Young's symmetrizer corresponding to a given standard Young tableau. For example, Young's symmetrizer corresponding to the Young tableau of Fig. 6.3(b) is $\{1 - (12)\} \times \{1 + (13)\}$, where (ij) is an exchanging operator of i and j.

Let us now consider a general linear transformation for a quantity $v = (v_1, v_2, \cdots, v_p)$ with p components:

$$v'_{t'} = \sum_{t=1}^{p} g_{t't} v_t \,, \tag{6.4.14}$$

where $\det(g) \neq 0$. The whole set of the transformations forms a group which is called a general linear transformation group $\mathrm{GL}(p)$. Consider a direct product of n quantities transforming as (6.4.14), then a "tensor" in the direct product space transforms as

$$v'_{t'_1 t'_2 \ldots t'_n} = \sum_{t_1 t_2 \ldots t_n} \left(\prod_{i=1}^{n} g_{t'_i t_i} \right) v_{t_1 t_2 \ldots t_n} \,. \tag{6.4.15}$$

The following properties are known for such a direct product space.

---

*) Young's symmetrizer may also be defined by $\mathcal{P}\mathcal{Q}$. But a mixed using of the two definitions must be avoided.

(1) The direct product space is completely reducible under a transformation of GL(p), i.e., it is decomposed into a direct sum of irreducible representation spaces of GL(p).
(2) Each irreducible representation space has one-one correspondence to a standard Young tableau in which the number of rows is at most p. Therefore the number of irreducible representation spaces of GL(p) contained in the direct product space is equal to that of all possible standard Young tableaus.*)
(3) The irreducible representations corresponding to various standard Young tableaus constructed from one of the Young diagrams are equivalent to one another. But the irreducible representation spaces corresponding to different Young diagrams are not equivalent to one another irrespective of the standard Young tableaus. That is, an irreducible representation is uniquely specified by a Young diagram.
(4) A tensor in the irreducible representation space corresponding to a given standard Young tableau is obtained by applying the Young symmetrizer, which is constructed from the standard tableau, to $v_{t_1t_2\ldots t_n}$.

In this operation, for example, the permutation of i and j means the exchange of the indices t_i and t_j. If the tensor obtained in this way is denoted as $(Yv)_{t_1t_2\ldots t_n}$, then its transformation is given by

$$(Yv)'_{t'_1t'_2\ldots t'_n} = \sum_{t_1t_2\ldots t_n}\left(\prod_{i=1}^{n} g_{t'_it_i}\right)(Yv)_{t_1t_2\ldots t_n} \,. \tag{6.4.16}$$

Our aim is to construct an irreducible representation of the rotation group from $u_{\xi_1\xi_2\ldots\xi_n}$ given in the previous section by the above general theory. As has been mentioned in §2.2, a representation of the rotation group is a representation of SU(2), which is a group formed by the whole set of 2×2 unitary matrices whose determinants are 1. But the unitarity is now not necessary to consider, because the unitarity plays an important role only in the consideration of products of $u = (u_1, u_2)$ with its complex conjugate $u^* = (u_1^*, u_2^*)$ or quantities transforming in the same manner. Therefore it is sufficient to consider GL(2) with a condition that the determinant of a transformation matrix is 1.

We shall first extract an irreducible representation of GL(2) and then attach the above condition to it. According to the general theory, in this case,

---

*)If the number of rows exceeds p, a tensor given by an operation of such a symmetrizer is anti-symmetric with respect to the indices larger than p. On the other hand, each index takes only p different values, thus the tensor vanishes identically. This is the reason for that the number of rows is at most p.

we consider a Young diagram whose number of rows is at most 2. Let r be the number of boxes in the second row (Fig. 6.4). An irreducible representation of GL(2) can be obtained from $u_{\xi_1\xi_2\ldots\xi_n}$ applied by the Young's symmetrizer Y corresponding to the standard Young tableau. From (6.4.13) the operation of the symmetrizer Y means, in this case, the operation of $\mathcal{Q}$ which anti-symmetrizes the r-pairs of indices after the operation of $\mathcal{P}$. Here, the anti-symmetric r-pairs of indices constructed by the operation $\mathcal{Q}$ have not any contribution to the rotation group as has been seen in (6.4.2).*) Therefore the irreducible representation of GL(2) given by the operation of Y on $u_{\xi_1\xi_2\ldots\xi_n}$ has the same transformation property in the rotation group as a quantity with $(n-2r)$ totally symmetric indices.

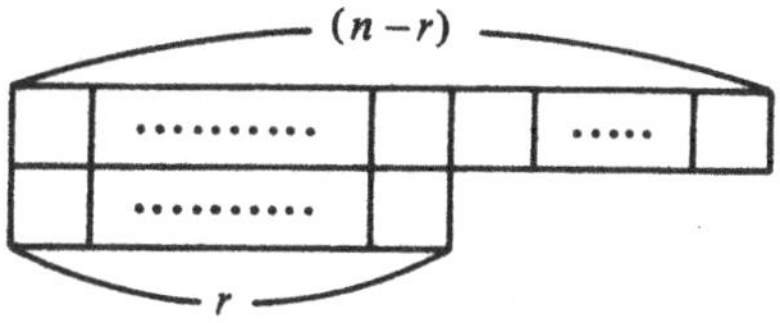

Fig. 6.4

According to the argument of the previous section, this representation is an irreducible representation of the rotation group, and belongs to the angular momentum of $\frac{n}{2} - r$. In other words, an irreducible representation space of GL(2) is an irreducible representation space of the rotation group. When a number n is given, irreducible representations of the rotation group given by different Young diagrams are not equivalent, while those given by the different standard Young tableaus of the same diagram are equivalent to one another. That is, when n is given, an irreducible representation of the rotation group is uniquely specified by a Young diagram with at most two rows. Thus, the number of irreducible representations with angular momentum $\frac{n}{2} - r$ obtained from $u_{\xi_1\xi_2\ldots\xi_n}$ is equal to that of standard Young tableaus constructed from the Young diagram of Fig. 6.4. From (6.4.8) this number is $n!(n-2r+1)/r!(n-r+1)!$ and is the number of irreducible representations with angular momentum $\frac{n}{2} - r$ obtained by the composition of n angular momenta of $1/2$.

When $n = 3$, for example, the Young diagrams to be considered are  and  . Only one standard Young tableau is associated to the former, and the operation of the corresponding symmetrizer on $u_{\xi_1\xi_2\xi_3}$ yields

*) At this stage we use the condition that the determinant of a transformation matrix is 1.

a totally symmetric quantity $u_{(\xi_1\xi_2\xi_3)}$. This belongs to the irreducible representation with angular momentum 3/2. While, two standard Young tableaus (Fig. 6.3) are associated to the later diagram, and the corresponding symmetrizers are $\{1-(13)\}\{(1+(12)\}$ and $\{1-(12)\}\{1+(13)\}$. Operating these symmetrizers on $u_{\xi_1\xi_2\xi_3}$ we get two states of angular momentum 1/2 with three indices:

$$u_{\xi_1\xi_2\xi_3} - u_{\xi_3\xi_2\xi_1} + u_{\xi_2\xi_1\xi_3} - u_{\xi_2\xi_3\xi_1} , \tag{6.4.17}$$

$$u_{\xi_1\xi_2\xi_3} - u_{\xi_2\xi_1\xi_3} + u_{\xi_3\xi_2\xi_1} - u_{\xi_3\xi_1\xi_2} . \tag{6.4.18}$$

In (6.4.17), the state of the eigenvalue 1/2 of the z-component of angular momentum, for example, is represented by $2u_{112} - u_{211} - u_{121} \neq 0$ and all the other components equal to 0. In (6.4.18), it is represented by $2u_{121} - u_{211} - u_{112} \neq 0$ and all the other components equal to 0.*)

In any case where the Young diagram (Fig. 6.4) is given the corresponding angular momentum is determined. We shall denote the wave function with such an angular momentum as $\phi^{(\pm)}_{\square}(\mathbf{k})$. This is obtained from $\phi^{(\pm)}_{\xi_1\xi_2\ldots\xi_n}(\mathbf{k})$ applied by a symmetrizer which is constructed from a Young diagram with the second row of length r. The indices $\xi_1, \xi_2, \ldots \xi_n$ are omitted in its expression. It is clear that the wave function transforms as

$$\phi^{(\pm)\prime}_{\square}(\mathbf{k}) = \left\{1 + i\left(\frac{1}{i}\mathbf{k} \times \frac{\partial}{\partial \mathbf{k}} + \frac{1}{2}\sum_{i=1}^{n} \sigma^{(i)}\right)\boldsymbol{\theta}\right\}\phi^{(\pm)}_{\square}(\mathbf{k})$$
$$(\theta\text{-transformation}) , \tag{6.4.19}$$

$$\phi^{(\pm)\prime}_{\square}(\mathbf{k}) = \left\{1 + k_0\left(\frac{1}{i}\frac{\partial}{\partial \mathbf{k}} - \frac{1}{2}\sum_{i=1}^{n} \frac{\mathbf{k} \times \sigma^{(i)}}{\omega_k(m+\omega_k)}\right)\boldsymbol{\tau}\right\}\phi^{(\pm)}_{\square}(\mathbf{k})$$
$$(\tau\text{-transformation}) . \tag{6.4.20}$$

Similarly to (6.3.5) and (6.3.6) we define a quantity with n indices $a_1, a_2, \ldots a_n$ from $\phi^{(\pm)}_{\square}(\mathbf{k})$. Then this quantity is given as a solution for

$$k_o\varphi_{\square}(\mathbf{k}) = \omega_k\beta^{(i)}\varphi_{\square}(\mathbf{k}) \qquad (i = 1, 2, \ldots n) , \tag{6.4.21}$$

---

*) These two states are linearly independent but are not orthogonal to each other. In general, the irreducible representation spaces corresponding to different standard Young tableaus constructed from the same Young diagram are linearly independent but their orthogonality does not always hold. Thus they must be orthogonalized by an appropriate superposition of them if necessary.

analogously to the argument of the previous section. Since the argument after this stage in the previous section is independent of the symmetry of indices, the equations also hold when the indices $(\ldots)_n$ are replaced by $\square$. Therefore, if $\psi_{\square}(x)$ is defined by

$$\psi_{\square}(\mathbf{k}) = \omega_k^{n/2} \prod_{j=1}^{n} U_F^{(j)}(\mathbf{k}) \varphi_{\square}(\mathbf{k}) \,, \tag{6.4.22}$$

$$\psi_{\square}(x) = \frac{1}{(2\pi)^{3/2}} \int \frac{d\mathbf{k}}{\omega_k} e^{i(\mathbf{k}\mathbf{x} - k_0 t)} \varphi_{\square}(\mathbf{k}) \,, \tag{6.4.23}$$

a set of equations

$$(\gamma_\mu^{(i)} \partial_\mu + m) \psi_{\square}(x) = 0 \quad (i = 1, 2, \ldots n) \tag{6.4.24}$$

is derived. Needless to say, k_0 in (6.4.23) takes values of ω_k and $-\omega_k$ for the positive and negative frequencies respectively. The Lorentz transformations are written as

$$\psi'_{\square}(x) = \Big(1 + \frac{i}{2} \sum_{i=1}^{n} \boldsymbol{\sigma}^{(i)} \boldsymbol{\theta}\Big) \psi_{\square}(\Lambda(\theta)^{-1} x) \quad (\theta\text{-transformation}) \,, \tag{6.4.25}$$

$$\psi'_{\square}(x) = \Big(1 - \frac{1}{2} \sum_{i=1}^{n} \boldsymbol{\alpha}^{(i)} \boldsymbol{\tau}\Big) \psi_{\square}(\Lambda(\tau)^{-1} x) \quad (\tau\text{-transformation}) \,. \tag{6.4.26}$$

Further, using

$$\bar{\psi}_{\square}(x) = \psi^*_{\square}(x) \prod_{i=1}^{n} \gamma_4^{(i)} \tag{6.4.27}$$

which is a generalization of (6.3.36), the covariant inner product is expressed as

$$\langle\langle 1 \,|\, 2 \rangle\rangle = \frac{1}{n |N|^2 m^{n-1}} \sum_{i=1}^{n} \int d\mathbf{x} \bar{\psi}_{\square}(x)_1 \gamma_4^{(i)} \psi_{\square}(x)_2 \,. \tag{6.4.28}$$

The spin of the particle described by Eqs. (6.4.24)–(6.4.28) corresponding to the Young diagram of Fig. 6.4 is clearly $\frac{n}{2} - r$. Since the factor $(\pm)^{n-1}$, which was omitted when the inner product was rewritten into the covariant inner product, can be written as $(\pm)^{2[(n/2)-r]-1}$, it may easily be recognized that (6.4.28) has the content of Table 6.1. Equations (6.4.24) are called generalized Bargmann-Wigner equations. The Bargmann-Wigner equations mentioned in the previous section are derived in the special case $r = 0$.

Since ψ has the indices $a_1, a_2, \dots a_n$ taking values from 1 to 4, we can formally consider Young diagrams with 3 or 4 rows corresponding to the indices. But if such ψ satisfies the Bargmann-Wigner type equations (6.4.24), it vanishes identically. This is recognized by noting that the equation of the type (6.4.21) for φ which is related to ψ by (6.4.22) has no non-zero solution corresponding to the Young diagram with 3 or more rows. In other words, for ψ satisfying a Bargmann-Wigner type equation it is sufficient to consider Young diagrams which have at most 2 rows.

§6.5 γ Matrices

The 4×4 matrices γ_μ defined by (6.2.42) in terms of α and β satisfy

$$\{\gamma_\mu, \gamma_\nu\} = 2\delta_{\mu\nu} \qquad (\mu, \nu = 1, 2, 3, 4) . \tag{6.5.1}$$

Since the explicit expressions for γ_μ written with α and β of (6.2.4) and (6.2.14) are not necessary for our general discussion, we shall start with (6.5.1) instead of the definition (6.2.42). In this case the following result can be derived: "The irreducible γ_μ ($\mu = 1, 2, 3, 4$) satisfying (6.5.1) are restricted to 4×4 matrices and every irreducible solution for (6.5.1) is equivalent to one another. That is, if γ_μ and γ'_μ are two irreducible solutions for (6.5.1), there exists a 4×4 matrix B such that

$$\gamma'_\mu = B\gamma_\mu B^{-1} \qquad (\mu = 1, 2, 3, 4) , \tag{6.5.2}$$

where $\det(B) \neq 0$."

Among the various methods to prove the theorem we shall adopt the theory of representation of a finite group. For this purpose we use the following well known properties of a finite group.

Let G be a finite group and f be the total number of its elements, i.e., the order of the group. Then, (1) the number of irreducible representations of G not equivalent to one another is equal to the number of classes of G. This number will be denoted as h. (2) Numbering the irreducible representations not equivalent to one another and letting d_i be the dimension of the i-th irreducible representation, we have

$$\sum_{i=1}^{h} d_i^2 = f . \tag{6.5.3}$$

We shall consider a group with 32 elements for G:

$$\left.\begin{array}{l} 1, \quad \gamma_\mu, \quad \gamma_\mu\gamma_\nu \ (\mu < \nu), \quad \gamma_\mu\gamma_5, \quad \gamma_5 , \\ -1, -\gamma_\mu, -\gamma_\mu\gamma_\nu \ (\mu < \nu), -\gamma_\mu\gamma_5, -\gamma_5 , \end{array}\right\} \tag{6.5.4}$$

where 1 denotes the identity matrix and γ_5 is

$$\gamma_5 = \gamma_1\gamma_2\gamma_3\gamma_4 \, . \tag{6.5.5}$$

It is easily seen in (6.5.1) that (6.5.4) form a group. This group has 17 classes which are given by,

$$\left.\begin{array}{l} 1, \ -1, \ (\gamma_\mu, -\gamma_\mu), \ (\gamma_\mu\gamma_\nu, -\gamma_\mu\gamma_\nu), \\ (\gamma_\mu\gamma_5, -\gamma_\mu\gamma_5), \ (\gamma_5, -\gamma_5) \, . \end{array}\right\} \tag{6.5.6}$$

Hence according to (1) the number of irreducible representations not equivalent to one another is 17. On the other hand, since $E = (1, -1)$ forms a normal subgroup (or an invariant subgroup) of this group, we can divide the group by E to have a quotient group (or a factor group) G/E which consists of 16 elements $E, \Gamma_\mu, \Gamma_{\mu\nu}$ $(\mu < \nu)$, $\Gamma_{\mu 5}$ and Γ_5:

$$\left.\begin{array}{l} E = (1, -1), \Gamma_\mu = (\gamma_\mu, -\gamma_\mu) \, , \\ \Gamma_{\mu\nu} = (\gamma_\mu\gamma_\nu, -\gamma_\mu\gamma_\nu) \quad (\mu < \nu) \, , \\ \Gamma_{\mu 5} = (\gamma_\mu\gamma_5, -\gamma_\mu\gamma_5) \, , \ \ \Gamma_5 = (\gamma_5, -\gamma_5) \, . \end{array}\right\} \tag{6.5.7}$$

Since this group is an Abelian group, its irreducible representation is a 1-dimensional representation. Therefore the representation of each Γ_μ is 1 or -1 in the irreducible representation, because E is 1 and

$$\Gamma_1^2 = \Gamma_2^2 = \Gamma_3^2 = \Gamma_4^2 = E \, . \tag{6.5.8}$$

The representations of the other elements can be obtained from the products of these Γ_μ's. The number of combinations in which Γ_μ take the value of 1 or -1 is $2^4 = 16$, hence we have 16 non-equivalent irreducible representations of this quotient group. On the other hand, since an irreducible representation of a quotient group is also an irreducible representation of the original group G, we have obtained 16 non-equivalent irreducible representations among 17 representations of G. Then if we put d to be the dimension of the remaining one irreducible representation, from (6.5.3) we have $d^2+16\times 1^2 = 32$ which has a solution $d = 4$. If irreducible γ_μ satisfying (6.5.1) are given, the irreducible representation of G is constructed from γ_μ, while none of the 16 1-dimensional representations obtained above satisfies (6.5.1). Therefore the remaining 4-dimensional representation must satisfy (6.5.1). In other words, the set of irreducible γ_μ satisfying (6.5.1) is that of 4×4 matrices and is unique apart from equivalent representations to it. (q.e.d.)

As a result, we can use (6.5.1) for the definition of γ_μ. We can hereafter restrict γ_μ to 4×4 Hermitian matrices for convenience without any loss of generality. Such a set of γ_μ is, of course, unitary equivalent to (6.2.42).

We shall here introduce 16 Hermitian matrices γ^A ($A = 1, 2, \ldots, 16$) which are given by

$$1, \gamma_\mu, \sigma_{\mu\nu}\ (\mu < \nu), i\gamma_\mu\gamma_5\ , \tag{6.5.9}$$

where

$$\sigma_{\mu\nu} = \frac{[\gamma_\mu, \gamma_\nu]}{2i}\ . \tag{6.5.10}$$

Using (6.5.1) we have

$$\mathrm{Tr}(\gamma^A\gamma^B) = 4\delta_{AB}\ , \tag{6.5.11}$$

and hence we find that the 16 γ's are linearly independent matrices. Since any 4×4 matrix M is determined if its 16 elements are given, M can be expanded uniquely in terms of γ^A:

$$M = \sum_{A=1}^{16} C_A\gamma^A\ . \tag{6.5.12}$$

By taking the trace of each side of (6.5.12) multiplied by γ^B and with the help of (6.5.11) the coefficients C_A are given by

$$C_A = \frac{1}{4}\mathrm{Tr}(\gamma^A M)\ . \tag{6.5.13}$$

When γ_μ satisfy (6.5.1), $-\gamma_\mu^{\mathrm{T}}$ also satisfy (6.5.1). Therefore there always exists a unitary matrix C such that

$$-\gamma_\mu^{\mathrm{T}} = C^{-1}\gamma_\mu C\ . \tag{6.5.14}$$

The matrix C satisfying (6.5.14) has an ambiguity of phase factor $e^{i\delta}$. This ambiguity, however, does not have any significant meaning because this factor can be included in the ununiqueness of the phase factor of wave function as will be seen later. The matrix C is called a charge conjugation matrix, and C has the following property. If we take the transposed matrix of each side of (6.5.14), we have $-\gamma_\mu = C^{\mathrm{T}}\gamma_\mu^{\mathrm{T}}(C^{\mathrm{T}})^{-1}$. Substituting (6.5.14) for this γ^{T} we find that $(C^{\mathrm{T}})^{-1}C$ is commutable with γ_μ. Hence it is a constant (by Schur's lemma), i.e., $C^{\mathrm{T}} = aC$. Further, taking the transposed matrix of each side of this equation and substituting them into the original equation we find that the possible value of a is either 1 or -1. Among these two possibilities a has to take the value -1, i.e.,

$$C^{\mathrm{T}} = -C\ . \tag{6.5.15}$$

If we assumed that $C^{\mathrm{T}} = C$, among the 16 linearly independent $\gamma^A C$, 10 γ^AC for which γ^A are γ_μ and $\sigma_{\mu\nu}$ become anti-symmetric $(\gamma^A C)^{\mathrm{T}} = -\gamma^A C$. But this is impossible since there are only 6 linearly independent anti-symmetric 4×4 matrices. Hence (6.5.15) must hold.

Using (6.5.1) we get a commutation relation for $\sigma_{\mu\nu}$:

$$[\sigma_{\mu\nu}, \sigma_{\lambda\rho}] = 2i(\delta_{\mu\lambda}\sigma_{\nu\rho} + \delta_{\mu\rho}\sigma_{\lambda\nu} + \delta_{\nu\rho}\sigma_{\mu\lambda} + \delta_{\nu\lambda}\sigma_{\rho\mu}) \ . \tag{6.5.16}$$

Comparing this equation with (2.3.15) we find that $\sigma_{\mu\nu}/2$ is a representation of $J_{[\mu\nu]}$, and from (2.3.17) we find that an infinitesimal transformation S_ω in this representation is given by

$$S_\omega = 1 + \frac{i}{4}\sigma_{\mu\nu}\omega_{[\mu\nu]} \ . \tag{6.5.17}$$

The representation $S(\Lambda)$ of a Lorentz transformation Λ is given by performing such infinitesimal transformation successively. A four component quantity transformed by $S(\Lambda)$ is called a Dirac spinor. With the help of (6.5.17) it is easily verified that $S(\Lambda)$ satisfies the following relations:

$$S(\Lambda)^\dagger\gamma_4 = \gamma_4 S(\Lambda)^{-1} \ , \tag{6.5.18}$$

$$C^{-1}S(\Lambda)^{-1} = S(\Lambda)^{\mathrm{T}}C^{-1} \ , \tag{6.5.19}$$

$$S(\Lambda)^{-1}\gamma_\mu S(\Lambda) = \Lambda_{\mu\nu}\gamma_\nu \ . \tag{6.5.20}$$

We introduced (6.2.42) for the explicit representation of γ_μ. That is called the Dirac representation and $\sigma_{\mu\nu}$ is given by

$$\sigma_{ij} = \sum_{k=1}^{3} \varepsilon_{ijk}\sigma_k \ , \tag{6.5.21}$$

$$\sigma_{j4} = \alpha_j \ . \tag{6.5.22}$$

In this case, the infinitesimal Lorentz transformation (6.5.17) is expressed in terms of the parameters $\boldsymbol{\theta}$ and $\boldsymbol{\tau}$ in the form

$$1 + \frac{i}{2}\boldsymbol{\sigma\theta} - \frac{1}{2}\boldsymbol{\alpha\tau} \ , \tag{6.5.23}$$

which is just the same as the transformation matrix acting on ψ of (6.2.29) and (6.2.30). In the Dirac representation the charge conjugation matrix is given by

$$C = \begin{pmatrix} \mathbf{0} & \sigma_2 \\ \sigma_2 & \mathbf{0} \end{pmatrix} \tag{6.5.24}$$

apart from the above mentioned ambiguous phase factor, where **0** in the above equation shows four 0's arranged in a 2×2 matrix. Such a shorthand notation will be used hereafter.

In addition, we frequently use another representation for γ_μ in which γ_5 is diagonal. In this representation, γ_μ are represented by

$$\gamma_j = \begin{pmatrix} \mathbf{0} & i\sigma_j \\ -i\sigma_j & \mathbf{0} \end{pmatrix} , \qquad \gamma_4 = \begin{pmatrix} \mathbf{0} & I \\ I & \mathbf{0} \end{pmatrix} , \tag{6.5.25}$$

and the diagonal γ_5 is

$$\gamma_5 = \begin{pmatrix} -I & \mathbf{0} \\ \mathbf{0} & I \end{pmatrix} , \tag{6.5.26}$$

where I is a 2×2 unit matrix. The generators of the Lorentz transformation are given by

$$\left.\begin{aligned} \sigma_{ij} &= \sum_{k=1}^{3} \varepsilon_{ijk}\sigma_k , \\ \sigma_{j4} &= \begin{pmatrix} \sigma_j & \mathbf{0} \\ \mathbf{0} & -\sigma_j \end{pmatrix} . \end{aligned}\right\} \tag{6.5.27}$$

As is seen in this equation, even if γ_μ are irreducible, the representation of the Lorentz transformation (6.5.17) constructed from them is not irreducible but it is a direct sum of (2.3.20) and (2.3.21). Therefore this representation is directly connected to the spinor representation mentioned in §2.1. The charge conjugation matrix is then given by

$$C = \begin{pmatrix} i\sigma_2 & \mathbf{0} \\ \mathbf{0} & -i\sigma_2 \end{pmatrix} . \tag{6.5.28}$$

Further we sometimes use the Majorana representation in which

$$C = -\gamma_4 . \tag{6.5.29}$$

In this representation, from (6.5.14) and (6.5.15) we have $\gamma_i^{\mathrm{T}} = \gamma_i$ $(i = 1, 2, 3)$ and $\gamma_4^{\mathrm{T}} = -\gamma_4$, or γ_i are real matrices and γ_4 is a purely imaginary matrix $(\gamma_4^* = -\gamma_4)$ from the Hermiticity of γ_μ. We shall give an example explicitly:

$$\left.\begin{aligned} \gamma_1 &= \begin{pmatrix} \mathbf{0} & \sigma_1 \\ \sigma_1 & \mathbf{0} \end{pmatrix} , \qquad \gamma_2 = \begin{pmatrix} I & \mathbf{0} \\ \mathbf{0} & -I \end{pmatrix} , \\ \gamma_3 &= \begin{pmatrix} \mathbf{0} & \sigma_3 \\ \sigma_3 & \mathbf{0} \end{pmatrix} , \qquad \gamma_4 = \begin{pmatrix} \mathbf{0} & \sigma_2 \\ \sigma_2 & \mathbf{0} \end{pmatrix} . \end{aligned}\right\} \tag{6.5.30}$$

In the Majorana representation, therefore, σ_{ij} $(i,j = 1,2,3)$ are purely imaginary matrices and σ_{i4} $(= -\sigma_{4i})$ are real matrices. Hence (6.5.17) becomes a real matrix and consequently $S(\Lambda)$ is also represented by a real matrix.

We shall give some additional properties of the γ matrices which are sometimes used.

Let γ_{ab} be the component of γ at the a-th row and the b-th column. Applying the well known orthogonality relation in a finite group to the element (6.5.4) of G, we have

$$2\{\delta_{ab}\delta_{cd} + (\gamma_\mu)_{ab}(\gamma_\mu)_{cd} + \sum_{\mu<\nu}(\gamma_\mu\gamma_\nu)_{ab}(\gamma_\nu\gamma_\mu)_{cd}$$
$$+ (\gamma_5\gamma_\mu)_{ab}(\gamma_\mu\gamma_5)_{cd} + (\gamma_5)_{ab}(\gamma_5)_{cd}\} = \frac{32}{4}\delta_{ad}\delta_{bc}\,, \qquad (6.5.31)$$

which leads to

$$\sum_{A=1}^{16}(\gamma^A)_{ab}(\gamma^A)_{cd} = 4\delta_{ad}\delta_{bc}\,. \qquad (6.5.32)$$

Further, the trace of a product of γ_μ's has the following property. If $F_{\mu_1\mu_2\ldots\mu_n}$ denotes $\mathrm{Tr}(\gamma_{\mu_1}\gamma_{\mu_2}\cdots\gamma_{\mu_n})$ then

$$F_{\mu_1\mu_2\ldots\mu_{2n+1}} = 0\,, \qquad (6.5.33)$$

because, from $\gamma_5^2 = 1$ and $\mathrm{Tr}(AB) = \mathrm{Tr}(BA)$ we obtain

$$F_{\mu_1\mu_2\ldots\mu_{2n+1}} = \mathrm{Tr}(\gamma_5\gamma_{\mu_1}\gamma_{\mu_2}\cdots\gamma_{\mu_{2n+1}}\gamma_5)\,. \qquad (6.5.34)$$

Using a relation $\gamma_5\gamma_\mu = -\gamma_\mu\gamma_5$, we move γ_5 at the left end in the trace of the above equation to the right end. Then we have $F_{\mu_1\mu_2\ldots\mu_{2n+1}} = -F_{\mu_1\mu_2\ldots\mu_{2n+1}}$, which leads to (6.5.33). Equation (6.5.33) is called Furry's theorem.

To get $F_{\mu_1\mu_2\ldots\mu_{2n}}$ we use a recursion formula. Since

$$F_{\mu_1\mu_2\ldots\mu_{2n}} = \mathrm{Tr}(\gamma_{\mu_{2n}}\gamma_{\mu_1}\gamma_{\mu_2}\cdots\gamma_{\mu_{2n-1}})\,, \qquad (6.5.35)$$

moving $\gamma_{\mu_{2n}}$ in the trace to the right end with the help of (6.5.1) and rearranging the equation we get

$$F_{\mu_1\mu_2\ldots\mu_{2n}} = \sum_{j=1}^{2n-1}(-1)^{j+1}\delta_{\mu_j\mu_{2n}}F_{\mu_1\mu_2\ldots\mu_{j-1}\mu_{j+1}\ldots\mu_{2n-1}}\,. \qquad (6.5.36)$$

On the other hand, from (6.5.1) $F_{\mu_1\mu_2}$ is given by

$$F_{\mu_1\mu_2} = 4\delta_{\mu_1\mu_2} \,. \tag{6.5.37}$$

To get

$$F^{(5)}_{\mu_1\mu_2\ldots\mu_{2n}} = \mathrm{Tr}(\gamma_5\gamma_{\mu_1}\gamma_{\mu_2}\cdots\gamma_{\mu_{2n}}) \tag{6.5.38}$$

we make use of an identity

$$\gamma_{\mu_1}\gamma_{\mu_2}\gamma_{\mu_3} = \varepsilon_{\mu_1\mu_2\mu_3\nu}\gamma_5\gamma_\nu + \delta_{\mu_1\mu_2}\gamma_{\mu_3} - \delta_{\mu_1\mu_3}\gamma_{\mu_2} + \delta_{\mu_2\mu_3}\gamma_{\mu_1} \,. \tag{6.5.39}$$

Multiplying this equation on the left by γ_5 and on the right by $\gamma_{\mu_4}\gamma_{\mu_5}\cdots\gamma_{\mu_{2n}}$, and taking its trace we have

$$\begin{aligned} F^{(5)}_{\mu_1\mu_2\ldots\mu_{2n}} &= \varepsilon_{\mu_1\mu_2\mu_3\nu}F_{\nu\mu_4\mu_5\ldots\mu_{2n}} + \delta_{\mu_1\mu_2}F^{(5)}_{\mu_3\mu_4\ldots\mu_{2n}} \\ &\quad - \delta_{\mu_1\mu_3}F^{(5)}_{\mu_2\mu_4\ldots\mu_{2n}} + \delta_{\mu_2\mu_3}F^{(5)}_{\mu_1\mu_2\ldots\mu_{2n}} \,. \end{aligned} \tag{6.5.40}$$

On the other hand, we have $F^{(5)}_{\mu_1\mu_2} = 0$, then from the above equation we get

$$F^{(5)}_{\mu_1\mu_2\mu_3\mu_4} = 4\varepsilon_{\mu_1\mu_2\mu_3\mu_4} \,. \tag{6.5.41}$$

Equation (6.5.40) together with (6.5.36) is a recursion formula for $F^{(5)}_{\mu_1\mu_2\ldots\mu_{2n}}$.

§6.6 Discrete Transformations

We constructed the irreducible representations of the Poincaré group that is a continuous group, and we studied the covariant formalisms which give equivalent descriptions to the irreducible representations (where the signs of the inner products were changed appropriately). We can find new invariants that were at first not assumed in the formalisms. For example, if $\psi_{\Box}(x)$ in §6.4 is written as $\psi_{\Box}(\mathbf{x}, t)$, the covariant equations (6.4.24) and (6.4.28) are invariant under a discrete transformation

$$\psi'_{\Box}(\mathbf{x}, t) = e^{i\delta}\prod_{j=1}^{n}(i\gamma_4^{(j)})\psi_{\Box}(-\mathbf{x}, t) \,, \tag{6.6.1}$$

where δ is a real number. This transformation is called a space reflection and is not contained in the Poincaré group which has been studied up to now. This fact implies that the irreducible representations of the Poincaré group are irreducible representations of a larger group which contains the space reflection and the like. In this section, we shall discuss such discrete transformations.

a) *Space reflection*

This is a transformation which reflects the three spatial coordinates, and is represented by an operator $\mathcal{R}$. According to §1.1, $\mathcal{R}$ is either unitary or anti-unitary if the theory is invariant under the space reflection. We shall here assume $\mathcal{R}$ to be a unitary operator defined in an irreducible representation space of the Poincaré group. This assumption is justified by the explicit construction of such an operator which will be done later.

If $T(a)$ is written as $T(a, a_0)$, by definition,

$$\mathcal{R}^{-1}T(\mathbf{a}, a_0)\mathcal{R} = T(-\mathbf{a}, a_0) \ . \tag{6.6.2}$$

Since the generator of space rotation $\mathbf{J}$ given by (5.1.21) is an axial vector, $\mathcal{R}$ must satisfy

$$\mathcal{R}^{-1}\mathbf{J}\mathcal{R} = \mathbf{J} \ . \tag{6.6.3}$$

On the other hand, the generator of τ-transformation $\mathbf{K}$ given by (5.1.22) must satisfy

$$\mathcal{R}^{-1}\mathbf{K}\mathcal{R} = -\mathbf{K} \ , \tag{6.6.4}$$

because the signs of spatial coordinates are changed by the space reflection. The commutation relations (2.3.14) are also satisfied if the sign of $\mathbf{K}$ is changed. Relations (6.6.2), (6.6.3) and (6.6.4) characterize $\mathcal{R}$. To construct $\mathcal{R}$ let

$$\mathcal{R} = qp \ , \tag{6.6.5}$$

where p is a unitary operator defined by

$$p\phi_\xi^{(\pm)}(\mathbf{k}) = \phi_\xi^{(\pm)}(-\mathbf{k}) \ . \tag{6.6.6}$$

As is easily seen in the explicit expressions of $\mathbf{J}$ and $\mathbf{K}$ given by (5.1.21), (5.1.22), and Eqs. (6.6.3), (6.6.4), (6.6.6), q commutes with $T(\mathbf{a}, a_0)$, $\mathbf{J}$ and $\mathbf{K}$. While, by assumption, $\mathcal{R}$ is a unitary operator in an irreducible representation space of the Poincaré group, and p is also unitary by (6.6.6). Then Schur's lemma implies that q is merely a number. Therefore we can write

$$\phi_\xi^{(\pm)'}(\mathbf{k}) = \mathcal{R}\phi_\xi^{(\pm)}(\mathbf{k}) = e^{i(\delta \pm c)}\phi_\xi^{(\pm)}(-\mathbf{k}) \ , \tag{6.6.7}$$

where δ and c are now arbitrary numbers, but they will be restricted to some extent if the covariant formalism is considered.

Let us consider a Dirac particle with spin $\frac{1}{2}$ for simplicity. Other cases are the straightforward extensions of this case similarly to the discussions up to now. Using $\varphi(\mathbf{k})$ of (6.2.2), Eq. (6.6.7) leads to

$$\varphi'(\mathbf{k}) = e^{i\delta}e^{i\beta c}\varphi(-\mathbf{k}) , \tag{6.6.8}$$

and with the help of (6.2.39), the space reflection of $\psi(\mathbf{k})$ is given by

$$\begin{aligned}\psi'(\mathbf{k}) &= e^{i\delta}U_F(\mathbf{k})e^{i\beta c}U_F(-\mathbf{k})^{-1}\psi(-\mathbf{k}) \\ &= e^{i\delta}\beta\Big(\frac{\beta m - \alpha\mathbf{k}}{\omega_k}\cos c + i\sin c\Big)\psi(-\mathbf{k}) \\ &= e^{i\delta}\beta\Big(\frac{k_0}{\omega_k}\cos c + i\sin c\Big)\psi(-\mathbf{k}) .\end{aligned} \tag{6.6.9}$$

If we apply the condition of covariance that the transformation coefficient is independent of k, we have $c = \pi/2$, and hence the right-hand side of (6.6.9) becomes $ie^{i\delta}\beta\psi(-\mathbf{k})$. We have used here the Dirac representation for γ_μ. In other representations we can use γ_4 instead of β. Converting this equation into the x-representation by (6.2.33) we obtain

$$\psi'(\mathbf{x}, t) = ie^{i\delta}\gamma_4\psi(-\mathbf{x}, t) . \tag{6.6.10}$$

It is clear that the extension of this result to the case of general spin leads to (6.6.1). In this way the space reflection has been introduced as a unitary operator in an irreducible representation space of the Poincaré group.

b) *Time reversal*

If the time axis is reflected, $T(\mathbf{a}, a_0)$ is transformed as

$$T(\mathbf{a}, a_0) \longrightarrow T(\mathbf{a}, -a_0) , \tag{6.6.11}$$

and consequently the exchange of the positive and negative frequency states occurs. In other words, it is impossible to perform the time reversal transformation without the exchange of the positive and negative frequency states. In this sense, we must use an anti-unitary transformation in order to leave the theory invariant under the transformation. As is seen in (1.1.11) we can use the complex conjugate of a wave function and in general the transformation is expressed as

$$\phi_\xi^{(\pm)}(\mathbf{k}) \longrightarrow [u^{(\pm)}\phi^{(\pm)*}(\mathbf{k})]_\xi . \tag{6.6.12}$$

Needless to say, the form of functions $\phi^{(\pm)}(\mathbf{k})$ on the right-hand side is the same as that on the left-hand side of (5.1.2), but they transform into $T(\mathbf{a}, -a_0)\phi_\xi^{(\pm)}(\mathbf{k})$ under the transformation of parallel translation. In (6.6.12), $u^{(\pm)}$ are unitary operators which leave the inner product invariant. We shall consider that they act only on the spin index ξ and do not involve $\mathbf{k}$. Applying the transformation (6.6.12) on the states $T(\mathbf{a}, -a_0)\phi_\xi^{(\pm)}(\mathbf{k})$ where the signs of frequency are exchanged by the transformation (6.6.11), we have

$$T(\mathbf{a}, -a_0)\phi_\xi^{(\pm)}(\mathbf{k}) \longrightarrow \sum_{\xi'} u_{\xi\xi'}^{(\pm)}[T(\mathbf{a}, -a_0)\phi_{\xi'}^{(\pm)}(\mathbf{k})]^*$$
$$= e^{-i\mathbf{k}\mathbf{a}\mp i\omega_k a_0} \sum_{\xi'} u_{\xi\xi'}^{(\pm)}\phi_{\xi'}^{(\pm)*}(\mathbf{k}) \; . \qquad (6.6.13)$$

In this equation we find that the momentum of the time reversed state $\sum_{\xi'} \times u_{\xi\xi'}^{(\pm)}\phi_{\xi'}^{(\pm)*}(\mathbf{k})$ changes its direction. This agrees with the classical picture of time reversal. Thus, if the wave functions in the time reversed world are denoted as $\phi_\xi^{(\pm)''}(\mathbf{k})$ in the momentum representation, we have

$$\phi_\xi^{(\pm)''}(\mathbf{k}) = \sum_{\xi'} u_{\xi\xi'}^{(\pm)}\phi_{\xi'}^{(\pm)*}(-\mathbf{k}) \; . \qquad (6.6.14)$$

Let us here consider the Lorentz transformation of $\phi_\xi^{(\pm)''}(\mathbf{k})$ to find $u^{(\pm)}$. First, substituting $-\mathbf{k}$ for $\mathbf{k}$ in (5.1.16) and taking the complex conjugate of each side we obtain

$$[\phi_\xi^{(\pm)'}(\mathbf{k})]'' = \sum_{\xi'} \left\{1 + i\left(\frac{1}{i}\mathbf{k} \times \frac{\partial}{\partial \mathbf{k}} - u^{(\pm)}\mathbf{S}^* u^{(\pm)^{-1}}\right)\right\}_{\xi\xi'} \phi_{\xi'}^{(\pm)''}(\mathbf{k}) \; . \quad (6.6.15)$$

If the theory is invariant under the time reversal, the above equation with $[\phi_\xi^{(\pm)'}(\mathbf{k})]'$ for $[\phi_\xi^{(\pm)''}(\mathbf{k})]''$ must be coincident with Eq. (5.1.16) with $\phi_\xi''(\mathbf{k})$ for $\phi_\xi(\mathbf{k})$. For this purpose we can put

$$u^{(\pm)}\mathbf{S}^* u^{(\pm)-1} = -\mathbf{S} \; . \qquad (6.6.16)$$

If we use the matrices given by (4.1.5) and (4.1.7) for $\mathbf{S}$, we have

$$S_1^* = S_1, \quad S_2^* = -S_2, \quad S_3^* = S_3 \; , \qquad (6.6.17)$$

since only S_2 is a purely imaginary matrix. Hence $u^{(\pm)}$ are operators for 180° rotation about the second axis. Of course, there is an ambiguity of phase factor also in this case and we can write generally

$$u^{(\pm)} = e^{i(\delta' \pm c')} e^{i\pi S_2} \; . \qquad (6.6.18)$$

Using these $u^{(\pm)}$ for the τ-transformation, from (5.1.22) and (5.1.24) we have

$$[\phi_{\xi}^{(\pm)'}(\mathbf{k})]'' = \sum_{\xi'} (1 + i\mathbf{K}\tau)_{\xi\xi'} \phi_{\xi'}^{(\pm)''}(\mathbf{k}) \,. \tag{6.6.19}$$

This equation has the opposite sign of $\mathbf{K}$ to that of (5.1.24), but this does not contradict the Lorentz invariance of the theory as has been noted in the discussion of space reflection. Therefore we can put $[\phi_{\xi}^{(\pm)'}(\mathbf{k})]'' = [\phi_{\xi}^{(\pm)''}(\mathbf{k})]'$ also in this case.

In order to see the connection to the covariant formalism we shall again consider the case of spin $\frac{1}{2}$. In this case we can put $\sigma_2/2$ for S_2, and using (6.6.18) we have the transformation of $\varphi(\mathbf{k})$ of (6.2.2) in the form

$$\varphi''(\mathbf{k}) = ie^{i\delta'} e^{ic'\beta} \sigma_2 \varphi^*(-\mathbf{k}) \,. \tag{6.6.20}$$

Hence

$$\begin{aligned} \psi''(\mathbf{k}) &= ie^{i\delta'} U_F(\mathbf{k}) e^{ic'\beta} \sigma_2 U_F^*(-\mathbf{k})^{-1} \psi(-\mathbf{K}) \\ &= ie^{i\delta'} \{\cos c' + iU_F^2(\mathbf{k})\beta \sin c'\} \sigma_2 \psi^*(-\mathbf{k}) \,. \end{aligned} \tag{6.6.21}$$

From the requirement of covariance we put $c' = 0$, then we can derive

$$\begin{aligned} \psi''(\mathbf{x}, t) &= \frac{i}{(2\pi)^{3/2}} e^{i\delta'} \int \frac{d\mathbf{k}}{\omega_k} e^{i(\mathbf{kx} - k_0 t)} \sigma_2 \psi^*(-\mathbf{k}) \\ &= ie^{i\delta'} \sigma_2 \psi^*(\mathbf{x}, -t) \end{aligned} \tag{6.6.22}$$

in the x-representation.

We have used here the Dirac representation for γ, and we must solve some problems to convert our expressions into the general representation of γ. The reason is that $\psi^*(x)$ does not satisfy the Dirac equation. In a general representation of γ, we shall derive the equation for $\bar{\psi}(x) = \psi^*(x)\gamma_4$ instead of $\psi^*(x)$. From (6.2.41) we get

$$(\gamma_\mu^{\mathrm{T}} \partial_\mu - m) \bar{\psi}(x) = 0 \,, \tag{6.6.23}$$

and using (6.5.14) we obtain the equation for

$$\psi^c(x) = C\bar{\psi}(x) \tag{6.6.24}$$

in the form

$$(\gamma_\mu \partial_\mu + m) \psi^c(x) = 0 \,, \tag{6.6.25}$$

which is the Dirac equation. The transformation which converts $\psi(x)$ into $\psi^c(x)$ is called a charge conjugation. Rewriting ψ^* of (6.6.22) into ψ^c in the Dirac representation by the use of (6.2.4) and (6.5.24), and using (6.2.42) we represent the 4×4 coefficient matrix in terms of γ matrices to get

$$\psi''(\mathbf{x},t) = ie^{i\delta'}\gamma_5\gamma_4\psi^c(\mathbf{x},-t) \ . \tag{6.6.26}$$

Although this equation has been obtained in the Dirac representation, it is valid in a general representation of γ because ψ^c satisfies the Dirac equation.

In the case of general spin we introduce $C^{(i)}$ such that $-\gamma_\mu^{(i)\mathrm{T}} = C^{(i)-1} \times \gamma_\mu^{(i)} C^{(i)}$ corresponding to each $\gamma_\mu^{(i)}$ of (6.4.24), and define

$$\psi^c_{\Box}(x) = \prod_{i=1}^{n} C^{(i)}\bar{\psi}_{\Box}(x) \ , \tag{6.6.27}$$

then the time reversal is given by

$$\psi''_{\Box}(\mathbf{x},t) = e^{i\delta'}\prod_{j=1}^{n}(i\gamma_5^{(j)}\gamma_4^{(j)})\psi^c_{\Box}(\mathbf{x},-t) \ . \tag{6.6.28}$$

Needless to say, the covariant inner product transforms as

$$\langle\langle 1 \,|\, 2 \rangle\rangle \longrightarrow \langle\langle 1'' \,|\, 2'' \rangle\rangle = \langle\langle 2 \,|\, 1 \rangle\rangle \tag{6.6.29}$$

under the time reversal.

c) *Charge conjugation*

The charge conjugation for a particle with arbitrary spin is defined by (6.6.27). From (6.5.14) and (6.5.15) we get

$$(\psi^c_{\Box}(x))^c = \prod_{i=1}^{n} C^{(i)}\overline{\psi^c_{\Box}}(x) = \psi_{\Box}(x) \ . \tag{6.6.30}$$

Since $\psi_{\Box}(x)$ and $\psi^c_{\Box}(x)$ satisfy the same equation (6.4.24), and since the transformation contains an operation of taking complex conjugate, we can regard the charge conjugation as a transformation that changes the direction of momentum $\mathbf{k}$ and at the same time exchanges the positive and negative frequency solutions. In this sense the charge conjugation cannot be defined in an irreducible representation space of the Poincaré group. But the theory has a symmetry between the positive and negative frequencies under the transformation. This is seen, for example, in the following way: In terms of $S(\Lambda)$ given in the previous section the Lorentz transform of $\psi_{\square}(x)$ is given by

$$\psi'_{\square}(x) = \prod_{j=1}^{n} S^{(j)}(\Lambda)\psi_{\square}(\Lambda^{-1}x) \,. \tag{6.6.31}$$

On the other hand, from (6.5.18) and the transposed matrix of (6.5.19) we have

$$\psi'^{c}_{\square}(x) = \prod_{j=1}^{n} S^{(j)}(\Lambda)\psi^{c}_{\square}(\Lambda^{-1}x) \,. \tag{6.6.32}$$

Therefore, putting $\psi'^{c}_{\square}(x) = \psi^{c'}_{\square}(x)$ we obtain the same transformation for $\psi^{c}_{\square}(x)$ with (6.6.31).

For the space reflection and the time reversal, from (6.6.1) and (6.6.28) we have

$$\psi'^{c}_{\square}(\mathbf{x},t) = e^{i\delta}\prod_{j=1}^{n}(i\gamma_4^{(j)})\varphi^{c}_{\square}(-\mathbf{x},t) \quad \text{(space reflection)}\,, \tag{6.6.33}$$

$$\psi''^{c}_{\square}(\mathbf{x},t) = e^{-i\delta'}\prod_{j=1}^{n}(i\gamma_5^{(j)}\gamma_4^{(j)})(\psi^{c}_{\square}(\mathbf{x},-t))^{c} \quad \text{(time reversal)}\,. \tag{6.6.34}$$

Therefore we can put $\psi'^{c}_{\square} = \psi^{c'}_{\square}$ and $\psi''^{c}_{\square} = \psi^{c''}_{\square}$ if $e^{-i\delta}$ and $e^{-i\delta'}$ are real.

The covariant inner product transforms as

$$\langle\langle\, 1 \,|\, 2 \,\rangle\rangle \longrightarrow \langle\langle\, 1^c \,|\, 2^c \,\rangle\rangle = (-1)^{n-1}\langle\langle\, 2 \,|\, 1 \,\rangle\rangle \tag{6.6.35}$$

under the charge conjugation, which has the different sign of coefficient corresponding to whether the spin is integral or half-integral. This is connected to the statistics of particles which will be discussed in Chapter 8.

In the Majorana representation the charge conjugation is represented in a simple form

$$\psi^{c}_{\square}(x) = \psi^{*}_{\square}(x) \,. \tag{6.6.36}$$

§6.7 Other Covariant Formalisms

We have discussed the relation between the irreducible representations of the Poincaré group corresponding to particles with finite masses and their covariant descriptions. We shall give a supplementary discussion for other covariant formalisms. They are, however, nothing but formal rewriting of those equations obtained already, and they do not include any essentially new information. But they are convenient in some cases.

a) *Spin 1 particle*

Let us regard the Bargmann-Wigner amplitude $\psi_{(ab)}(x)$ as the element of a 4×4 symmetric matrix $\psi(x)$ at the a-th row and the b-th column. Then Eqs. (6.3.25) are written as

$$\left.\begin{aligned} (\gamma_\mu \partial_\mu + m)\psi(x) = 0 \,, \\ \psi(x)(\gamma_\mu^{\mathrm{T}} \overleftarrow{\partial}_\mu + m) = 0 \,. \end{aligned}\right\} \tag{6.7.1}$$

where $\psi(x)\overleftarrow{\partial}_\mu$ means $\partial_\mu \psi(x)$. The second equation is easily derived from the first one because the matrix $\psi(x)$ is symmetric. Therefore it is sufficient to consider the former equation only. Since $\psi(x)$ can be expanded uniquely in terms of the 10 linearly independent matrices $\gamma_\mu C$ and $\sigma_{\mu\nu} C$, putting $\psi(x)C^{-1} = \chi(x)$ we have

$$\chi(x) = V_\mu(x)\gamma_\mu - \frac{i}{2} T_{[\mu\nu]}(x)\sigma_{\mu\nu} \,. \tag{6.7.2}$$

On the other hand, since $\psi(x)$ is transformed into $S(\Lambda)\psi(\Lambda^{-1}x)S(\Lambda)^{\mathrm{T}}$ by a Lorentz transformation, with the help of (6.5.19) the transformation of $\chi(x)$ is given by

$$\begin{aligned} \chi'(x) &= S(\Lambda)\chi(\Lambda^{-1}x)S(\Lambda)^{-1} \\ &= V_\mu(\Lambda^{-1}x)S(\Lambda)\gamma_\mu S(\Lambda)^{-1} - \frac{i}{2} T_{[\mu\nu]}(\Lambda^{-1}x)S(\Lambda)\sigma_{\mu\nu}S(\Lambda)^{-1} \\ &= \Lambda_{\nu\mu} V_\mu(\Lambda^{-1}x)\gamma_\nu - \frac{i}{2} \Lambda_{\rho\mu}\Lambda_{\sigma\nu} T_{[\mu\nu]}(\Lambda^{-1}x)\sigma_{\rho\sigma} \,, \end{aligned} \tag{6.7.3}$$

where (6.5.20) has been used to derive the right-hand side. Consequently $V_\mu(x)$ transforms as a vector and $T_{[\mu\nu]}(x)$ transforms as an anti-symmetric tensor. Let us write the equation of motion in terms of these quantities. For this purpose, we substitute (6.7.2) into the first equation of (6.7.1), and rearrange it with the help of relations

$$\gamma_\mu \gamma_\nu = \delta_{\mu\nu} + i\sigma_{\mu\nu} \tag{6.7.4}$$

and

$$i\gamma_\mu \sigma_{\lambda\rho} = \varepsilon_{\mu\lambda\rho\sigma}\gamma_5\gamma_\sigma + \delta_{\mu\lambda}\gamma_\rho - \delta_{\mu\rho}\gamma_\lambda \tag{6.7.5}$$

which is obtained from (6.5.39). Then we have

$$\begin{aligned}\partial_\mu V_\mu(x) &- \{\partial_\mu T_{[\mu\nu]}(x) - mV_\nu(x)\}\gamma_\nu \\ &+ \frac{i}{2}\{\partial_\mu V_\nu(x) - \partial_\nu V_\mu(x) - mT_{[\mu\nu]}(x)\}\sigma_{\mu\nu} \\ &- \frac{i}{2}\varepsilon_{\mu\nu\lambda\rho}\partial_\mu T_{[\nu\lambda]}(x)\gamma_5\gamma_\rho = 0 \,, \end{aligned} \tag{6.7.6}$$

from which we immediately obtain

$$mT_{[\mu\nu]}(x) = \partial_\mu V_\nu(x) - \partial_\nu V_\mu(x) \,, \tag{6.7.7}$$

$$mV_\nu(x) = \partial_\mu T_{[\mu\nu]}(x) \,, \tag{6.7.8}$$

$$\partial_\mu V_\mu(x) = 0 \,, \tag{6.7.9}$$

$$\varepsilon_{\mu\nu\lambda\rho}\partial_\nu T_{[\mu\lambda]}(x) = 0 \,. \tag{6.7.10}$$

Needless to say, these equations are equivalent to the Bargmann-Wigner equations (6.7.1). These four equations are, however, not independent and, for example, Eqs. (6.7.9) and (6.7.10) are derived from (6.7.8) and (6.7.7) respectively. A set of equations (6.7.7) and (6.7.8) is called a Proca equation and is one of covariant formalisms which describe a spin 1 particle.

On the other hand, if we use (6.7.7) to eliminate $T_{[\mu\nu]}(x)$ from the other equations, we have a set of equations

$$(\partial_\mu^2 - m^2)V_\mu(x) = 0 \,, \tag{6.7.11}$$

$$\partial_\mu V_\mu = 0 \,, \tag{6.7.12}$$

which is equivalent to (6.7.7)–(6.7.10). This formalism is frequently used because of the following merit: It describes a particle by the 4-component vector $V_\mu(x)$ while the Bargmann-Wigner amplitude $\psi_{(ab)}(x)$ has 10 components, and $V_\mu(x)$ and $T_{[\mu\nu]}(x)$ in the Proca equation have also 10 components.

Next, we shall represent the covariant inner product in terms of $V_\mu(x)$ and $T_{[\mu\nu]}(x)$. If we take the transposed matrix of $\psi(x) = \chi(x)C$, we have $\psi(x) = C^{\mathrm{T}}\chi^{\mathrm{T}}(x)$ since $\psi(x)$ is a symmetric matrix, and hence we can write $\psi^*(x) = C^{-1}\chi^\dagger(x)$, where $\chi^\dagger(x)$ is the Hermitian conjugate matrix of $\chi(x)$. As a result, from $\bar{\psi}(x) = \gamma_4^{\mathrm{T}}\psi^*(x)\gamma_4$ we have

$$\bar{\psi}(x) = -C^{-1}\gamma_4\chi^\dagger(x)\gamma_4 \,. \tag{6.7.13}$$

Hence, from (6.3.39) we obtain

$$\langle\langle 1|2\rangle\rangle = \frac{1}{2|N|^2 m}\int dx\Big(\sum_{a_1 a_1' a_2} \bar{\psi}_{(a_1 a_2)}(x)_1(\gamma_4)_{a_1 a_1'}\psi_{(a_1' a_2)}(x)_2$$

$$+\sum_{a_1 a_2 a_2'} \bar{\psi}_{(a_1 a_2)}(x)_1(\gamma_4)_{a_2 a_2'}\psi_{(a_1 a_2')}(x)_2\Big)$$

$$= \frac{1}{|N|^2 m}\int dx \mathrm{Tr}(\bar{\psi}(x)_1\gamma_4\psi(x)_2)$$

$$= \frac{-1}{|N|^2 m}\int dx \mathrm{Tr}(\gamma_4\chi^\dagger(x)_1\chi(x)_2)\ . \tag{6.7.14}$$

Introduce $V_\mu(x)_h$ and $T_{[\mu\nu]}(x)_h$ corresponding to $\chi(x)_h$ $(h = 1, 2)$ with the help of (6.7.2), and calculate the right-hand side of (6.7.14) by the method of trace mentioned in §6.5. Then, putting $T_{[4\mu]}(x)_h = iT_{[0\mu]}(x)_h$ we have

$$\langle\langle 1|2\rangle\rangle = \frac{4i}{|N|^2 m}\sum_{j=1}^{3}\int dx(T^*_{[0j]}(x)_1 V_j(x)_2 - V_j^*(x)_1 T_{[0j]}(x)_2)$$

$$= \frac{4}{|N|^2 m^2 i}\int dx\Big\{\Big(\frac{\partial \mathbf{V}^*(x)_1}{\partial t} + \nabla V_0(x)_1^*\Big)\mathbf{V}(x)_2$$

$$- \mathbf{V}^*(x)_1\Big(\frac{\partial \mathbf{V}(x)_2}{\partial t} + \nabla V_0(x)_2\Big)\Big\}$$

$$= \frac{4}{|N|^2 m i}\int dx\Big\{\Big(\frac{\partial \mathbf{V}^*(x)_1}{\partial t}\mathbf{V}(x)_2 - \frac{\partial V_0^*(x)_1}{\partial t}V_0(x)_2\Big)$$

$$- \Big(\mathbf{V}^*(x)_1\frac{\partial \mathbf{V}(x)_2}{\partial t} - V_0^*(x)_1\frac{\partial V_0(x)_2}{\partial t}\Big)\Big\}\ . \tag{6.7.15}$$

The last expression has been derived by integrating by parts and with the help of (6.7.12). We usually put $|N|^2 = 8/m^2$ in the above equation.

b) *Spin 3/2 particle*

Another example is given if a similar method is applied to the Bargmann-Wigner amplitude for a spin 3/2 particle. The amplitude $\psi_{(abc)}(x)$ with three symmetric indices has 20 components. If it is expanded similarly to the case of a spin 1 particle, the symmetry for the indices a, b and c leads to

$$\sum_{b'}\psi_{(ab'c)}(x)(C^{-1})_{b'b} = (\gamma_\mu)_{ab}\psi_{\mu,c}(x) - \frac{i}{2}(\sigma_{\mu\nu})_{ab}\psi_{[\mu\nu],c}(x)$$

$$= (\gamma_\mu)_{cb}\psi_{\mu,a}(x) - \frac{i}{2}(\sigma_{\mu\nu})_{cb}\psi_{[\mu\nu],a}(x)\ , \tag{6.7.16}$$

where $\psi_{\mu,a}(x)$ and $\psi_{[\mu\nu],a}(x)$ transform under a Lorentz transformation like a direct product of a vector and a Dirac spinor, and that of an anti-symmetric tensor and a Dirac spinor respectively. They are, however, not linearly independent as is seen in the above equation. Let us find the relation between them. Multiplying (6.7.16) by $(\gamma^A)_{bc}$ and summing up with respect to b and c, we have

$$\begin{aligned}\gamma_\mu\gamma^A\psi_\mu(x) &- \frac{i}{2}\sigma_{\mu\nu}\gamma^A\psi_{[\mu,\nu]}(x) \\ &= \mathrm{Tr}(\gamma_\mu\gamma^A)\psi_\mu(x) - \frac{i}{2}\mathrm{Tr}(\gamma_{\mu\nu}\gamma^A)\psi_{[\mu\nu]}(x)\,,\end{aligned} \tag{6.7.17}$$

where we have omitted the indices of Dirac spinors. Substituting γ^A of (6.5.9) and using (6.5.11) we obtain the following relations:

$$\gamma_\mu\psi_\mu(x) - \frac{i}{2}\sigma_{\mu\nu}\psi_{[\mu\nu]}(x) = 0\,, \tag{6.7.18}$$

$$\gamma_\mu\gamma_\lambda\psi_\mu(x) - \frac{i}{2}\sigma_{\mu\nu}\gamma_\lambda\psi_{[\mu\nu]}(x) = 4\psi_\lambda(x)\,, \tag{6.7.19}$$

$$\gamma_\mu\sigma_{\lambda\rho}\psi_\mu(x) - \frac{i}{2}\sigma_{\mu\nu}\sigma_{\lambda\rho}\psi_{[\mu\nu]}(x) = -4i\psi_{[\lambda\rho]}(x)\,, \tag{6.7.20}$$

$$\gamma_\mu\gamma_\lambda\gamma_5\psi_\mu(x) - \frac{i}{2}\sigma_{\mu\nu}\gamma_\lambda\gamma_5\psi_{[\mu\nu]}(x) = 0\,, \tag{6.7.21}$$

$$\gamma_\mu\gamma_5\psi_\mu(x) - \frac{i}{2}\sigma_{\mu\nu}\gamma_5\psi_{[\mu\nu]}(x) = 0\,. \tag{6.7.22}$$

Comparing (6.7.18) with (6.7.22) multiplied by γ_5 we get

$$\gamma_\mu\psi_\mu(x) = 0\,, \tag{6.7.23}$$

$$\sigma_{\mu\nu}\psi_{[\mu\nu]}(x) = 0\,, \tag{6.7.24}$$

and from (6.7.21) multiplied by γ_5 we obtain

$$\begin{aligned}\{\gamma_\mu,\gamma_\lambda\}\psi_\mu(x) - \gamma_\lambda\gamma_\mu\psi_\mu(x) + \frac{i}{2}[\sigma_{\mu\nu},\gamma_\lambda]\psi_{[\mu\nu]}(x) + \frac{i}{2}\gamma_\lambda\sigma_{\mu\nu}\psi_{[\mu\nu]}(x) \\ = 2\psi_\lambda(x) + 2\gamma_\mu\psi_{[\mu\nu]}(x) = 0\,,\end{aligned} \tag{6.7.25}$$

where (6.7.23) and (6.7.24) have been used. Hence we have

$$\psi_\nu(x) = -\gamma_\mu\psi_{[\mu\nu]}(x)\,. \tag{6.7.26}$$

A similar argument is applicable to (6.7.19) and (6.7.20). But we do not get any new relation from them. We can now regard (6.7.23) and (6.7.26) as conditions for $\psi_\mu(x)$ and $\psi_{[\mu\nu]}(x)$ because (6.7.24) is derived from (6.7.23) and (6.7.26). As $\psi_\mu(x)$ and $\psi_{[\mu\nu]}(x)$ have a total of 40 components, and the number of conditions is 20, we have 20 independent components whose number is just that of components of $\psi_{(abc)}(x)$. In other words, (6.7.23) and (6.7.26) are equivalent to the condition that the indices of $\psi_{(abc)}(x)$ are totally symmetric.

The equations of motion for $\psi_\mu(x)$ and $\psi_{[\mu\nu]}(x)$ are derived from the Bargmann-Wigner equations, that is,

$$\sum_{b'c'}(\gamma_\mu\partial_\mu + m)_{cc'}\psi_{(ab'c')}(C^{-1})_{b'b} = (\gamma_\mu)_{ab}\{(\gamma_\lambda\partial_\lambda + m)\psi_\mu(x)\}_c - \frac{i}{2}(\sigma_{\mu\nu})_{ab}\{(\gamma_\lambda\partial_\lambda + m)\psi_{[\mu\nu]}(x)\}_c = 0 \tag{6.7.27}$$

gives

$$(\gamma_\lambda\partial_\lambda + m)\psi_\mu(x) = 0 \ , \tag{6.7.28}$$

$$(\gamma_\lambda\partial_\lambda + m)\psi_{[\mu\nu]} = 0 \ . \tag{6.7.29}$$

Multiplying (6.7.28) by γ_μ and using (6.7.23) we get

$$\partial_\mu\psi_\mu(x) = 0 \ , \tag{6.7.30}$$

and from (6.7.29) multiplied by γ_μ, (6.7.26) and (6.7.28) we obtain

$$m\psi_\mu(x) = \partial_\lambda\psi_{[\lambda\mu]}(x) \ . \tag{6.7.31}$$

Next, another equation in the Bargmann-Wigner equations

$$\sum_{a'b'}(\gamma_\lambda\partial_\lambda + m)_{aa'}\psi_{(a'b'c)}(x)(C^{-1})_{b'b} = \{(\gamma_\lambda\partial_\lambda + m)\gamma_\mu\}_{ab}\psi_{\mu,c}(x) - \frac{i}{2}\{(\gamma_\lambda\partial_\lambda + m)\sigma_{\mu\nu}\}_{ab}\psi_{[\mu\nu],c}(x) = 0 \tag{6.7.32}$$
*)

---

*) Since (6.7.23) and (6.7.26) are equivalent to the condition that indices of $\psi_{(abc)}$ are totally symmetric, (6.7.32) must also be derived from (6.7.23), (6.7.26) and (6.7.28). In fact, tracing the above argument from (6.7.23) and (6.7.26) conversely, we have (6.7.18)–(6.7.22) from which we get (6.7.17). Multiplying each side of it by $[(\gamma_\mu\partial_\mu + m)\gamma^A]_{ab}$ and summing up with respect to A, we obtain (6.7.32) with the help of (6.5.31).

multiplied by $(\gamma^A)_{ba}$ and summed up with respect to a and b gives

$$m\psi_{[\mu\nu]}(x) = \partial_\mu \psi_\nu(x) - \partial_\nu \psi_\mu(x) \,, \tag{6.7.33}$$

when $\gamma^A = \sigma_{\rho\sigma}$ (No new equation is derived for other γ^A). It is, however, easily found that all the equations obtained above are derived from (6.7.23) and (6.7.28). Here, we can interpret (6.7.23) as the definition of $\psi_{[\mu\nu]}(x)$ in terms of $\psi_\mu(x)$. Namely we can use the following set of equations for a spin $\frac{3}{2}$ particle with finite mass:

$$\left.\begin{aligned} (\gamma_\lambda \partial_\lambda + m)\psi_\mu(x) = 0 \,, \\ \gamma_\mu \psi_\mu(x) = 0 \,. \end{aligned}\right\} \tag{6.7.34}$$

This set of equations is called a Rarita-Schwinger equation for a spin $\frac{3}{2}$ particle.

The covariant inner product is obtained from (6.3.39) and (6.7.16) and is shown to be expressed as

$$\begin{aligned} \langle\langle 1 | 2 \rangle\rangle &= \frac{4i}{|N|^2 m} \sum_{j=1}^{3} \int dx (\psi^*_{[0j]}(x)_1 \gamma_4 \psi_j(x)_2 - \psi^*_j(x)_1 \gamma_4 \psi_{[0j]}(x)_2) \\ &= \frac{8}{|N|^2 m^2} \int dx \Big(\sum_{j=1}^{3} \psi^*_j(x)_1 \psi_j(x)_2 - \psi^*_0(x)_1 \psi_0(x)_2 \Big) \,. \end{aligned} \tag{6.7.35}$$

c) *Generalization*

We rewrote the Bargmann-Wigner amplitudes for the cases of spin 1 and $\frac{3}{2}$, and obtained several covariant formalisms which look different. That long discussion was intended to show the explicit relation between the Bargmann-Wigner amplitudes and other types of covariant formalism. It is also useful as an exercise for treating the γ matrices, and finally is connected to the argument of §7.3. We must, however, perform some complicated calculations when we apply our method to particles with larger spin, because we must calculate many component quantities, i.e., the Bargmann-Wigner amplitudes. Therefore, we avoid such calculations, and we make use of new covariant amplitudes directly connected to the irreducible representations of the Poincaré group in order to generalize the results of a) and b). For this purpose we shall summarize the argument in the case of spin 1 particle.

According to (6.3.17) we can write

$$\psi(\mathbf{k}) = \omega_k U_F(\mathbf{k})\varphi(\mathbf{k})U_F(\mathbf{k})^{\mathrm{T}} \tag{6.7.36}$$

for a spin 1 particle, where $\psi(\mathbf{k})$ and $\varphi(\mathbf{k})$ are 4×4 symmetric matrices. The operator $U_F(\mathbf{k})$ given by (6.2.11) is expressed in terms of γ matrices of (6.2.42) as

$$U_F(\mathbf{k}) = \frac{\omega_k + m - i\boldsymbol{\gamma}\mathbf{k}}{\sqrt{2\omega_k(\omega_k + m)}}\,, \tag{6.7.37}$$

where $\boldsymbol{\gamma} = (\gamma_1, \gamma_2, \gamma_3)$ and they are, of course, γ matrices in the Dirac representation. We shall proceed to use this representation hereafter. The Fourier component $\chi(\mathbf{k})$ of the left-hand side of (6.7.2) is given by

$$\chi(\mathbf{k}) = \psi(\mathbf{k})C^{-1} = \omega_k U_F(\mathbf{k})\varphi(\mathbf{k})C^{-1}U_F(\mathbf{k})^{-1}\,. \tag{6.7.38}$$

Here, use (6.3.5), (6.3.6) (with $n = 2$) and (6.5.24) for $\varphi(\mathbf{k})C^{-1}$, and define $\boldsymbol{\phi}^{(\pm)}(\mathbf{k})$ $(= (\phi_1^{(\pm)}(\mathbf{k}),\ \phi_2^{(\pm)}(\mathbf{k}), \phi_3^{(\pm)}(\mathbf{k}))$ by

$$\phi^{(\pm)}(\mathbf{k})\sigma_2 = \mp 2i\sum_{j=1}^{3}\phi_j^{(\pm)}(\mathbf{k})\sigma_j = \mp 2i\boldsymbol{\phi}^{(\pm)}(\mathbf{k})\boldsymbol{\sigma}\,. \tag{6.7.39}$$ *)

Since

$$\varphi(\mathbf{k}) = N\begin{pmatrix} \phi^{(+)}(\mathbf{k}) & \mathrm{O} \\ \mathrm{O} & \phi^{(-)}(\mathbf{k}) \end{pmatrix}\,,$$

we have

$$\begin{aligned} \varphi(\mathbf{k})C^{-1} &= N\begin{pmatrix} \mathrm{O} & -2i\boldsymbol{\phi}^{(+)}(\mathbf{k})\boldsymbol{\sigma} \\ 2i\boldsymbol{\phi}^{(-)}(\mathbf{k})\boldsymbol{\sigma} & \mathrm{O} \end{pmatrix} \\ &= N\{(1+\gamma_4)\boldsymbol{\gamma}\boldsymbol{\phi}^{(+)}(\mathbf{k}) + (1-\gamma_4)\boldsymbol{\gamma}\boldsymbol{\phi}^{(-)}(\mathbf{k})\}\,. \end{aligned} \tag{6.7.40}$$

---

*) Note that $\phi^{(\pm)}(\mathbf{k})$ on the left-hand side of (6.7.39) are 2×2 matrices, while $\phi_j^{(\pm)}(\mathbf{k})$ on the right-hand side are $\pm i\mathrm{Tr}(\phi^{(\pm)}(\mathbf{k})\sigma_2\sigma_j)/4$ which are merely complex numbers.

Hence, from (6.7.2) and (6.7.38) we get

$$\begin{aligned}V_\mu(\mathbf{k}) &= \frac{1}{4}\mathrm{Tr}(\gamma_\mu\chi(\mathbf{k}))\\ &= \frac{N}{8(\omega_k+m)}\sum_{j=1}^{3}[\mathrm{Tr}\{\gamma_\mu(m+\omega_k-i\gamma\mathbf{k})(1+\gamma_4)\\ &\quad\times\gamma_j(m+\omega_k+i\gamma\mathbf{k})\}\phi_j^{(+)}(\mathbf{k})\\ &\quad+\mathrm{Tr}\{\gamma_\mu(m+\omega_k-i\gamma\mathbf{k})(1-\gamma_\mu)\gamma_j(m+\omega_k+i\gamma\mathbf{k})\}\phi_j^{(-)}(\mathbf{k})]\\ &= N\sum_{j=1}^{3}F_{\mu j}(\mathbf{k})\phi_j(\mathbf{k})\ ,\end{aligned} \tag{6.7.41}$$

where $F_{\mu j}$ and $\phi_j(\mathbf{k})$ are

$$F_{\mu j} = m\delta_{\mu j} + \frac{\sum_{i=1}^{3}\delta_{\mu i}k_i}{\omega_k+m}k_j + \frac{k_4}{\omega_k}\delta_{\mu 4}k_j\ , \tag{6.7.42}$$

$$\phi_j(\mathbf{k}) = \phi_j^{(+)}(\mathbf{k}) + \phi_j^{(-)}(\mathbf{k}) \quad (j=1,2,3)\ . \tag{6.7.43}$$

From (6.7.41) and (6.7.42) we get easily

$$k_\mu V_\mu(\mathbf{k}) = 0\ , \tag{6.7.44}$$

which is nothing but (6.7.12). Therefore $V_i(\mathbf{k})$ $(i=1,2,3)$ can be considered as independent components and are expressed in terms of $\phi_j(\mathbf{k})$ as

$$V_i(\mathbf{k}) = N\sum_{j=1}^{3}F_{ij}\phi_j(\mathbf{k})\ . \tag{6.7.45}$$

Conversely, when $V_i(\mathbf{k})$ are given, $\phi_j(\mathbf{k})$ are expressed as

$$\left.\begin{aligned}\phi_i(\mathbf{k}) &= \frac{1}{N}\sum_{j=1}^{3}\tilde{F}_{ij}\phi_j(\mathbf{k})\ ,\\ \tilde{F}_{ij} &= \frac{1}{m}\left(\delta_{ij}-\frac{k_ik_j}{\omega_k(\omega_k+m)}\right)\ ,\end{aligned}\right\} \tag{6.7.46}$$

In fact, $\sum_{l=1}^{3}\tilde{F}_{il}F_{lj} = \delta_{ij}$ is easily verified by a direct calculation. This implies that the three components $\phi_j(\mathbf{k})$ are equivalent to the four components $V_\mu(\mathbf{k})$ satisfying (6.7.44).

Since $\phi^{(\pm)}(\mathbf{k})$ transform as (6.3.4) (with $n = 2$) under an infinitesimal spatial rotation, we find with the help of (6.7.39) that $\phi_i(\mathbf{k})$ transform as a 3-dimensional vector. Hence, from (5.1.14) and (5.1.15) we get

$$\boldsymbol{\phi}'(\mathbf{k}) = \boldsymbol{\phi}(\mathbf{k} - \mathbf{k}\times\boldsymbol{\theta}) - \boldsymbol{\theta}\times\boldsymbol{\phi}(\mathbf{k}) \quad (\theta\text{-transformation}), \tag{6.7.47}$$

$$\boldsymbol{\phi}'(\mathbf{k}) = \boldsymbol{\phi}(\mathbf{k} + \boldsymbol{\tau} k_0) - \frac{k_0}{\omega_k(m+\omega_k)}(\mathbf{k}\times\boldsymbol{\tau})\times\boldsymbol{\phi}(\mathbf{k}) \quad (\tau\text{-transformation}) . \tag{6.7.48}$$

Using these equations (6.7.41), (6.7.42) and (6.7.44) we can verify that $V_\mu(\mathbf{k})$ transform as a 4-dimensional vector under a Lorentz transformation. The calculation is left as a reader's exercise.

The three independent components of $\boldsymbol{\phi}(\mathbf{k})$ satisfy

$$k_0\phi_j^{(\pm)}(\mathbf{k}) = \pm\omega_k\phi_j^{(\pm)}(\mathbf{k}) . \tag{6.7.49}$$

Therefore, from this equation and (6.7.44) we find that

$$V_\mu(x) = \frac{1}{(2\pi)^{3/2}}\int\frac{d\mathbf{k}}{\omega_k}e^{i(\mathbf{k}\mathbf{x}-k_0t)}V_\mu(\mathbf{k}) \tag{6.7.50}$$

satisfies (6.7.11) and (6.7.12). Conversely, from $V_\mu(x)$ satisfying those equations we get (6.7.46), and we can trace the above argument back to obtain $\phi_i(\mathbf{k})$ satisfying (6.7.47)–(6.7.49), thus we find that these descriptions are equivalent. In this way the covariant description with $V_\mu(x)$ is directly connected by (6.7.46) to an irreducible representation of the little group. We note, by the way, that $F_{\mu j}$ satisfy

$$\sum_{j=1}^{3}F_{\mu j}F_{\nu j} = m^2\left(\delta_{\mu\nu} + \frac{k_\mu k_\nu}{m^2}\right) . \tag{6.7.51$^{*)}$}$$

In order to generalize the above argument to the case of integral spin n we proceed as follows. First, construct a direct product of n 3-dimensional vectors, and let $u_{(i_1,i_2,\ldots i_n)}$ $(i_1, i_2, \ldots i_n = 1,2,3)$ be a tensor of the n-th rank in which all indices are symmetrized. This direct product space consists of irreducible representations of the rotation group with angular momenta

---

*) When $k_\mu^2 + m^2 = 0$ and $R_{\mu\nu}(k) = \delta_{\mu\nu} + k_\mu k_\nu/m^2$ we have $R_{\mu\nu}(k)R_{\nu\lambda}(k) = R_{\mu\lambda}(k)$ and $k_\mu R_{\mu\nu}(k) = 0$. Therefore $R_{\mu\nu}(k)$ is a projection operator which extracts the part satisfying (6.7.44) from an arbitrary vector $V_\mu(k)$ (where $k_\mu^2 + m^2 = 0$).

$n, n-1, n-2, \ldots$. To extract an angular momentum we shall set subsidiary conditions

$$\sum_{j=1}^{3} u_{(jji_3\ldots i_n)} = 0 \; . \tag{6.7.52}$$

Since the indices of $u_{(i_1\ldots i_n)}$ are symmetric, (6.7.52) stands for the sum with respect to arbitrary two indices, and it is invariant under a 3-dimensional rotation. On the other hand, a vector $\mathbf{u} = (u_1, u_2, u_3)$ belongs to an irreducible representation space with angular momentum 1, and the state vector, for which $u_1 + iu_2 \neq 0$ in this space and all the other components vanish, belongs to angular momentum 1 around the third axis (cf. (4.3.4)). Therefore a direct product of n such vectors is a state where the angular momentum around the third axis is n, and its components are linear combinations of the symmetric tensor $u_{(i_1,i_2,\ldots i_n)}$. It does not contradict (6.7.52) because the indices do not take the value 3. In other words, the space formed by the symmetric tensors of the n-th rank satisfying (6.7.52) contains the irreducible representation space of angular momentum n. Now, the number of components of $u_{(i_1,i_2\ldots i_n)}$ is the number of repeated combinations of integers 1,2,3 taken n at a time, and is $(n+2)(n+1)/2$. While the number of the conditions (6.7.52) is $n(n-1)/2$ which is the number of combinations of integers 1,2,3 taken $n-2$ at a time. Hence the number of the independent components of the symmetric tensor satisfying (6.7.52) becomes $(n+2)(n+1)/2 - n(n-1)/2 = 2n+1$. This is nothing but the dimension of the irreducible representation space with angular momentum n and consequently the irreducible representation with angular momentum n is extracted by (6.7.52).

In a spin n irreducible representation space of the Poincaré group let us consider $\phi_{(i_1i_2\ldots i_n)}(\mathbf{k})$ $(= \phi^{(+)}_{(i_1i_2\ldots i_n)}(\mathbf{k}) + \phi^{(-)}_{(i_1i_2\ldots i_n)}(\mathbf{k}))$ corresponding to the above $u_{(i_1i_2\ldots i_n)}$. Of course,

$$k_0\phi^{(\pm)}_{(i_1i_2\ldots i_n)}(\mathbf{k}) = \pm\omega_k\phi^{(\pm)}_{(i_1i_2\ldots i_n)}(\mathbf{k}) \; , \tag{6.7.53}$$

then if we define

$$U_{\mu_1\mu_2\ldots\mu_n}(\mathbf{k}) = \sum_{i_1i_2\ldots i_n=1,2,3} \left(\prod_{l=1}^{n} F_{\mu_l i_l}\right)\phi_{(i_1i_2\ldots i_n)}(\mathbf{k}) \; , \tag{6.7.54}$$

we can easily verify that the symmetric tensor $U_{\mu_1\mu_2\ldots\mu_n}(\mathbf{k})$ of the n-th rank satisfies

$$k_\mu U_{\mu_1\mu_2\ldots\mu_n}(\mathbf{k}) = 0 \; . \tag{6.7.55}$$

Further, from (6.7.42) we have

$$F_{\mu i}F_{\mu j} = m^2\delta_{ij} + \left(1 - \frac{k_0^2}{\omega_k^2}\right)k_i k_j = m^2\delta_{ij}\ , \tag{6.7.56}$$

and with the help of (6.7.52) and (6.7.53) we get

$$U_{\mu\mu\mu_3\ldots\mu_n}(\mathbf{k}) = 0\ . \tag{6.7.57}$$

Equations (6.7.55) and (6.7.57) are equivalent to (6.7.52). There is no more condition for the symmetric tensor $U_{\mu_1\mu_2\ldots\mu_n}(\mathbf{k})$. If we introduce

$$U_{\mu_1\mu_2\ldots\mu_n}(x) = \frac{1}{(2\pi)^{3/2}}\int \frac{d\mathbf{k}}{\omega_k}e^{i(\mathbf{k}\mathbf{x}-k_0 t)}U_{\mu_1\mu_2\ldots\mu_n}(\mathbf{k})\ , \tag{6.7.58}$$

we obtain a covariant formalism for a spin n particle:

$$(\partial_\mu^2 - m^2)U_{\mu_1\mu_2\ldots\mu_n}(x) = 0\ , \tag{6.7.59}$$

$$\partial_\mu U_{\mu\mu_2\ldots\mu_n}(x) = 0\ , \tag{6.7.60}$$

$$U_{\mu\mu\mu_3\ldots\mu_n}(x) = 0\ . \tag{6.7.61}$$

The set of equations (6.7.59)–(6.7.61) is called a Fierz-Pauli equation. The inner product is expressed, by the use of (5.1.7) and (6.7.54), in the form

$$\begin{aligned}\langle 1, \pm \,|\, 2, \pm \rangle &= \frac{1}{|N'|^2}\sum_{i_1,i_2\ldots i_n=1,2,3}\int \frac{d\mathbf{k}}{\omega_k}\phi^{(\pm)*}_{(i_1i_2\ldots i_n)}(\mathbf{k})_1\phi^{(\pm)}_{(i_1i_2\ldots i_n)}(\mathbf{k})_2 \\ &= \frac{1}{|N'|^2}\sum_{\substack{i_1\ldots i_n\\ j_1\ldots j_n\\ j_1'\ldots j_n'}}\int \frac{d\mathbf{k}}{\omega_k}U^{(\pm)*}_{j_1\ldots j_n}(\mathbf{k})_1\prod_{l=1}^{n}(\tilde{F}_{i_lj_l}\tilde{F}_{i_lj_l'})U^{(\pm)}_{j_1'\ldots j_n'}(\mathbf{k})_2\end{aligned} \tag{6.7.62}$$

where N' is an appropriate normalization constant. Using the relation

$$\sum_{i=1,2,3}\tilde{F}_{ij}\tilde{F}_{ij'} = \frac{1}{m^2}\left(\delta_{jj'} - \frac{k_jk_{j'}}{\omega_k^2}\right) \tag{6.7.63}$$

which is derived from (6.7.46), and with the help of (6.7.55) we obtain

$$\begin{aligned}&\sum_{i_l,j_l,j_l'}U^{(\pm)*}_{j_1\ldots j_{l-1}j_lj_{l+1}\ldots j_n}(\mathbf{k})_1\tilde{F}_{i_lj_l}\tilde{F}_{i_lj_l'}U^{(\pm)}_{j_1'\ldots j_{l-1}'j_l'j_{l+1}'\ldots j_n'}(\mathbf{k})_2 \\ &\quad = \frac{1}{m^2}U^{(\pm)*}_{j_1\ldots j_{l-1}\mu j_{l+1}\ldots j_n}(\mathbf{k})_1U^{(\pm)}_{j_1'\ldots j_{l-1}'\mu j_{l+1}'\ldots j_n'}(\mathbf{k})_2\ ,\end{aligned} \tag{6.7.64}$$

where $U^{(\pm)*}_{j_1\ldots\mu\ldots j_n}$ is the complex conjugate of $U^{(\pm)}_{j_1\ldots\mu\ldots j_n}$ when $\mu = 1,2,3$, and is the complex conjugate of $U^{(\pm)}_{j_1\ldots 0\ldots j_n}$ multiplied by i when $\mu = 4$. In general, when a indices among the indices of tensor $U_{\mu_1\ldots\mu_n}$ take the value 4, we define $U^*_{\mu_1\ldots\mu_n}$ in the following manner: Replace all such indices by 0 after multiplying by $(-i)^a$ and then multiply the complex conjugate of it by $(i)^a$. This is not the simple complex conjugate of $U_{\mu_1\ldots\mu_n}$ (which is denoted as $(U_{\mu_1\ldots\mu_n})^*$), but it transforms as a tensor of the n-th rank. Such a usage of the symbol * is applicable only to tensors, and for other quantities it is used to represent the usual complex conjugate.

We shall now note that (6.7.64) is rewritten in another form. Since (6.7.63) is expressed as

$$\sum_{i=1}^{3} \tilde{F}_{ij}\tilde{F}_{ij'} = \frac{1}{2m^2\omega_k}\left(2\delta_{jj'}\omega_k - \frac{k_j}{\omega_k}k_{j'} - k_j\frac{k_{j'}}{\omega_k}\right), \tag{6.7.65}$$

applying (6.7.53) and (6.7.55) to the part in the parentheses of this equation we have

$$\begin{aligned}
(6.7.64) = \frac{\pm 1}{2m^2\omega_k}\Big\{&\sum_j 2k_0 U^{(\pm)*}_{j_1\ldots j_{l-1}jj_{l+1}\ldots j_n}(\mathbf{k})_1 U^{(\pm)}_{j'_1\ldots j'_{l-1}jj'_{l+1}\ldots j'_n}(\mathbf{k})_2 \\
&-\sum_j k_j\big(U^{(\pm)*}_{j_1\ldots j_{l-1}jj_{l+1}\ldots j_n}(\mathbf{k})_1 U^{(\pm)}_{j'_1\ldots j'_{l-1}0j'_{l+1}\ldots j'_n}(\mathbf{k})_2 \\
&+U^{(\pm)*}_{j_1\ldots j_{l-1}0j_{l+1}\ldots j_n}(\mathbf{k})_1 U^{(\pm)}_{j'_1\ldots j'_{l-1}j'j'_{l+1}\ldots j'_n}(\mathbf{k})_2\big)\Big\} \\
= \frac{\pm 1}{2m^2\omega_k}\sum_j &\big\{\big(k_0 U^{(\pm)*}_{j_1\ldots j_{l-1}jj_{l+1}\ldots j_n}(\mathbf{k})_1 - k_j U^{(\pm)*}_{j_1\ldots j_{l-1}0j_{l+1}\ldots j_n}(\mathbf{k})_1 \\
&\times U^{(\pm)}_{j'_1\ldots j'_{l-1}jj'_{l+1}\ldots j'_n}(\mathbf{k})_2 + U^{(\pm)*}_{j_1\ldots j_{l-1}jj_{l+1}\ldots j_n}(\mathbf{k})_1 \\
&\times \big(k_0 U^{(\pm)}_{j'_1\ldots j'_{l-1}jj'_{l+1}\ldots j'_n}(\mathbf{k})_2 - k_j U^{(\pm)}_{j'_1\ldots j'_{l-1}0j'_{l+1}\ldots j'_n}(\mathbf{k})_2\big)\big\}\,.
\end{aligned} \tag{6.7.66}$$

Using (6.7.64) for the indices j_l and j'_l $(l \geq 2)$ on the right-hand side of

(6.7.62), and using (6.7.66) for j_1 and j_1', we can derive the inner product

$$\langle 1,\pm\,|\,2,\pm\rangle = \frac{\pm 1}{2m^{2n}|\,N'\,|^2}\sum_j\int\frac{d\mathbf{k}}{\omega_k^2}\{(k_0U^{(\pm)*}_{j\mu_2\ldots j_n}(\mathbf{k})_1$$

$$- k_jU^{(\pm)*}_{0\mu_2\ldots\mu_n}(\mathbf{k})_1)U^{(\pm)}_{j\mu_2\ldots\mu_n}(\mathbf{k})_2 + U^{(\pm)*}_{j\mu_2\ldots\mu_n}(\mathbf{k})_1$$

$$\times\,(k_0U^{(\pm)}_{j\mu_2\ldots\mu_n}(\mathbf{k})_2 - k_jU^{(\pm)}_{0\mu_2\ldots\mu_n}(\mathbf{k})_2)\}\ . \tag{6.7.67}$$

The covariant inner product is connected to (6.7.67) as $\langle\langle\,1,\pm\,|\,2,\pm\,\rangle\rangle = \pm\,\langle\,1,\pm\,|\,2,\pm\,\rangle$ which is one of the properties of an integral spin, and hence from (6.7.58) we get in the x-representation

$$\begin{aligned}\langle\langle\,1\,|\,2\,\rangle\rangle &= \frac{1}{2im^{2n}|\,N'\,|^2}\sum_j\int d\mathbf{x}\Big\{\Big(\frac{\partial}{\partial t}U^*_{j\mu_2\ldots\mu_n}(x)_1 + \frac{\partial}{\partial x_j}U^*_{0\mu_2\ldots\mu_n}(x)_1\Big)\\ &\quad\times U_{j\mu_2\ldots\mu_n}(x)_2 - U^*_{j\mu_2\ldots\mu_n}(x)_1\Big(\frac{\partial}{\partial t}U_{j\mu_2\ldots\mu_n}(x)_2\\ &\quad+\frac{\partial}{\partial x_j}U_{0\mu_2\ldots\mu_n}(x)_2\Big)\Big\}\\ &= \frac{1}{2m^{2n-1}|\,N'\,|^2}\int d\mathbf{x}(U^*_{[4\mu_1]\mu_2\ldots\mu_n}(x)_1U_{\mu_1\mu_2\ldots\mu_n}(x)_2\\ &\quad- U^*_{\mu_1\mu_2\ldots\mu_n}(x)_1U_{[4\mu_1]\mu_2\ldots\mu_n}(x)_2)\\ &= \frac{1}{2im^{2n}|\,N'\,|}\int d\mathbf{x}\Big(\frac{\partial}{\partial t}U^*_{\mu_1\mu_2\ldots\mu_n}(x)_1U_{\mu_1\mu_2\ldots\mu_n}(x)_2\\ &\quad- U^*_{\mu_1\mu_2\ldots\mu_n}(x)_1\frac{\partial}{\partial t}U_{\mu_1\mu_2\ldots\mu_n}(x)_2\Big)\ ,\end{aligned} \tag{6.7.68}$$

where we have used

$$mU_{[\nu\lambda]\mu_2\ldots\mu_n}(x) = \partial_\nu U_{\lambda\mu_2\ldots\mu_n}(x) - \partial_\lambda U_{\nu\mu_2\ldots\mu_n}(x)\ . \tag{6.7.69}$$

Equation (6.7.68) is nothing but the generalization of (6.7.15).

Next, we shall consider a particle with half-integral spin. Its angular momentum $n + 1/2$ is constructed by a direct product $v_{(i_1i_2\ldots i_n),\xi}$ of $u_{(i_1i_2\ldots i_n)}$ satisfying (6.7.52) and a spinor u_ξ ($\xi = 1, 2$). According to the composition law of angular momenta this direct product space consists of two kinds of angular momenta $n+1/2$ and $n-1/2$, and so we must eliminate the part of $n-1/2$ by setting subsidiary conditions on $v_{(i_1 i_2 \ldots i_n),\xi}$. We shall set

$$\sum_{j=1}^{3} \sigma_j v_{(j i_2 \ldots i_n)} = 0 , \tag{6.7.70}$$

where $v_{(i_1 i_2 \ldots i_n)}$ is a brief account of $v_{(i_1 i_2 \ldots i_n),\xi}$. Noting that the indices $i_1, i_2, \ldots i_n$ of $v_{(i_1 i_2 \ldots i_n)}$ satisfying (6.7.70) are symmetric we apply $\sigma_i \sigma_j = \delta_{ij} + [\sigma_i, \sigma_j]/2$ on $v_{(i j i_3 \ldots i_n)}$ and sum up with respect to i and j to derive

$$\sum_{j=1}^{3} v_{(j j i_3 \ldots i_n)} = 0 . \tag{6.7.71}$$

That is, (6.7.71) is a result from (6.7.70) and is not an independent condition. It is easily verified in the following way that $v_{(i_1 i_2 \ldots i_n)}$ describes a state of angular momentum $n+1/2$ when it satisfies (6.7.70). Since the number of components of $v_{(i_1 i_2 \ldots i_n)}$ is $(n+1)(n+2)$ and that of the conditions (6.7.70) is $n(n+1)$, then the number of independent components of $v_{(i_1 i_2 \ldots i_n)}$ satisfying (6.7.70) is $(n+1)(n+2) - n(n+1) = 2(n+1/2)+1$. This is equal to the dimension of an irreducible representation space with angular momentum $n+1/2$. On the other hand, $\sum_j \sigma_j v_{(j i_2 \ldots i_n)}$ is a direct product of a tensor of the $(n-1)$-th rank and a spinor in 3-dimensional space, and it has angular momenta of at most $n-1/2$. Among the angular momenta of $v_{(i_1 \ldots i_n)}$, therefore, the part of $n+1/2$ is never eliminated by (6.7.70). In other words, $v_{(i_1 \ldots i_n)}$ satisfying (6.7.70) contains an irreducible representation with angular momentum $n+1/2$. Since its dimension is, as has been seen, just $2(n+1/2)+1$, thus we can conclude that such $v_{(i_1 \ldots i_n)}$ gives an irreducible representation space with angular momentum $n+1/2$.

Corresponding to $v_{(i_1 \ldots i_n)}$ a wave function $\phi^{(\pm)}_{(i_1 \ldots i_n),\xi}(\mathbf{k})$ describing a spin $n+1/2$ particle in an irreducible representation of the Poincaré group is obtained. As usual, we shall introduce a 4-component function $\varphi_{(i_1 \ldots i_n)}(\mathbf{k})$ $(a = 1,2,3,4)$ for given $i_1, \ldots i_n$:

$$\varphi_{(i_1 \ldots i_n)}(\mathbf{k}) = \begin{pmatrix} \phi^{(+)}_{(i_1 \ldots i_n)}(\mathbf{k}) \\ \phi^{(-)}_{(i_1 \ldots i_n)}(\mathbf{k}) \end{pmatrix} . \tag{6.7.72}$$

In the following the index a will not be written explicitly. Then

$$k_0 \varphi_{(i_1 \ldots i_n)}(\mathbf{k}) = \beta \omega_k \varphi_{(i_1 \ldots i_n)}(\mathbf{k}) , \tag{6.7.73}$$

and from (6.7.70) we obtain

$$\sum_{j=1}^{3} \sigma_j \varphi_{(ji_2\ldots i_n)}(\mathbf{k}) = 0 \ . \tag{6.7.74}$$

Combining the above general theory for integral spin and the argument for a Dirac particle in §6.2, we find immediately that the covariant amplitude

$$\psi_{\mu_1\mu_2\ldots\mu_n}(\mathbf{k}) = \omega_k^{1/2} \sum_{i_1\ldots i_n} \left(\prod_{l=1}^{n} F_{\mu_l i_l}\right) U_F(\mathbf{k}) \varphi_{(i_1\ldots i_n)}(\mathbf{k}) \ , \tag{6.7.75}$$

for a spin $\frac{1}{2}$ particle symmetric with respect to $\mu_1, \mu_2 \ldots \mu_n$ satisfies the following relations:

$$(i\gamma_\mu k_\mu + m)\psi_{\mu_1\mu_2\ldots\mu_n}(\mathbf{k}) = 0 \ , \tag{6.7.76}$$

$$\psi_{\mu\mu\mu_3\ldots\mu_n}(\mathbf{k}) = 0 \ , \tag{6.7.77}$$

$$k_\mu \psi_{\mu\mu_2\ldots\mu_n}(\mathbf{k}) = 0 \ . \tag{6.7.78}$$

Next, to find conditions corresponding to (6.7.74) we shall calculate $\gamma_\mu \psi_{\mu\mu_2\ldots\mu_n}(\mathbf{k})$. From (6.7.42) we get

$$\gamma_\mu F_{\mu j} = m\gamma_j + \frac{\gamma_\mu k_\mu}{\omega_k + m} k_j + \frac{m\gamma_4 k_4}{\omega_k(\omega_k + m)} k_j \ . \tag{6.7.79}$$

Noting here

$$(i\gamma_\mu k_\mu + m)U_F(\mathbf{k})\varphi_{(i_1\ldots i_n)}(\mathbf{k}) = 0 \ ,$$

from (6.7.78) and (6.7.79) we obtain

$$\begin{aligned} \gamma_\mu \psi_{\mu\mu_2\ldots\mu_n}(\mathbf{k}) = \omega_k^{1/2} m \sum_j \left(\gamma_j + \frac{ik_j}{\omega_k + m} + \frac{ik_0\beta}{\omega_k(\omega_k + m)} k_j\right) U_F(\mathbf{k}) \\ \times \sum_{i_2\ldots i_n} \left(\prod_{l=2}^{n} F_{\mu_l i_l}\right) \varphi_{(ji_2\ldots i_n)}(\mathbf{k}) \ . \end{aligned} \tag{6.7.80}$$

On the other hand, from (6.7.37) we get

$$\begin{aligned} \gamma_j U_F(\mathbf{k}) &= U_F(\mathbf{k})^{-1}\gamma_j - \frac{2ik_j}{\sqrt{2\omega_k(\omega_k + m)}} \\ &= U_F(\mathbf{k})^{-1}\gamma_j - \frac{ik_j}{\omega_k + m}(U_F(\mathbf{k}) + U_F(\mathbf{k})^{-1}) \ . \end{aligned} \tag{6.7.81}$$

Since

$$k_0\beta U_F(\mathbf{k})\varphi_{(ji_2\ldots i_n)}(\mathbf{k}) = U_F(\mathbf{k})^{-1}k_0\beta\varphi_{(ji_2\ldots i_n)}(\mathbf{k}) \\ = \omega_k U_F(\mathbf{k})^{-1}\varphi_{(ji_2\ldots i_n)}(\mathbf{k}) \ , \tag{6.7.82}$$

substituting these equations into (6.7.80) we obtain

$$\gamma_\mu\psi_{\mu\mu_2\ldots\mu_n}(\mathbf{k}) = \omega_k^{1/2} m U_F(\mathbf{k})^{-1} \sum_{i_2\ldots i_n} \left(\prod_{l=2}^{n} F_{\mu_l i_l}\right) \sum_j \gamma_j \varphi_{(ji_2\ldots i_n)}(\mathbf{k}) \ . \tag{6.7.83}$$

Now γ_j in the Dirac representation is

$$\gamma_j = \begin{pmatrix} \mathbf{0} & -iI \\ iI & \mathbf{0} \end{pmatrix} \sigma_j \ .$$

From Eqs. (6.7.74) and (6.7.83) we therefore obtain

$$\gamma_\mu\psi_{\mu\mu_2\ldots\mu_n}(\mathbf{k}) = 0 \ . \tag{6.7.84}$$

Conversely, if this equation holds, (6.7.74) is derived through (6.7.83).

Since (6.7.77) is derived from (6.7.84), and since (6.7.78) is derived from (6.7.76) and (6.7.84), we obtain the following equations which describe a spin $n + 1/2$ particle with finite mass:

$$(\gamma_\mu\partial_\mu + m)\psi_{\mu_1\mu_2\ldots\mu_n}(x) = 0 \ , \tag{6.7.85}$$

$$\gamma_\mu\psi_{\mu\mu_2\ldots\mu_n}(x) = 0 \ , \tag{6.7.86}$$

where

$$\psi_{\mu_1\mu_2\ldots\mu_n}(x) = \frac{1}{(2\pi)^{3/2}} \int \frac{d\mathbf{k}}{\omega_k} e^{i(\mathbf{k}\mathbf{x}-k_0 t)} \psi_{\mu_1\mu_2\ldots\mu_n}(\mathbf{k}) \ . \tag{6.7.87}$$

The set of equations (6.7.85) and (6.7.86) is called a Rarita-Schwinger equation for a spin $n+1/2$ particle, and is a generalization of the result in b). Further, using the relation (6.7.64) the covariant inner product is given by

$$\langle\langle 1 \,|\, 2 \rangle\rangle = \langle 1 \,|\, 2 \rangle \\ = \frac{1}{m^{2n} |\, N' \,|^2} \int d\mathbf{x}\, \psi^*_{\mu_1\mu_2\ldots\mu_n}(x)_1 \psi_{\mu_1\mu_2\ldots\mu_n}(x)_2 \ . \tag{6.7.88}$$

This is also a generalization of (6.7.35).

Chapter 7

Covariant Formalism II — Massless Particles

§7.1 Particles with Discrete Spin

The irreducible representations of the Poincaré group corresponding to massless particles are, as has been mentioned in §5.2, classified into two kinds: The representations with discrete spin and with continuous spin. In this section we shall find covariant descriptions of the former.

The Lorentz transformations of the wave function of such a particle have been given by (5.2.21) and (5.2.22). Let us consider the simplest case of spin 0, i.e., the case $S = 0$. In this case (5.2.21) and (5.2.22) are just the same equations as (6.1.1) and (6.1.2) with $m = 0$, and hence the covariant formalism can be obtained from (6.1.8)–(6.1.10) with $m = 0$. The result is

$$\partial_\mu^2 U(x) = 0 \ , \tag{7.1.1}$$

$$U'(x) = U(\Lambda^{-1}x) \ , \tag{7.1.2}$$

$$\langle\langle 1 \,|\, 2 \rangle\rangle = \frac{1}{2i} \int d\mathbf{x} \Big(\frac{\partial U^*(x)_1}{\partial t} U(x)_2 - U^*(x)_1 \frac{\partial U(x)_2}{\partial t} \Big) \ . \tag{7.1.3}$$

The equations corresponding to those after (6.1.11) are

$$U_\mu(x) = \partial_\mu U(x) \ , \tag{7.1.4}$$

$$\partial_\mu U_\mu(x) = 0 \ , \tag{7.1.5}$$

and the covariant inner product is easily obtained from (6.1.13) with $m = 0$. The Kemmer type equation is given by (6.1.15) with m replaced by a 5×5 diagonal matrix M whose element at p-th row and q-th column is $(M)_{pq} = \delta_{pq} - \delta_{p5}\delta_{q5}$ $(p, q = 1, \ldots, 5)$. The parameter m in (6.1.17) should be dropped. That is, in the case of spin 0, the covariant formalism is nothing but the result of simple deformations of equations in §6.1, and in this sense we can say that the peculiarity of a massless particle appears in the case $|S| \geq \frac{1}{2}$. Thus we shall first consider the case of spin $\frac{1}{2}$, i.e., the case $|S| = \frac{1}{2}$, similarly to the case of particle with finite mass.

The infinitesimal Lorentz transformations of the wave functions $\phi^{(\pm)}_{1/2}(\mathbf{k})$ and $\phi^{(\pm)}_{-1/2}(\mathbf{k})$ corresponding to $S = 1/2$ and $-1/2$ respectively are given, using (5.2.23)–(5.2.26), by

$$\phi^{(\pm)'}_{1/2}(\mathbf{k}) = \left[1 + i\left\{\mathbf{k} \times \left(\frac{1}{i}\frac{\partial}{\partial \mathbf{k}} + \frac{1}{2}\frac{\mathbf{n} \times \mathbf{k}}{|\mathbf{k}|(|\mathbf{k}| + k_3)}\right) + \frac{1}{2}\frac{\mathbf{k}}{|\mathbf{k}|}\right\}\boldsymbol{\theta}\right] \phi^{(\pm)}_{1/2}(\mathbf{k}) \,, \tag{7.1.6}$$

$$\phi^{(\pm)'}_{-1/2}(\mathbf{k}) = \left[1 + i\left\{\mathbf{k} \times \left(\frac{1}{i}\frac{\partial}{\partial \mathbf{k}} - \frac{1}{2}\frac{\mathbf{n} \times \mathbf{k}}{|\mathbf{k}|(|\mathbf{k}| + k_3)}\right) - \frac{1}{2}\frac{\mathbf{k}}{|\mathbf{k}|}\right\}\boldsymbol{\theta}\right] \phi^{(\pm)}_{-1/2}(\mathbf{k}) \tag{7.1.7}$$

for θ-transformation, and

$$\phi^{(\pm)'}_{1/2}(\mathbf{k}) = \left\{1 + ik_0\left(\frac{1}{i}\frac{\partial}{\partial \mathbf{k}} + \frac{1}{2}\frac{\mathbf{n} \times \mathbf{k}}{|\mathbf{k}|(|\mathbf{k}| + k_3)}\right)\boldsymbol{\tau}\right\}\phi^{(\pm)}_{1/2}(\mathbf{k}) \,, \tag{7.1.8}$$

$$\phi^{(\pm)'}_{-1/2}(\mathbf{k}) = \left\{1 + ik_0\left(\frac{1}{i}\frac{\partial}{\partial \mathbf{k}} - \frac{1}{2}\frac{\mathbf{n} \times \mathbf{k}}{|\mathbf{k}|(|\mathbf{k}| + k_3)}\right)\boldsymbol{\tau}\right\}\phi^{(\pm)}_{-1/2}(\mathbf{k}) \tag{7.1.9}$$

for τ-transformation, where $\mathbf{n}$ is a unit vector directing to the third axis, i.e.,

$$\mathbf{n} = (0, 0, 1) \,. \tag{7.1.10}$$

Similarly to (6.2.2) in the case of finite mass it is convenient to consider appropriate combinations of positive and negative frequency states to get a covariant formalism. For this purpose we define

$$\varphi^{R(+)}(\mathbf{k}) = \begin{pmatrix} \phi^{(+)}_{1/2}(\mathbf{k}) \\ 0 \end{pmatrix}, \qquad \varphi^{R(-)}(\mathbf{k}) = \begin{pmatrix} 0 \\ \phi^{(-)}_{-1/2}(\mathbf{k}) \end{pmatrix}, \tag{7.1.11}$$

$$\varphi^{L(+)}(\mathbf{k}) = \begin{pmatrix} 0 \\ \phi^{(+)}_{-1/2}(\mathbf{k}) \end{pmatrix}, \qquad \varphi^{L(-)}(\mathbf{k}) = \begin{pmatrix} \phi^{(-)}_{1/2}(\mathbf{k}) \\ 0 \end{pmatrix}, \tag{7.1.12}$$

and we introduce two-component state vectors with positive and negative frequencies

$$\varphi^R(\mathbf{k}) = \varphi^{R(+)}(\mathbf{k}) + \varphi^{R(-)}(\mathbf{k}) \, , \tag{7.1.13}$$

$$\varphi^L(\mathbf{k}) = \varphi^{L(+)}(\mathbf{k}) + \varphi^{L(-)}(\mathbf{k}) \, . \tag{7.1.14}$$

From (7.1.6)–(7.1.9) the Lorentz transformations of $\varphi^R(\mathbf{k})$ and $\varphi^L(\mathbf{k})$ are given by

$$\varphi^{R,L'}(\mathbf{k}) = \left[1 + i\left\{\mathbf{k} \times \left(\frac{1}{i}\frac{\partial}{\partial \mathbf{k}} + \frac{\sigma_3}{2}\frac{(\mathbf{n} \times \mathbf{k})}{|\mathbf{k}|(|\mathbf{k}| + k_3)}\right) + \frac{\sigma_3}{2}\frac{\mathbf{k}}{|\mathbf{k}|}\right\}\boldsymbol{\theta}\right]\varphi^{R,L}(\mathbf{k})$$

$$(\theta\text{-transformation}) \, , \tag{7.1.15}$$

$$\varphi^{R,L'}(\mathbf{k}) = \left\{1 + ik_0\left(\frac{1}{i}\frac{\partial}{\partial \mathbf{k}} + \frac{\sigma_3(\mathbf{n} \times \mathbf{k})}{2|\mathbf{k}|(|\mathbf{k}| + k_3)}\right)\boldsymbol{\tau}\right\}\varphi^{R,L}(\mathbf{k})$$

$$(\tau\text{-transformation}) \, , \tag{7.1.16}$$

and (5.2.27) can be rewritten as

$$k_0\varphi^R(\mathbf{k}) = |\mathbf{k}|\sigma_3\varphi^R(\mathbf{k}) \, , \tag{7.1.17}$$

$$k_0\varphi^L(\mathbf{k}) = -|\mathbf{k}|\sigma_3\varphi^L(\mathbf{k}) \, . \tag{7.1.18}$$

However, since the third axis plays a special role in these equations, we must remove such a speciality of the axis in order to get a covariant formalism. Introducing a unitary operator $U(\mathbf{k})$ such that

$$U(\mathbf{k})\sigma_3 U(\mathbf{k})^{-1} = \frac{(\boldsymbol{\sigma}\mathbf{k})}{|\mathbf{k}|} \, , \tag{7.1.19}$$

we define

$$\chi^{R,L}(\mathbf{k}) = U(\mathbf{k})\varphi^{R,L}(\mathbf{k}) \, , \tag{7.1.20}$$

then Eqs. (7.1.17) and (7.1.18) are rewritten by (7.1.19) as

$$k_0\chi^R(\mathbf{k}) = (\boldsymbol{\sigma}\mathbf{k})\chi^R(\mathbf{k}) \, , \tag{7.1.21}$$

$$k_0\chi^L(\mathbf{k}) = -(\boldsymbol{\sigma}\mathbf{k})\chi^L(\mathbf{k}) \, . \tag{7.1.22}$$

According to (7.1.19), $U(\mathbf{k})$ is considered to be a rotation generator with an angle θ ($k_3 = |\mathbf{k}|\cos\theta$) about the axis through the origin and perpendicular to the plane formed by $\mathbf{k}$ and $\mathbf{n}$. Hence we have

$$\begin{aligned} U(\mathbf{k}) &= \exp\left(i\frac{\theta}{2}\frac{\boldsymbol{\sigma}(\mathbf{k} \times \mathbf{n})}{|\mathbf{k} \times \mathbf{n}|}\right) = \cos\frac{\theta}{2} + i\frac{\boldsymbol{\theta}(\mathbf{k} \times \mathbf{n})}{|\mathbf{k} \times \mathbf{n}|}\sin\frac{\theta}{2} \\ &= \frac{1}{\sqrt{2(\mathbf{k}^2 + (\mathbf{n}\mathbf{k})|\mathbf{k}|)}}\{|\mathbf{k}| + (\mathbf{n}\mathbf{k}) + i\boldsymbol{\sigma}(\mathbf{k} \times \mathbf{n})\} \, . \end{aligned} \tag{7.1.23}$$

In order to get the Lorentz transformations of $\chi^{R,L}(\mathbf{k})$, as is seen in (7.1.15) and (7.1.16), we must calculate $U(\mathbf{k})(\partial/\partial\mathbf{k})U(\mathbf{k})^{-1}$, and with the help of (7.1.23) we find

$$\begin{aligned}\frac{1}{i}U(\mathbf{k})\Big(\frac{\partial}{\partial\mathbf{k}}\Big)U(\mathbf{k})^{-1} &= \frac{1}{i}\frac{\partial}{\partial\mathbf{k}} + U(\mathbf{k})\frac{\partial U(\mathbf{k})^{-1}}{\partial\mathbf{k}} \\ &= \frac{1}{i}\frac{\partial}{\partial\mathbf{k}} + \frac{1}{2}\frac{(\boldsymbol{\sigma}\times\mathbf{k})}{\mathbf{k}^2} - \frac{1}{2}\frac{\mathbf{n}\times\mathbf{k}}{|\mathbf{k}|^2(|\mathbf{k}|+k_3)}(\boldsymbol{\sigma}\mathbf{k})\,. \end{aligned} \tag{7.1.24}$$

In deriving this expression we have used the following relations:

$$\mathbf{A}\times(\mathbf{B}\times\mathbf{C}) = \mathbf{B}(\mathbf{AC}) - (\mathbf{AB})\mathbf{C} - \sum_{j=1}^{3}[\mathbf{B}, A_j]C_j\,,$$

$$\boldsymbol{\sigma}\times\boldsymbol{\sigma} = 2i\boldsymbol{\sigma}\,,$$

$$\boldsymbol{\sigma}(\mathbf{A}\boldsymbol{\sigma}) = i(\mathbf{A}\times\boldsymbol{\sigma}) + \mathbf{A}\,.$$

As a result, considering (7.1.19) we have

$$U(\mathbf{k})\Big(\frac{1}{i}\frac{\partial}{\partial\mathbf{k}} + \frac{\sigma_3}{2}\frac{\mathbf{n}\times\mathbf{k}}{|\mathbf{k}|(|\mathbf{k}|+k_3)}\Big)U(\mathbf{k})^{-1} = \frac{1}{i}\frac{\partial}{\partial\mathbf{k}} + \frac{1}{2}\frac{\boldsymbol{\sigma}\times\mathbf{k}}{\mathbf{k}^2}\,, \tag{7.1.25}$$

and hence the transformations of $\chi^{R,L}(\mathbf{k})$ are given by

$$\chi^{R,L'}(\mathbf{k}) = \Big[1 + i\Big\{\frac{1}{i}\Big(\mathbf{k}\times\frac{\partial}{\partial\mathbf{k}}\Big) + \frac{\boldsymbol{\sigma}}{2}\Big\}\boldsymbol{\theta}\Big]\chi^{R,L}(\mathbf{k})$$
$$(\theta\text{-transformation})\,, \tag{7.1.26}$$

$$\chi^{R,L'}(\mathbf{k}) = \Big\{1 + ik_0\Big(\frac{1}{i}\frac{\partial}{\partial\mathbf{k}} + \frac{1}{2}\frac{\boldsymbol{\sigma}\times\mathbf{k}}{\mathbf{k}^2}\Big)\boldsymbol{\tau}\Big\}\chi^{R,L}(\mathbf{k})$$
$$(\tau\text{-transformation})\,. \tag{7.1.27}$$

The right-hand side of (7.1.27) is now rewritten by the use of an identity

$$\frac{\boldsymbol{\sigma}\times\mathbf{k}}{2\mathbf{k}^2} = \frac{1}{2\mathbf{k}^2 i}(-\boldsymbol{\sigma}(\mathbf{k}\boldsymbol{\sigma}) + \mathbf{k})\,,$$

(7.1.21), and (7.1.22) as

$$(7.1.27) = \Big\{1 + \Big(k_0\frac{\partial}{\partial\mathbf{k}} \mp \frac{\boldsymbol{\sigma}}{2} + \frac{\mathbf{k}}{2k_0}\Big)\boldsymbol{\tau}\Big\}\chi^{R,L}(\mathbf{k})\,. \tag{7.1.28}$$

Here it is easily seen that $\mathbf{k}/2k_0$ can be eliminated by putting $\psi^{R,L}(\mathbf{k}) = |\mathbf{k}|^{1/2}\chi^{R,L}(\mathbf{k})$ from the transformation equations of $\psi^{R,L}(\mathbf{k})$ similarly to the case of finite mass. The signs $\mp$ in (7.1.28) correspond to R and L respectively, and this notation will be used also hereafter.

The above discussion is summarized as follows. We begin with (7.1.13) and (7.1.14), and put

$$\psi^{R,L}(\mathbf{k}) = |\,\mathbf{k}\,|^{1/2} U(\mathbf{k}) \phi^{R,L}(\mathbf{k}) \,, \tag{7.1.29}$$

then $\psi^R(\mathbf{k})$ and $\psi^L(\mathbf{k})$ transform as

$$\psi^{R,L'}(\mathbf{k}) = \left(1 + \frac{i}{2}\boldsymbol{\theta}\boldsymbol{\sigma}\right)\psi^{R,L}(\mathbf{k} - \mathbf{k}\times\boldsymbol{\theta}) \quad (\theta\text{-transformation}) \,, \tag{7.1.30}$$

$$\psi^{R,L'}(\mathbf{k}) = \left(1 \mp \frac{1}{2}\boldsymbol{\tau}\boldsymbol{\sigma}\right)\psi^{R,L}(\mathbf{k} + \boldsymbol{\tau} k_0) \quad (\tau\text{-transformation}) \,, \tag{7.1.31}$$

and they satisfy

$$k_0 \psi^R(\mathbf{k}) = (\mathbf{k}\boldsymbol{\sigma})\psi^R(\mathbf{k}) \,, \tag{7.1.32}$$

$$k_0 \psi^L(\mathbf{k}) = -(\mathbf{k}\boldsymbol{\sigma})\psi^L(\mathbf{k}) \,. \tag{7.1.33}$$

Therefore, if we define their amplitude in the x-representation by

$$\psi^{R,L}(x) = \frac{1}{(2\pi)^{3/2}} \int \frac{d\mathbf{k}}{|\,\mathbf{k}\,|}\, e^{i(\mathbf{k}\mathbf{x} - k_0 t)} \psi^{R,L}(\mathbf{k}) \,, \tag{7.1.34}$$

as usual, we have

$$\left(\frac{\partial}{\partial t} + \boldsymbol{\sigma}\boldsymbol{\nabla}\right)\psi^R(x) = 0 \,, \tag{7.1.35}$$

$$\left(\frac{\partial}{\partial t} - \boldsymbol{\sigma}\boldsymbol{\nabla}\right)\psi^L(x) = 0 \,, \tag{7.1.36}$$

and the Lorentz transformations are

$$\psi^{R,L'}(x) = \left(1 + \frac{i}{2}\boldsymbol{\theta}\boldsymbol{\sigma}\right)\psi^{R,L}(\Lambda(\theta)^{-1}x) \qquad (\theta\text{-transformation}) \,, \tag{7.1.37}$$

$$\psi^{R,L'}(x) = \left(1 \mp \frac{1}{2}\boldsymbol{\tau}\boldsymbol{\sigma}\right)\psi^{R,L}(\Lambda(\tau)^{-1}x) \qquad (\tau\text{-transformation}) \,. \tag{7.1.38}$$

Further it is clear that (5.2.28) gives the inner products

$$\langle\, 1\,|\,2\,\rangle^{R,L} = \langle\langle\, 1\,|\,2\,\rangle\rangle = \int d\mathbf{x} \psi^{R,L*}(x)_1 \psi^{R,L}(x)_2 \tag{7.1.39}$$

for R and L. In this way we have obtained the covariant formalism for a particle with $|S| = \frac{1}{2}$. Equations (7.1.35) and (7.1.36) are called Weyl equations and describe massless particles with spin $\frac{1}{2}$.

Combining $\psi^R(x)$ and $\psi^L(x)$ we define

$$\psi(x) = \begin{pmatrix} \psi^R(x) \\ \psi^L(x) \end{pmatrix} , \tag{7.1.40}$$

and we find from (7.1.35) and (7.1.36) that $\psi(x)$ satisfies

$$\left\{ \begin{pmatrix} \mathrm{O} & i\sigma \\ -i\sigma & \mathrm{O} \end{pmatrix} \nabla + \begin{pmatrix} \mathrm{O} & I \\ I & \mathrm{O} \end{pmatrix} \partial_4 \right\} \psi(x) = 0 . \tag{7.1.41}$$

This is a Dirac equation for $m = 0$ in the representation with diagonal γ_5, which has been given in (6.5.25). Of course, the independent components of the solutions for this equation do not form an irreducible representation space of the Poincaré group, i.e., are reducible, so we must multiply $\psi(x)$ by projection operator $\frac{1}{2}(1-\gamma_5)$ or $\frac{1}{2}(1+\gamma_5)$ to extract the function corresponding to $\psi^R(x)$ or $\psi^L(x)$. Then we have

$$\frac{1}{2}\gamma_\mu \partial_\mu (1 - \gamma_5) \psi(x) = 0 , \tag{7.1.42}$$

$$\frac{1}{2}\gamma_\mu \partial_\mu (1 + \gamma_5) \psi(x) = 0 , \tag{7.1.43}$$

corresponding to (7.1.35) and (7.1.36) respectively. In this case the inner products are given by

$$\langle\langle 1 | 2 \rangle\rangle^{R,L} = \frac{1}{2} \int d\mathbf{x} \psi^*(x)_1 (1 \mp \gamma_5) \psi(x)_2 . \tag{7.1.44}$$

The extension of the above argument to the case $|S| = n/2$ ($n > 1$) can be carried out in a similar manner to the case of finite mass in §6.3.

First, supposing that $\rho_1, \rho_2, \ldots, \rho_n$ take the value 1 or 2, we define totally symmetric functions $\varphi^{R(\pm)}_{(\rho_1 \cdots \rho_n)}(\mathbf{k})$ and $\varphi^{L(\pm)}_{(\rho_1 \cdots \rho_n)}(\mathbf{k})$ in place of $\phi^{(\pm)}_{n/2}(\mathbf{k})$ and $\phi^{(\pm)}_{-n/2}(\mathbf{k})$ as follows:

$$\varphi^{R(+)}_{(\rho_1 \cdots \rho_n)}(\mathbf{k}) = \begin{cases} \phi^{(+)}_{n/2}(\mathbf{k}) & (\text{for } \rho_1 = \rho_2 = \ldots = \rho_n = 1) , \\ 0 & (\text{in other cases}) , \end{cases} \tag{7.1.45}$$

$$\varphi^{R(-)}_{(\rho_1\ldots\rho_n)}(\mathbf{k}) = \begin{cases} \phi^{(-)}_{-n/2}(\mathbf{k}) & (\text{for } \rho_1 = \rho_2 = \ldots = \rho_n = 2)\ , \\ 0 & (\text{in other cases})\ , \end{cases} \tag{7.1.46}$$

$$\varphi^{L(+)}_{(\rho_1\ldots\rho_n)}(\mathbf{k}) = \begin{cases} \phi^{(+)}_{-n/2}(\mathbf{k}) & (\text{for } \rho_1 = \rho_2 = \ldots = \rho_n = 2)\ , \\ 0 & (\text{in other cases})\ , \end{cases} \tag{7.1.47}$$

$$\varphi^{L(-)}_{(\rho_1\ldots\rho_n)}(\mathbf{k}) = \begin{cases} \phi^{(-)}_{n/2}(\mathbf{k}) & (\text{for } \rho_1 = \rho_2 = \ldots = \rho_n = 1)\ , \\ 0 & (\text{in other cases})\ . \end{cases} \tag{7.1.48}$$

Further we define

$$\varphi^{R,L}_{(\rho_1\ldots\rho_n)}(\mathbf{k}) = \varphi^{R,L(+)}_{(\rho_1\ldots\rho_n)}(\mathbf{k}) + \varphi^{R,L(-)}_{(\rho_1\ldots\rho_n)}(\mathbf{k})\ . \tag{7.1.49}$$

We denote hereafter the indices $(\rho_1 \ldots \rho_n)$ as $(\ldots)_n$ for simplicity. Since we can easily find that (7.1.45) and (7.1.46) are nothing but the solutions for

$$k_0\varphi^R_{(\ldots)_n}(\mathbf{k}) = |\,\mathbf{k}\,|\sigma_3^{(i)}\varphi^R_{(\ldots)_n}(\mathbf{k}) \qquad (i = 1, 2, \ldots n)\ , \tag{7.1.50}$$

we can use the above equations instead of (7.1.45) and (7.1.46). Similarly we can use

$$k_0\varphi^L_{(\ldots)_n}(\mathbf{k}) = -|\,\mathbf{k}\,|\sigma_3^{(i)}\varphi^L_{(\ldots)_n}(\mathbf{k}) \qquad (i = 1, 2, \ldots n) \tag{7.1.51}$$

instead of (7.1.47) and (7.1.48).

According to (5.2.21) and (5.2.22) we can consider that $\varphi^R_{(\ldots)_n}(\mathbf{k})$ and $\varphi^L_{(\ldots)_n}(\mathbf{k})$ transform under a Lorentz transformation like a direct product of n $\varphi^R(\mathbf{k})$'s and n $\varphi^L(\mathbf{k})$'s in (7.1.13) and (7.1.14) respectively, and we have

$$\begin{aligned}\varphi^{R,L\prime}_{(\ldots)_n}(\mathbf{k}) = \Bigg[1 + i\Bigg\{&\mathbf{k}\times\left(\frac{1}{i}\frac{\partial}{\partial\mathbf{k}} + \sum_{i=1}^{n}\frac{\sigma_3^{(i)}}{2}\frac{\mathbf{n}\times\mathbf{k}}{|\,\mathbf{k}\,|(|\,\mathbf{k}\,| + k_3)}\right) \\ &+ \sum_{i=1}^{n}\frac{\sigma_3^{(i)}}{2}\frac{\mathbf{k}}{|\,\mathbf{k}\,|}\Bigg\}\boldsymbol{\theta}\Bigg]\varphi^{R,L}_{(\ldots)_n}(\mathbf{k}) \\ &(\theta\text{-transformation})\ ,\end{aligned} \tag{7.1.52}$$

$$\begin{aligned}\varphi^{R,L\prime}_{(\ldots)_n}(\mathbf{k}) = \left\{1 + ik_0\left(\frac{1}{i}\frac{\partial}{\partial\mathbf{k}} + \sum_{i=1}^{n}\frac{\sigma_3^{(i)}}{2}\frac{\mathbf{n}\times\mathbf{k}}{|\,\mathbf{k}\,|(|\,\mathbf{k}\,| + k_3)}\right)\boldsymbol{\tau}\right\}\varphi^{R,L}_{(\ldots)_n}(\mathbf{k}) \\ (\tau\text{-transformation})\ .\end{aligned} \tag{7.1.53}$$

If $U^{(i)}(\mathbf{k})$ denotes the unitary operator (7.1.23) acting on the index ρ_i and if (7.1.29) is generalized to

$$\psi^{R,L}_{(\ldots)_n}(\mathbf{k}) = |\mathbf{k}|^{n/2} \prod_{i=1}^{n} U^{(i)}(\mathbf{k}) \varphi^{R,L}_{(\ldots)_n}(\mathbf{k}) , \tag{7.1.54}$$

a similar argument to the case $|S| = \frac{1}{2}$ gives

$$\psi^{R,L'}_{(\ldots)_n}(\mathbf{k}) = \Big(1 + \frac{i}{2}\boldsymbol{\theta} \sum_{i=1}^{n} \boldsymbol{\sigma}^{(i)}\Big) \psi^{R,L}_{(\ldots)_n}(\mathbf{k} - \mathbf{k} \times \boldsymbol{\theta}) \quad (\theta\text{-transformation}) , \tag{7.1.55}$$

$$\psi^{R,L'}_{(\ldots)_n}(\mathbf{k}) = \Big(1 \mp \frac{1}{2}\boldsymbol{\tau} \sum_{i=1}^{n} \boldsymbol{\sigma}^{(i)}\Big) \psi^{R,L}_{(\ldots)_n}(\mathbf{k} + \boldsymbol{\tau} k_0) \quad (\tau\text{-transformation}) . \tag{7.1.56}$$

From (7.1.50) and (7.1.51) we get

$$k_0 \psi^{R}_{(\ldots)_n}(\mathbf{k}) = (\mathbf{k}\boldsymbol{\sigma}^{(i)}) \psi^{R}_{(\ldots)_n}(\mathbf{k}) \quad (i = 1, 2, \ldots n) , \tag{7.1.57}$$

$$k_0 \psi^{L}_{(\ldots)_n}(\mathbf{k}) = -(\mathbf{k}\boldsymbol{\sigma}^{(i)}) \psi^{L}_{(\ldots)_n}(\mathbf{k}) \quad (i = 1, 2, \ldots n) . \tag{7.1.58}$$

Therefore, introducing amplitudes in the x-representation by

$$\psi^{R,L}_{(\ldots)_n}(x) = \frac{1}{(2\pi)^{3/2}} \int \frac{d\mathbf{k}}{|\mathbf{k}|} e^{i(\mathbf{k}\mathbf{x} - k_0 t)} \psi^{R,L}_{(\ldots)_n}(\mathbf{k}) , \tag{7.1.59}$$

we obtain immediately

$$\psi^{R,L'}_{(\ldots)_n}(x) = \Big(1 + \frac{i}{2}\boldsymbol{\theta} \sum_{i=1}^{n} \boldsymbol{\sigma}^{(i)}\Big) \psi^{R,L}_{(\ldots)_n}(\Lambda(\theta)^{-1} x) \quad (\theta\text{-transformation}) , \tag{7.1.60}$$

$$\psi^{R,L'}_{(\ldots)_n}(x) = \Big(1 \mp \frac{1}{2}\boldsymbol{\tau} \sum_{i=1}^{n} \boldsymbol{\sigma}^{(i)}\Big) \psi^{R,L}_{(\ldots)_n}(\Lambda(\tau)^{-1} x) \quad (\tau\text{-transformation}) , \tag{7.1.61}$$

and the equations of motion are

$$\Big(\frac{\partial}{\partial t} + \boldsymbol{\sigma}^{(i)} \nabla\Big) \psi^{R}_{(\ldots)_n}(x) = 0 \quad (i = 1, 2, \ldots n) , \tag{7.1.62}$$

$$\Big(\frac{\partial}{\partial t} - \boldsymbol{\sigma}^{(i)} \nabla\Big) \psi^{L}_{(\ldots)_n}(x) = 0 \quad (i = 1, 2, \ldots n) . \tag{7.1.63}$$

Equations (7.1.62) and (7.1.63) are the Bargmann-Wigner equations for a massless particle with spin $n/2$. The symbols R and L denote the right-handed and left-handed polarizations respectively. For example, when $n = 2$, (7.1.62) and (7.1.63) become the equations for a right-handed photon and a left-handed photon respectively (cf. §7.3). If we regard the indices $a_1, a_2, \ldots, a_n$ of $\psi_{(a_1 \ldots a_n)}(x)$ as the indices of Dirac spinor, which take the values from 1 to 4, we can write (7.1.62) and (7.1.63) in the following forms:

$$\gamma_\mu^{(i)} \partial_\mu \Big[\prod_{j=1}^{n} \frac{1}{2}(1 - \gamma_5)^{(j)} \Big] \psi_{(\ldots)_n}(x) = 0 \, , \tag{7.1.64}$$

$$\gamma_\mu^{(i)} \partial_\mu \Big[\prod_{j=1}^{n} \frac{1}{2}(1 + \gamma_5)^{(j)} \Big] \psi_{(\ldots)_n}(x) = 0 \, . \tag{7.1.65}$$

The covariant inner products may, analogously to the argument in the case of particle with finite mass, be written by the use of (5.2.28) and (7.1.54) as

$$\langle\langle\, 1 \,|\, 2 \,\rangle\rangle^{R,L} = \frac{1}{|\,N\,|^2} \sum_{\rho_1 \rho_2 \ldots \rho_n = 1,2} \int \frac{d\mathbf{k}}{|\,\mathbf{k}\,|^2} \frac{\psi^{R,L*}_{(\rho_1 \ldots \rho_n)}(\mathbf{k})_1 \psi^{R,L}_{(\rho_1 \ldots \rho_n)}(\mathbf{k})_2}{k_0^{n-1}} \, . \tag{7.1.66}$$

These covariant inner products are related to (5.2.28) as

$$\left. \begin{aligned} \langle\, n/2; 1, \pm \,|\, n/2; 2, \pm \,\rangle &= (\pm 1)^{n-1} \langle\langle\, 1, \pm \,|\, 2, \pm \,\rangle\rangle^{R,L} \, , \\ \langle\, -n/2; 1, \pm \,|\, -n/2; 2, \pm \,\rangle &= (\pm 1)^{n-1} \langle\langle\, 1, \pm \,|\, 2, \pm \,\rangle\rangle^{L,R} \, , \end{aligned} \right\} \tag{7.1.67}$$

apart from the normalization constant $1/|\,N\,|^2$. In the case $n \geq 2$, however, the right-hand side of (7.1.66) cannot be expressed in the x-representation by the use of (7.1.59), because, when the mass is zero the extra factor $1/k_0^{n-1}$ cannot be eliminated by the equation of motion in an analogous manner to the case of (6.3.35). A further study is therefore necessary to express the covariant inner products in the x-representation. It will be given in §7.3. But before proceeding we shall consider the discrete transformations.

§7.2 Discrete Transformations

Let us first consider the space reflection. We assume that there exists a unitary operator $\mathcal{R}$ satisfying (6.6.2)–(6.6.4). In this case the operators $\mathbf{J}$ and $\mathbf{K}$ are given by (5.2.25) and (5.2.26), and they are denoted as $\mathbf{J}(S)$ and $\mathbf{K}(S)$ hereafter in order to clarify their dependences on S. Now, similarly to (6.6.5), we put $\mathcal{R} = qp$ and

$$p\phi_S^{(\pm)}(\mathbf{k}) = \phi_S^{(\pm)}(-\mathbf{k}) \, , \tag{7.2.1}$$

then we find from (6.6.2) that q is a unitary operator commuting with $\mathbf{k}$, and that it is a function of $\mathbf{k}$ only. Thus we denote it as $q(\mathbf{k})$. By definition we get, from (6.6.3) and (6.6.4), the following equations for $q(\mathbf{k})$:

$$\left.\begin{aligned} \mathcal{R}^{-1}J_1(S)\mathcal{R} &= q(-\mathbf{k})^{-1}\Big\{\frac{1}{i}\Big(\mathbf{k}\times\frac{\partial}{\partial\mathbf{k}}\Big)_1 - S\frac{k_1}{|\mathbf{k}|-k_3}\Big\}q(-\mathbf{k}) = J_1(S)\,,\\ \mathcal{R}^{-1}J_2(S)\mathcal{R} &= q(-\mathbf{k})^{-1}\Big\{\frac{1}{i}\Big(\mathbf{k}\times\frac{\partial}{\partial\mathbf{k}}\Big)_2 - S\frac{k_2}{|\mathbf{k}|-k_3}\Big\}q(-\mathbf{k}) = J_2(S)\,,\\ \mathcal{R}^{-1}J_3(S)\mathcal{R} &= q(-\mathbf{k})^{-1}\Big\{\frac{1}{i}\Big(\mathbf{k}\times\frac{\partial}{\partial\mathbf{k}}\Big)_3 + S\Big\}q(-\mathbf{k}) = J_3(S)\,, \end{aligned}\right\} \tag{7.2.2}$$

and

$$\left.\begin{aligned} \mathcal{R}^{-1}K_1(S)\mathcal{R} &= q(-\mathbf{k})^{-1}\Big\{k_0\Big(\frac{1}{i}\frac{\partial}{\partial k_1} - S\frac{k_2}{|\mathbf{k}|(|\mathbf{k}|-k_3)}\Big)\Big\}q(-\mathbf{k}) = -K_1(S),\\ \mathcal{R}^{-1}K_2(S)\mathcal{R} &= q(-\mathbf{k})^{-1}\Big\{k_0\Big(\frac{1}{i}\frac{\partial}{\partial k_2} + S\frac{k_1}{|\mathbf{k}|(|\mathbf{k}|-k_3)}\Big)\Big\}q(-\mathbf{k}) = -K_2(S),\\ \mathcal{R}^{-1}K_3(S)\mathcal{R} &= q(-\mathbf{k})^{-1}\Big(\frac{1}{i}k_0\frac{\partial}{\partial k_3}\Big)q(-\mathbf{k}) = -K_3(S)\,. \end{aligned}\right\} \tag{7.2.3}$$

However, it is easily shown in the following way that $q(\mathbf{k})$, which is defined in an irreducible representation space of the Poincaré group and which satisfies the above equations, does not exist except for the case $S = 0$. Putting $q(-\mathbf{k}) = u_1(\mathbf{k})u_2(\mathbf{k})$ where $u_1(\mathbf{k})$ is defined by

$$u_1(\mathbf{k}) = \exp\Big\{-2iS\tan^{-1}\Big(\frac{k_2}{k_1}\Big)\Big\}\,, \tag{7.2.4}$$

we can derive from (7.2.2) and (7.2.3)

$$\mathcal{R}^{-1}\mathbf{J}(S)\mathcal{R} = u_2(\mathbf{k})^{-1}\mathbf{J}(-S)u_2(\mathbf{k}) = \mathbf{J}(S)\,, \tag{7.2.5}$$
$$\mathcal{R}^{-1}\mathbf{K}(S)\mathcal{R} = -u_2(\mathbf{k})^{-1}\mathbf{K}(-S)u_2(\mathbf{k}) = -\mathbf{K}(S)\,. \tag{7.2.6}$$

That is, $u_2(\mathbf{k})$ must be a unitary operator which transforms $\mathbf{J}(-S)$ into $\mathbf{J}(S)$ and $\mathbf{K}(-S)$ into $\mathbf{K}(S)$. But, when $S \neq 0$, the operators with S of different sign belong to different irreducible representations of the Poincaré group. We find therefore that such $u_2(\mathbf{k})$ and the unitary space reflection operator $\mathcal{R}$ do not exist in a single irreducible representation space of the Poincaré group.

Therefore, in order to define the space reflection we must use an anti-unitary operator or we must construct our theory in a direct sum space of two irreducible representations of S and $-S$.

In the former case, when the complex conjugates of each side of (5.2.23) and (5.2.24) are taken, the signs of S are changed. Combining this fact with the above argument we can certainly obtain a definition of the space reflection, but such a definition is not desirable because the positive and negative frequencies are exchanged when the complex conjugates are taken.

On the other hand, in the latter case, as is seen in the above calculation, the transformations of state vectors are generally given by

$$\phi_S^{(\pm)'}(\mathbf{k}) = \exp\left(i\delta_S^{(\pm)}\right)\exp\left\{-2iS\tan^{-1}\left(\frac{k_2}{k_1}\right)\right\}\phi_{-S}^{(\pm)}(-\mathbf{k}) , \qquad (7.2.7)$$

where $\delta_S^{(\pm)}$ are real constants.

When $S = 0$, putting $\delta_0^{(+)} = \delta_0^{(-)} = \delta$ we can easily verify that $U(x)$ in (7.1.1) transforms under the space reflection as

$$U'(\mathbf{x}, t) = e^{i\delta}U(-\mathbf{x}, t) . \qquad (7.2.8)$$

When $|S| \geq \frac{1}{2}$, similarly to the case of finite mass, the condition of $|S| = \frac{1}{2}$ is automatically extended to the case of general S. Thus we put $|S| = \frac{1}{2}$, and from (7.1.11)–(7.1.14) we have

$$\varphi^{R'}(\mathbf{k}) = e^{i(\delta+\sigma_3 c)}\exp\left\{-i\sigma_3\tan^{-1}\left(\frac{k_2}{k_1}\right)\right\}\sigma_1\varphi^L(-\mathbf{k}) , \qquad (7.2.9)$$

where $\delta = (\delta_{1/2}^{(+)} + \delta_{-1/2}^{(-)})/2$ and $c = (\delta_{1/2}^{(+)} - \delta_{-1/2}^{(-)})/2$. Calculating by the use of (7.1.29) and the explicit form of $U(\mathbf{k})$ in (7.1.23) we obtain

$$\begin{aligned}\psi^{R'}(\mathbf{k}) &= \left[U(\mathbf{k})e^{i(\delta+\sigma_3 c)}\exp\left\{-i\sigma_3\tan^{-1}\left(\frac{k_2}{k_1}\right)\right\}U(\mathbf{k})^{-1}U(\mathbf{k})\sigma_1 U(-\mathbf{k})^{-1}\right] \\ &\qquad \times \psi^L(-\mathbf{k}) \\ &= e^{i\delta}\left(i\sin c + \frac{(\mathbf{k}\sigma)}{|\mathbf{k}|}\cos c\right)\psi^L(-\mathbf{k}) . \end{aligned} \qquad (7.2.10)$$

The condition of covariance leads to $c = \pi/2$. Thus in the x-representation we have

$$\psi^{R'}(\mathbf{x}, t) = ie^{i\delta}\psi^L(-\mathbf{x}, t) , \qquad (7.2.11)$$

and similarly

$$\psi^{L'}(\mathbf{x}, t) = ie^{i\hat{\delta}}\psi^R(-\mathbf{x}, t) , \qquad (7.2.12)$$

where we have put $(\delta^{(-)}_{1/2} + \delta^{(+)}_{-1/2})/2 = \hat{\delta}$ and $(\delta^{(-)}_{1/2} - \delta^{(+)}_{-1/2})/2 = \pi/2$. Generalizing these equations to the case $|S| \geq \frac{1}{2}$ we immediately obtain

$$\begin{aligned} \psi^{R'}_{(\dots)_n}(\mathbf{x}, t) &= e^{i\delta}\psi^{L}_{(\dots)_n}(-\mathbf{x}, t) , \\ \psi^{L'}_{(\dots)_n}(\mathbf{x}, t) &= e^{i\hat{\delta}}\psi^{R}_{(\dots)_n}(-\mathbf{x}, t) . \end{aligned} \tag{7.2.13}$$

In this way, if we require the invariance of the theory under the space reflection in which the positive and negative frequencies are not exchanged, we find that it becomes possible only when both equations (7.1.62) and (7.1.63) exist together for $|S| \geq \frac{1}{2}$.

Therefore, if only one of $\psi^{R}_{(\dots)_n}(x)$ and $\psi^{L}_{(\dots)_n}(x)$ exists in the world, there is no space reflection which leaves the description of such a particle invariant. A typical example is a neutrino ($|S| = 1/2$) which has the L-amplitude only and has no R-amplitude as has been well known in experiments. That is, we can construct no space reflection which leaves the theory invariant. As has been mentioned in §1.2, in the framework of special relativity there is originally no requirement that the theory must be invariant under discrete transformations. Thus the above situation itself is not inconsistent theoretically. While a photon ($|S| = 1$) with zero mass has, in contrast to the neutrino, both R- and L-amplitudes, i.e., right-handed and left-handed polarizations. Further it is well known that there are photons in the elliptic or linear polarizations which are the superpositions of R- and L-amplitudes. It may be a convention that we do not regard R- and L-amplitudes as different kinds of photons but consider as different states of a photon in this way. At least in practice we characterize the free particle picture of a photon by an irreducible representation of a larger group consisting of the Poincaré group and the space reflection. Such a treatment, in which we use only the Poincaré group for a neutrino and the larger group for a photon, seems to have a lack of consistency, but this situation is in fact connected to the problem of their interactions, i.e., the problem of the emission and absorption of photons and neutrinos. In this book we cannot afford to discuss the details of the problem, but within the case of free particles we can realize, from the expressions of the covariant inner products given in the next section, that it is natural to consider both R and L together for a photon.

We shall next discuss the time reversal. By an analogous argument as in §6.6 we denote the time reversed amplitudes of $\phi^{(\pm)}_S(\mathbf{k})$ as $\phi^{(\pm)''}_S(\mathbf{k})$, and we write

$$\phi^{(\pm)''}_S(\mathbf{k}) = \exp(i\delta^{(\pm)'}_S)\exp\left\{-i2S\tan^{-1}\left(\frac{k_2}{k_1}\right)\right\}\phi^{(\pm)*}_S(-\mathbf{k}) , \tag{7.2.14}$$

where $\phi_S^{(\pm)}(\mathbf{k})$ on the right-hand side transform, like $\phi_\xi^{(\pm)}(\mathbf{k})$ in (6.6.12), as $T(\mathbf{a}, -a_0)\phi_S^{(\pm)}(\mathbf{k})$ under a translation of the coordinates. It is easy to rewrite (7.2.14) into transformation equations for the amplitudes in the covariant formalism.

For $S = 0$, putting $\delta_0^{(\pm)'} = \delta'$ we have

$$U''(\mathbf{x}, t) = e^{i\delta'} U^*(\mathbf{x}, -t) , \tag{7.2.15}$$

and for $|S| = 1/2$ putting $(\delta_{1/2}^{(+)'} + \delta_{-1/2}^{(-)'})/2 = \delta_R'$ and $(\delta_{1/2}^{(+)'} - \delta_{-1/2}^{(-)'})/2 = c_R' + \pi/2$, from (7.2.14), (7.1.11) and (7.1.13) we get

$$\begin{aligned}\varphi^{R''}(\mathbf{k}) &= \exp\left\{i\left(\delta_R' + \sigma_3 c_R' + \frac{\sigma_3\pi}{2}\right)\right\} \exp\left\{-i\sigma_3 \tan^{-1}\left(\frac{k_2}{k_1}\right)\right\} \varphi^{R*}(-\mathbf{k}) \\ &= \exp\{i(\delta_R' + \sigma_3 c_R')\} \exp\left\{-i\sigma_3 \tan^{-1}\left(\frac{k_2}{k_1}\right)\right\} \sigma_1\sigma_2 \varphi^{R*}(-\mathbf{k}) .\end{aligned} \tag{7.2.16}$$

If we use a relation $\sigma_2 U^*(-\mathbf{k})^{-1}\sigma_2 = U(-\mathbf{k})^{-1}$ from (7.1.29) we get

$$\begin{aligned}\psi^{R''}(\mathbf{k}) = \Big[&U(\mathbf{k}) \exp\{i(\delta_R' + \sigma_3 c_R')\} \exp\left\{-i\sigma_3 \tan^{-1}\left(\frac{k_2}{k_1}\right)\right\} U(\mathbf{k})^{-1} \\ &\times U(\mathbf{k})\sigma_1 U(\mathbf{k})^{-1}\Big] \sigma_2 \psi^{R*}(-\mathbf{k}) .\end{aligned} \tag{7.2.17}$$

Since the part of $[\ldots]$ in the above equation has the same form as part of $[\ldots]$ in (7.2.10), putting $c_R' = \pi/2$ we have

$$\psi^{R''}(\mathbf{k}) = i e^{i\delta_R'} \sigma_2 \psi^{R*}(-\mathbf{k}) , \tag{7.2.18}$$

and in the x-representation we obtain

$$\psi^{R''}(\mathbf{x}, t) = i e^{i\delta_R'} \sigma_2 \psi^{R*}(\mathbf{x}, -t) . \tag{7.2.19}$$

Similarly putting $(\delta_{1/2}^{(-)'} + \delta_{-1/2}^{(+)'})/2 = \delta_L'$ and $(\delta_{1/2}^{(-)'} - \delta_{-1/2}^{(+)'})/2 = \pi$ we get

$$\psi^{L''}(\mathbf{x}, t) = i e^{i\delta_L'} \sigma_2 \psi^{L*}(\mathbf{x}, -t) . \tag{7.2.20}$$

The extensions of these equations to the case $|S| \geq 1/2$ are clearly given by

$$\psi_{(\ldots)_n}^{R,L''}(\mathbf{x}, t) = e^{i\delta_{R,L}'} \prod_{i=1}^{n} \sigma_2^{(i)} \psi_{(\ldots)_n}^{R,L*}(\mathbf{x}, -t) . \tag{7.2.21}$$

As is seen in the above argument the time reversal does not require the coexistence of ψ^R and ψ^L in contrast to the space reflection, and can be defined for each of them by the use of an anti-unitary transformation.

In the case $m = 0$, when $|S| \geq 1/2$ the charge conjugation cannot be defined for each ψ^R and ψ^L separately. This is because, in the process of this transformation we must take the complex conjugates of the amplitudes leaving x_μ as they are, and as has been noted, this operation changes S into $-S$. In other words, the charge conjugation transformation becomes possible only when both ψ^R and ψ^L exist together.

§7.3 Covariant Inner Products

When the value of spin is 1 or greater than 1, eliminating the extra factor $1/k_0^{n-1}$ on the right-hand side of (7.1.66) we shall try to express the covariant inner products in terms of the amplitudes in the x-representation. To do this we must rewrite $\sum_{\rho_1\rho_2\ldots\rho_n=1,2}\psi^{R,L*}_{(\rho_1\rho_2\ldots\rho_n)}(\mathbf{k})_1\psi^{R,L}_{(\rho_1\rho_2\ldots\rho_n)}(\mathbf{k})_2$ in the numerator into a form proportional to k_0^{n-1}. But, since it is impossible for the Bargmann-Wigner amplitudes, we shall first convert them into amplitudes with tensor indices from which we shall extract k_0 as we have done in §6.7. It becomes, however, a slightly complicated calculation for a particle with general spin, and we shall discuss here only the cases of spin 1 and 3/2.

a) *Spin 1 particle*

Put $n = 2$ in (7.1.64) and (7.1.65). Applying $\prod_{i=1,2}\left\{\frac{1}{2}(1-\gamma_5)^{(i)}\right\}$ and $\prod_{i=1,2}\left\{\frac{1}{2}(1+\gamma_5)^{(i)}\right\}$ on $\psi_{(a_1,a_2)}(x)$ $(a_1, a_2 = 1,2,3,4)$ we have R- and L-amplitudes respectively, and regarding $\psi_{(a_1,a_2)}(x)$ as the element of the 4×4 symmetric matrix $\psi(x)$ at a_1-th row and a_2-th column we can write them as $\frac{1-\gamma_5}{2}\psi(x)\frac{1-\gamma_5^T}{2}$ and $\frac{1+\gamma_5}{2}\psi(x)\frac{1+\gamma_5^T}{2}$. Multiplying these amplitudes on the right by C^{-1} and putting $\psi(x)C^{-1} = \chi(x)$, we obtain from (7.1.64) the equations for the R-amplitude

$$\gamma_\mu\partial_\mu\frac{1-\gamma_5}{2}\chi(x)\frac{1-\gamma_5}{2} = 0\,, \tag{7.3.1}$$

$$\frac{1-\gamma_5}{2}\chi(x)\frac{1-\gamma_5}{2}\gamma_\mu\overleftarrow{\partial}_\mu = 0\,, \tag{7.3.2}$$

where a relation $C\gamma_5^TC^{-1} = \gamma_5$ has been used. The equations for the L-amplitude are also given by the above expressions with γ_5 replaced by $-\gamma_5$. Since one of these two equations is derived from the other, we shall hereafter consider (7.3.1) only. As a result of the fact that $\psi(x)$ is a symmetric matrix,

$\chi(x)$ is expanded in the form (6.7.2). If it is multiplied on both sides by $(1-\gamma_5)/2$, $V_\mu(x)$ is dropped because γ_μ are anti-commutable with γ_5, and hence substitution of it into (7.3.1) leads to the following equation for $T_{[\mu\nu]}(x)$:

$$\gamma_\mu \sigma_{\lambda\rho}(1-\gamma_5)\partial_\mu T_{[\lambda\rho]}(x) = 0 \ . \tag{7.3.3}$$

If the product of γ matrices on the left-hand side is rewritten by the use of (6.7.5), Eq. (7.3.3) becomes

$$\partial_\lambda T^R_{[\lambda\rho]}(x) = 0 \ , \tag{7.3.4}$$

$$T^R_{[\lambda\rho]}(x) = T_{[\lambda\rho]}(x) + \frac{\varepsilon_{\lambda\rho\mu\nu}}{2} T_{[\mu\nu]}(x) \ . \tag{7.3.5}$$

Similarly, for the L-amplitude the following equations are easily derived from (7.3.3) with γ_5 replaced by $-\gamma_5$:

$$\partial_\lambda T^L_{[\lambda\rho]}(x) = 0 \ , \tag{7.3.6}$$

$$T^L_{[\lambda\rho]}(x) = T_{[\lambda\rho]}(x) - \frac{\varepsilon_{\lambda\rho\mu\nu}}{2} T_{[\mu\nu]}(x) \ . \tag{7.3.7}$$

Equations (7.3.4)–(7.3.7) are nothing but the Maxwell equations. If we put here $\sum_{i,j=1,2,3} \varepsilon_{ijk} T_{[ij]}(x)/2 = H_k(x)$ and $T_{[4k]} = iE_k(x)$ $(k = 1,2,3)$, (7.3.4) and (7.3.6) become the equations for $\mathbf{H}(x) - i\mathbf{E}(x)$ and $\mathbf{H}(x) + i\mathbf{E}(x)$ respectively, i.e., the equations for photons in the right-handed and left-handed polarizations. In this case, although $\mathbf{H}(x)$ and $\mathbf{E}(x)$ satisfy the Maxwell equations, since they are obtained from the wave functions which are probability waves, it must be noted that they are neither the magnetic nor the electric fields themselves in the real space-time, but are complex quantities quite different conceptually.

In order to get the covariant inner products we convert the amplitudes into those in the momentum representation by

$$\chi(x) = \frac{1}{(2\pi)^{3/2}} \int \frac{d\mathbf{k}}{|\mathbf{k}|} e^{i(\mathbf{k}\mathbf{x} - k_0 t)} \chi(\mathbf{k}) \ , \tag{7.3.8}$$

$$T_{[\mu\nu]}(x) = \frac{1}{(2\pi)^{3/2}} \int \frac{d\mathbf{k}}{|\mathbf{k}|} e^{i(\mathbf{k}\mathbf{x} - k_0 t)} T_{[\mu\nu]}(\mathbf{k}) \ . \tag{7.3.9}$$

Then (7.1.66) becomes, with the help of $(1 \mp \gamma_5)\chi(\mathbf{k})(1 \mp \gamma_5) = -iT_{[\mu\nu]}(\mathbf{k}) \times \sigma_{\mu\nu}(1 \mp \gamma_5)$,

$$\begin{aligned} \langle\langle 1 | 2 \rangle\rangle &= \frac{1}{|N|^2} \int \frac{d\mathbf{k}}{8k_0 |\mathbf{k}|^2} \mathrm{Tr}\{(1 \mp \gamma_5)\chi^\dagger(\mathbf{k})_1 (1 \mp \gamma_5)\chi(\mathbf{k})_2 (1 \mp \gamma_5)\} \\ &= \frac{1}{|N|^2} \int \frac{d\mathbf{k}}{8k_0 |\mathbf{k}|^2} (T_{[\mu\nu]}(\mathbf{k})_1)^* T_{[\lambda\rho]}(\mathbf{k})_2 \mathrm{Tr}\{\sigma_{\mu\nu}\sigma_{\lambda\rho}(1 \mp \gamma_5)\} \ , \end{aligned} \tag{7.3.10}$$

where $\chi^\dagger(\mathbf{k})$ is the Hermitian conjugate matrix of $\chi(\mathbf{k})$. In analogy to electromagnetism we here express $T_{[\mu\nu]}(\mathbf{k})$ in terms of vector potentials $A_\mu(\mathbf{k})$ in the form

$$T_{[\mu\nu]}(\mathbf{k}) = i(k_\mu A_\nu(\mathbf{k}) - k_\nu A_\mu(\mathbf{k})) \,. \tag{7.3.11}$$

On the other hand, since the trace in (7.3.10) is calculated according to the argument of §6.5, as

$$\mathrm{Tr}\{\sigma_{\mu\nu}\sigma_{\lambda\rho}(1 \mp \gamma_5)\} = 4(\delta_{\mu\lambda}\delta_{\nu\rho} - \delta_{\mu\rho}\delta_{\nu\lambda} \pm \varepsilon_{\mu\nu\lambda\rho}) \,, \tag{7.3.12}$$

the numerator on the right-hand side of (7.3.10) becomes

$$\begin{aligned} &16\{(\mathbf{k}^2 + k_0^2)(\mathbf{A}^*(\mathbf{k})_1\mathbf{A}(\mathbf{k})_2 + A_0^*(\mathbf{k})_1 A_0(\mathbf{k})_2) - (\mathbf{k}\mathbf{A}(\mathbf{k})_1 + k_0 A_0(\mathbf{k})_1)^* \\ &\qquad \times (\mathbf{k}\mathbf{A}(\mathbf{k})_2 + k_0 A_0(\mathbf{k})_2) \mp 2ik_0(\mathbf{k}[\mathbf{A}^*(\mathbf{k})_1 \times \mathbf{A}(\mathbf{k})_2])\} \\ &= 32k_0\{k_0(\mathbf{A}^*(\mathbf{k})_1\mathbf{A}(\mathbf{k})_2 - A_0^*(\mathbf{k})_1 A_0(\mathbf{k})_2) \mp i\,\mathbf{k}[\mathbf{A}^*(\mathbf{k})_1 \times \mathbf{A}(\mathbf{k})_2]\} \,. \end{aligned} \tag{7.3.13}$$

In this way, the common factor k_0 has been extracted, where we have used the Lorentz condition for the vector potentials

$$k_\mu A_\mu(\mathbf{k}) = 0 \tag{7.3.14}$$

and a relation $\mathbf{k}^2 = k_0^2$. As a result, if we let the normalization constant $1/|N|^2$ in (7.3.10) be 1/8, and if we define

$$A_\mu(x) = \frac{1}{(2\pi)^{3/2}} \int \frac{d\mathbf{k}}{|\mathbf{k}|} e^{i(\mathbf{k}\mathbf{x} - k_0 t)} A_\mu(\mathbf{k}) \,, \tag{7.3.15}$$

we have

$$\begin{aligned} \langle\langle 1|2\rangle\rangle^{R,L} &= \frac{1}{4i}\int d\mathbf{x}\Big(\frac{\partial \mathbf{A}^*(x)_1}{\partial t}\mathbf{A}(x)_2 - \mathbf{A}^*(x)_1\frac{\partial \mathbf{A}(x)_2}{\partial t} - \frac{\partial A_0^*(x)_1}{\partial t}A_0(x)_2 \\ &\quad + A_0^*(x)_1\frac{\partial A_0(x)_2}{\partial t}\Big) \pm \frac{1}{4}\int d\mathbf{x}\{(\nabla \times \mathbf{A}^*(x)_1)\mathbf{A}(x)_2 \\ &\quad + \mathbf{A}^*(x)_1(\nabla \times \mathbf{A}(x)_2)\} \\ &= \frac{1}{4}\int d\mathbf{x}\Big\{\frac{\partial A_\mu^*(x)_1}{\partial x_4}A_\mu(x)_2 - A_\mu^*(x)_1\frac{\partial A_\mu(x)_2}{\partial x_4} \\ &\quad \pm ((\nabla \times \mathbf{A}^*(x)_1)\mathbf{A}(x)_2 + \mathbf{A}^*(x)_1(\nabla \times \mathbf{A}(x)_2))\Big\} \,. \end{aligned} \tag{7.3.16}$$

The meaning of the symbol * in $A_\mu^*(x)$ has already been given after (6.7.64). Needless to say about the symbol ± on the right-hand side, the upper sign corresponds to R and the lower sign corresponds to L. A similar notation will be used also in the next paragraph.

In this way the covariant inner products have been expressed in terms of the amplitudes in the x-representation. As is seen in the definition, however, it must be noted that the amplitudes $A_\mu(x)$ used here properly belong to neither R nor L, but they can be introduced only in the case where both R- and L-amplitudes exist together. We gave in (7.3.16) the covariant inner products for each R and L, but in the above sense it is rather natural to define the covariant inner product in a direct product space of the R- and L-representation spaces (where the space reflection and the charge conjugation can be defined by the argument of the previous section). Of course, it is given by

$$\begin{aligned}\langle\langle 1|2\rangle\rangle &= \langle\langle 1|2\rangle\rangle^R + \langle\langle 1|2\rangle\rangle^L \\ &= \frac{1}{2}\int dx\left(\frac{\partial A_\mu^*(x)_1}{\partial x_4}A_\mu(x)_2 - A_\mu^*(x)_1\frac{\partial A_\mu(x)_2}{\partial x_4}\right) . \end{aligned} \tag{7.3.17}$$

Incidentally, it is well known that $A_\mu(x)$ are not uniquely determined even if $T_{[\mu\nu]}(x)$ are given, and have the freedom of a gauge transformation. That is, $T_{[\mu\nu]}(x)$ do not change when $A_\mu(x)+\partial_\mu\Lambda(x)(\partial_\mu^2\Lambda(x)=0)$ are used in place of $A_\mu(x)$, and the equations for $A_\mu(x)$ are also left invariant. The invariances of (7.3.16) and (7.3.17) under this transformation are easily verified by the use of the amplitudes in the momentum representation. The integrands of (7.3.16) and (7.3.17) are, however, not gauge invariant.

b) *Spin 3/2 Particle and Others*

The R- and L-amplitudes are given by

$$\begin{aligned}&\frac{1}{8}\sum_{a'b'c'}(1\mp\gamma_5)_{aa'}(1\mp\gamma_5)_{bb'}(1\mp\gamma_5)_{cc'}\psi_{(a'b'c')}(x) \\ &= \frac{1}{8}\sum_{a'b'c'}(1\mp\gamma_5)_{aa'}(1\mp\gamma_5)_{cc'}\psi_{(a'b'c')}(x)(1\mp\gamma_5^T)_{b'b} \end{aligned} \tag{7.3.18}$$

Multiplying these expressions on the right by C^{-1} we have

$$\frac{1}{8}\sum_{a'b'c'}(1\mp\gamma_5)_{aa'}(1\mp\gamma_5)_{cc'}\Big(\sum_{b''}\psi_{(a'b''c')}(x)(C^{-1})_{b''b'}\Big)(1\mp\gamma_5)_{b'b} . \tag{7.3.19}$$

Since $\sum_{b''}\psi_{(a'b''c')}(x)(C^{-1})_{b''b'}$ is expanded in a similar manner to (6.7.16), substituting it into the above expression we obtain

$$(7.3.19) = -\frac{i}{4}(\sigma_{\mu\nu})_{ab}\psi^{R,L}_{[\mu\nu],c}(x) = -\frac{i}{4}(\sigma_{\mu\nu})_{cb}\psi^{R,L}_{[\mu\nu],a}(x) , \tag{7.3.20}$$

where

$$\psi^{R,L}_{[\mu\nu]}(x) = \frac{1}{2}\left(1 \mp \gamma_5)(\psi_{[\mu\nu]}(x) \pm \frac{\varepsilon_{\mu\nu\lambda\rho}}{2}\psi_{[\lambda\rho]}(x)\right) , \tag{7.3.21}$$

and we have used the relation $\sigma_{\mu\nu}\gamma_5 = -\varepsilon_{\mu\nu\lambda\rho}\sigma_{\lambda\rho}/2$ to derive (7.3.20). As is easily seen in the definition, $\psi^{R,L}_{[\mu\nu]}(x)$ satisfy

$$\frac{1}{2}\varepsilon_{\mu\nu\lambda\rho}\psi^{R,L}_{[\lambda\rho]}(x) = \pm\psi^{R,L}_{[\mu\nu]}(x) , \tag{7.3.22}$$

$$\gamma_5\psi^{R,L}_{[\mu\nu]}(x) = \mp\psi^{R,L}_{[\mu\nu]}(x) . \tag{7.3.23}$$

Let us multiply (7.3.20) by $(\gamma^A)_{ba}$ and sum up with respect to a and b to get other relations for $\psi^{R,L}_{[\mu\nu]}(x)$. It is sufficient to consider the cases where γ^A is 1, γ_5 or $\sigma_{\lambda\rho}$ because both sides of the equations will vanish when γ^A is γ_λ or $\gamma_\lambda\gamma_5$. Now, if we let γ^A be 1 or γ_5 we have

$$\sigma_{\mu\nu}\psi^{R,L}_{[\mu\nu]}(x) = 0 . \tag{7.3.24}$$

Next, putting $\gamma^A = \sigma_{\lambda\rho}$ we get, from (7.3.20),

$$\mathrm{Tr}(\sigma_{\mu\nu}\sigma_{\lambda\rho})\psi^{R,L}_{[\mu\nu]}(x) = \sigma_{\mu\nu}\sigma_{\lambda\rho}\psi^{R,L}_{[\mu\nu]}(x) . \tag{7.3.25}$$

The right-hand side becomes $[\sigma_{\mu\nu}, \sigma_{\lambda\rho}]\psi^{R,L}_{[\mu\nu]}(x)$ with the help of (7.3.24), and the commutator can be rewritten by the use of (6.4.16). On the other hand, according to the argument about the traces of the products of γ matrices in §6.5, $\mathrm{Tr}(\sigma_{\mu\nu}\sigma_{\lambda\rho})$ on the left-hand side becomes $4(\delta_{\mu\lambda}\delta_{\nu\rho} - \delta_{\mu\rho}\delta_{\nu\lambda})$, and (7.3.25) becomes finally

$$\psi^{R,L}_{[\lambda\rho]}(x) = \frac{i}{2}(\sigma_{\mu\nu}\psi^{R,L}_{[\mu\rho]}(x) - \sigma_{\mu\rho}\psi^{R,L}_{[\mu\lambda]}(x)) . \tag{7.3.26}$$

In this way we have obtained the relations (7.3.24) and (7.3.26) for $\psi^{R,L}_{[\mu\nu]}(x)$, and we can further simplify them in the following way. Multiplying (7.3.26) by γ_λ we have

$$\gamma_\lambda\psi^{R,L}_{[\lambda\rho]}(x) = -\frac{1}{2}(2\gamma_\mu\psi^{R,L}_{[\mu\rho]}(x) + \varepsilon_{\lambda\mu\rho\tau}\gamma_5\gamma_\tau\psi^{R,L}_{[\mu\lambda]}(x)) , \tag{7.3.27}$$

where we have used the relations $\gamma_\lambda\sigma_{\mu\lambda} = 3i\gamma_\mu$ and $\gamma_\lambda\sigma_{\mu\rho} = -i(\varepsilon_{\lambda\mu\rho\tau}\gamma_5\gamma_\tau + \delta_{\lambda\mu}\gamma_\rho - \delta_{\lambda\rho}\gamma_\mu)$ which is derived from (6.5.39). The second term in the parentheses of (7.3.27) becomes $2\gamma_\tau\psi^{R,L}_{[\tau\rho]}(x)$ with the help of (7.3.22) and (7.3.23), and hence the right-hand side becomes $-2\gamma_\mu\psi^{R,L}_{[\mu\rho]}(x)$, which leads to

$$\gamma_\mu\psi^{R,L}_{[\mu\nu]}(x) = 0 . \tag{7.3.28}$$

Equations (7.3.24) and (7.3.26) are easily derived from (7.3.28). We can therefore conclude that (7.3.28) are sufficient conditions for $\psi^{R,L}_{[\mu\nu]}(x)$.

The equations of motion for $\psi^{R,L}_{[\mu\nu]}(x)$ are obtained from (7.1.64) and (7.1.65) as

$$\gamma_\mu \partial_\mu \psi^{R,L}_{[\lambda\rho]}(x) = 0 \ . \tag{7.3.29}$$

Multiplying these equations by γ_λ and using (7.3.28) we get

$$\partial_\mu \psi^{R,L}_{[\mu\nu]}(x) = 0 \ . \tag{7.3.30}$$

We want to write $\psi_{[\mu\nu]}(x)$ in (7.3.21) as

$$\psi_{[\mu\nu]}(x) = \partial_\mu \psi_\nu(x) - \partial_\nu \psi_\mu(x) \tag{7.3.31}$$

similarly to the case of spin 1, but since we cannot get $\psi_{[\mu\nu]}(x)$ from (7.3.21) only, we shall impose equations

$$(1 \pm \gamma_5)\left(\psi_{[\mu\nu]}(x) \pm \frac{\varepsilon_{\mu\nu\lambda\rho}}{2}\psi_{[\lambda\rho]}(x)\right) = 0 \ , \tag{7.3.32}$$

which give no restriction on $\psi^{R,L}_{[\mu\nu]}(x)$. Then we have $\psi_{[\mu\nu]}(x) = \psi^{R}_{[\mu\nu]}(x) + \psi^{L}_{[\mu\nu]}(x)$, and from (7.3.28) and (7.3.29) the equations for $\psi_{[\mu\nu]}(x)$ are

$$\gamma_\mu \partial_\mu \psi_{[\lambda\rho]}(x) = 0 \ , \tag{7.3.33}$$

$$\gamma_\mu \psi_{[\mu\nu]}(x) = 0 \ . \tag{7.3.34}$$

Conversely, if $\psi^{R,L}_{[\mu\nu]}(x)$ are given by (7.3.21), the above equations lead to (7.3.28), (7.3.29) and (7.3.32) as will be verified in the following way. First, (7.3.29) is immediately verified to hold. Next, from (6.5.39) we have a relation

$$\frac{i}{2}\gamma_5(\gamma_\lambda \sigma_{\mu\nu}\gamma_\rho - \gamma_\rho \sigma_{\mu\nu}\gamma_\lambda) = \varepsilon_{\mu\nu\lambda\rho} + \gamma_5(\delta_{\mu\lambda}\delta_{\nu\rho} - \delta_{\mu\rho}\delta_{\nu\lambda}) \ . \tag{7.3.35}$$

This equation is also derived by calculating the 16 coefficients of γ^A, with the help of the formula for the trace in §6.5, by which the left-hand side of the above equation is expanded. If (7.3.35) acts on $\psi_{[\lambda\rho]}(x)$, (7.3.34) leads to

$$\frac{1}{2}\varepsilon_{\mu\nu\lambda\rho}\psi_{[\lambda\rho]}(x) = -\gamma_5 \psi_{[\mu\nu]}(x) \ . \tag{7.3.36}$$

Thus (7.3.28) and (7.3.32) are easily verified by the definition (7.3.21) of $\psi^{R,L}_{[\mu\nu]}(x)$. Equations (7.3.33) and (7.3.34) consequently become equivalent to (7.3.28), (7.3.29) and (7.3.32), and we can use $\psi_{[\mu\nu]}(x)$ instead of $\psi^{R,L}_{[\mu\nu]}(x)$.

Introducing $\psi_{[\mu\nu]}(x)$ in this way, it can be expressed as (7.3.31). Here, if we assume that $\psi_\mu(x)$ satisfy

$$\gamma_\mu \partial_\mu \psi_\nu(x) = 0 \,, \tag{7.3.37}$$
$$\gamma_\mu \psi_\mu(x) = 0 \,, \tag{7.3.38}$$

we can derive (7.3.33) and (7.3.34). The function $\psi_{[\mu\nu]}(x)$, and Eqs. (7.3.37) and (7.3.38) are invariant under the transformation

$$\psi_\mu(x) \longrightarrow \psi_\mu(x) + \partial_\mu \psi(x) \,, \tag{7.3.39}$$
$$\gamma_\mu \partial_\mu \psi(x) = 0 \,. \tag{7.3.40}$$

Equation (7.3.39) is a gauge transformation for a spin $\frac{3}{2}$ particle, and (7.3.37) and (7.3.38) have two independent solutions, which are not affected by this transformation, for the positive and negative frequencies respectively. They are the solutions corresponding to the R- and L-amplitudes.

Now, the covariant inner products are obtained from (7.1.66) and (7.3.21) in terms of the amplitudes in the momentum representation:

$$\begin{aligned}\langle\langle 1|2\rangle\rangle^{R,L} &= \frac{1}{16|N|^2}\int \frac{d\mathbf{k}}{|\mathbf{k}|^2}\frac{(\psi^{R,L}_{[\mu\nu]}(\mathbf{k})_1)^*\psi^{R,L}_{[\mu\nu]}(\mathbf{k})_2 \mathrm{Tr}(\sigma_{\mu\nu}\sigma_{\lambda\rho})}{k_0^2} \\ &= \frac{1}{2|N|^2}\int \frac{d\mathbf{k}}{|\mathbf{k}|^2}\frac{(\psi^{R,L}_{[\mu\nu]}(\mathbf{k})_1)^*\psi^{R,L}_{[\mu\nu]}(\mathbf{k})_2}{k_0^2} \,.\end{aligned} \tag{7.3.41}$$

Since we get $\psi^{R,L}_{[\mu\nu]}(x) = \frac{1}{2}(1 \mp \gamma_5)\psi_{[\mu\nu]}(x)$ from (7.3.21) and (7.3.36), we obtain, by the use of (7.3.31),

$$\begin{aligned}\langle\langle 1|2\rangle\rangle^{R,L} &= \frac{1}{|N|^2}\int \frac{d\mathbf{k}}{|\mathbf{k}|^2}\frac{(\psi_{[\mu\nu]}(\mathbf{k})_1)^*(1\mp\gamma_5)\psi_{[\mu\nu]}(\mathbf{k})}{k_0^2} \\ &= \frac{4}{|N|^2}\int \frac{d\mathbf{k}}{|\mathbf{k}|^2}(\boldsymbol{\psi}^*(\mathbf{k})_1(1\mp\gamma_5)\boldsymbol{\psi}(\mathbf{k})_2 - \psi_0^*(\mathbf{k})_1(1\mp\gamma_5)\psi_0(\mathbf{k})_2) \\ &= \frac{4}{|N|^2}\int d\mathbf{x}\psi_\mu^*(x)_1(1\mp\gamma_5)\psi_\mu(x)_2 \,.\end{aligned} \tag{7.3.42}$$

In the derivation of this expression we have used $k_\mu\psi_\mu(x) = 0$ which is derived from (7.3.37) and (7.3.38). Needless to say, $\boldsymbol{\psi}(x)$ and $\psi_0(x)$ are the space- and time-components of $\psi_\mu(x)$, and $\psi_\mu(x)$ and $\psi_\mu(\mathbf{k})$ are connected by the relation

$$\psi_\mu(x) = \frac{1}{(2\pi)^{3/2}}\int \frac{d\mathbf{k}}{|\mathbf{k}|}e^{i(\mathbf{k}\mathbf{x}-k_0t)}\psi_\mu(\mathbf{k}) \,. \tag{7.3.43}$$

In contrast to the case of spin 1, $\frac{1}{2}(1\mp\gamma_5)\psi_\mu(x)$ are amplitudes connected to $\psi^{R,L}_{[\mu\nu]}(x)$ by (7.3.31) and (7.3.36). That is, if we write

$$\psi^{R,L}_\mu(x)=\frac{1}{2}(1\mp\gamma_5)\psi_\mu(x)\ , \tag{7.3.44}$$

we have

$$\psi^{R,L}_{[\mu\nu]}(x)=\partial_\mu\psi^{R,L}_\nu(x)-\partial_\nu\psi^{R,L}_\mu(x)\ . \tag{7.3.45}$$

Hence, putting $|N|^2=8$ we obtain

$$\langle\langle 1\,|\,2\rangle\rangle^{R,L}=\int d\mathbf{x}\psi^{R,L*}_\mu(x)_1\psi^{R,L}_\mu(x)_2\ . \tag{7.3.46}$$

In this way the covariant inner products of the R- and L-amplitudes are expressed by the corresponding amplitudes $\psi^R_\mu(x)$ and $\psi^L_\mu(x)$ respectively. It must be noted that this point is quite different from the case of spin 1 in the previous paragraph. Therefore, as far as this point is concerned, the Poincaré group characterizes sufficiently the particle picture of massless particles with spin $\frac{3}{2}$, and we need no irreducible representation of a larger group including the space reflection. However, the fact that the amplitudes themselves and the integrands in the expressions of covariant inner products in the x-representation are not gauge invariant, is a common property with the case of spin 1 which was discussed in the previous section. This is also true for the cases of higher spins.

The method in this section will be summarized as follows: We rewrote the spin indices of the Bargmann-Wigner amplitudes for a massless particle into the tensor indices as much as possible, next we expressed the obtained tensors in terms of the differentials of tensors of lower rank, and by the use of k_0 in the expression we eliminated k_0 in the denominator of (7.1.66). This method needs somewhat tedious calculation, but is applicable to particles with higher spins. In the application, however, we cannot avoid introducing gauge non-invariant amplitudes such as $A_\mu(x)$ and $\psi^{R,L}_\mu(x)$. In this case, such gauge non-invariant quantities are considered to be not just the convenient quantities in the covariant descriptions of relativistic quantum mechanics, but rather to have a more fundamental meaning.

As has been seen in the above two examples and in the cases of spin 0 and $\frac{1}{2}$, there are relations between the original inner products in an irreducible representation space of the Poincaré group and the covariant inner products for massless particles with discrete spin, similar to those for the integral and half-integral spins in the case of finite mass. The covariant norms are classified in the Table 6.1. This can be regarded as a property of particles with finite spin freedom. In the cases of infinite spin freedom, i.e., for massless particles with continuous spin, we shall show in the next section that a quite different result will appear.

Another method in which the covariant inner products are obtained without the use of the Bargmann-Wigner amplitudes mentioned in c) of §6.7 is also possible, but it may be not necessarily simple.

§7.4 Particles with Continuous Spin

According to the discussion in §5.2, a massless particle with continuous spin belongs to either the single-valued or the double-valued irreducible representation of the Poincaré group. In this section we shall study the covariant descriptions of such particles.

a) *Particles in a Single-Valued Representation*

The behavior of this kind of particles is given by (5.2.30)–(5.2.37). Now, let ξ_1 and ξ_2 be the first and the second components of a 3-dimensional vector $\boldsymbol{\xi}$ and denote the wave function as $\phi^{(\pm)}(\mathbf{k},\boldsymbol{\xi})$. Since the variables of the wave function in (5.2.29) are $\mathbf{k}, \xi_1$ and ξ_2, we can write $\phi^{(\pm)}(\mathbf{k},\boldsymbol{\xi}) = \delta(\xi_3)\phi^{(\pm)}(\mathbf{k}, \xi_1, \xi_2)$ or

$$\xi_3\phi^{(\pm)}(\mathbf{k},\boldsymbol{\xi}) = 0 \,. \tag{7.4.1}$$

The Lorentz transformations of $\phi^{(\pm)}(\mathbf{k},\boldsymbol{\xi})$ are the same as those of $\phi^{(\pm)}(\mathbf{k}, \xi_1, \xi_2)$. It is easily verified by rewriting $\phi^{(\pm)}(\mathbf{k}, \xi_1, \xi_2)$ into $\phi^{(\pm)}(\mathbf{k},\boldsymbol{\xi})$ after multiplying each side of (5.2.30) and (5.2.31) by $\delta(\xi_3)$. We shall hereafter use

$$\phi(\mathbf{k},\boldsymbol{\xi}) = \phi^{(+)}(\mathbf{k},\boldsymbol{\xi}) + \phi^{(-)}(\mathbf{k},\boldsymbol{\xi}) \,. \tag{7.4.2}$$

Then (5.2.34) is expressed as

$$(\mathbf{k}^2 - k_0^2)\phi(\mathbf{k},\boldsymbol{\xi}) = 0 \,. \tag{7.4.3}$$

Needless to say, $\phi^{(\pm)}(\mathbf{k},\boldsymbol{\xi})$ are the solutions of the above equation for $k_0 \gtrless 0$ respectively.

The generators of the Lorentz transformation expressed in terms of the unit vector (7.1.10) directing the third axis are obtained from (5.2.30) and (5.2.31) in the form

$$\mathbf{J} = \mathbf{K} \times \left(\frac{1}{i}\frac{\partial}{\partial \mathbf{k}} + \hat{l}_3 \frac{\mathbf{n}\times\mathbf{k}}{|\mathbf{k}|(|\mathbf{k}| + k_3)}\right) + \hat{l}_3 \frac{\mathbf{k}}{|\mathbf{k}|} \,, \tag{7.4.4}$$

$$\mathbf{K} = -k_0\left(\frac{1}{i}\frac{\partial}{\partial \mathbf{k}} + \hat{l}_3 \frac{\mathbf{n}\times\mathbf{k}}{|\mathbf{k}|(|\mathbf{k}| + k_3)}\right) - \mathbf{F}\frac{k_0}{\mathbf{k}^2} \,, \tag{7.4.5}$$

where $\hat{l}_3$ is the third component of the following vector $\hat{\mathbf{l}} = (\hat{l}_1, \hat{l}_2, \hat{l}_3)$:

$$\hat{\mathbf{l}} = \frac{1}{i}\left(\boldsymbol{\xi} \times \frac{\partial}{\partial \boldsymbol{\xi}}\right) , \tag{7.4.6}$$

further $\mathbf{F} = (F_1, F_2, F_3)$ is given by

$$\left.\begin{aligned} F_1 &= \xi_1 - \frac{k_1\xi_1 + k_2\xi_2}{|\mathbf{k}|(|\mathbf{k}| + k_3)} k_1 , \\ F_2 &= \xi_2 - \frac{k_1\xi_1 + k_2\xi_2}{|\mathbf{k}|(|\mathbf{k}| + k_3)} k_2 , \\ F_3 &= -\frac{(k_1\xi_1 + k_2\xi_2)}{|\mathbf{k}|} , \end{aligned}\right\} \tag{7.4.7}$$

and it is connected to $\mathcal{F}_\tau(\xi_1, \xi_2)$ in (5.2.33) by a relation $\tau\mathbf{F} = \mathcal{F}_\tau(\xi_1, \xi_2)$. Since the third axis plays a special role in (7.4.1), we introduce a unitary operator U such that

$$|\mathbf{k}|U\xi_3 U^{-1} = \mathbf{k}\boldsymbol{\xi} \tag{7.4.8}$$

like in the case of spin $\frac{1}{2}$ in §7.1, and we denote the wave function transformed by this operator by

$$\chi(\mathbf{k}, \boldsymbol{\xi}) = U\phi(\mathbf{k}, \boldsymbol{\xi}) . \tag{7.4.9}$$

Explicitly U is given by

$$U = \exp\left(i\theta \frac{\hat{\mathbf{l}}(\mathbf{k} \times \mathbf{n})}{|\mathbf{k} \times \mathbf{n}|}\right) = \exp\{i\theta(\hat{l}_1 \sin\phi - \hat{l}_2 \cos\phi)\} , \tag{7.4.10}$$

where θ and ϕ are the angles of the polar coordinates, i.e., $\mathbf{k}/|\mathbf{k}| = (\sin\theta\cos\phi, \sin\theta\sin\phi, \cos\theta)$. We can easily verify that U satisfies (7.4.8) with the help of the following well known relation:

$$\begin{aligned} e^{-A}Be^{A} &= \sum_{n=0}^{\infty} \frac{B_n}{n!} , \\ B_n &= [B_{n-1}, A], \quad B_0 = B , \end{aligned} \tag{7.4.11}$$

and $[\xi_i, \hat{l}_j] = i\sum_{k=1}^{3} \varepsilon_{ijk}\xi_k$. Therefore (7.4.1) and (5.2.35) become

$$(\mathbf{k}\boldsymbol{\xi})\chi(\mathbf{k}, \boldsymbol{\xi}) = 0 , \tag{7.4.12}$$

$$\left(\boldsymbol{\xi}^2 - \frac{(\mathbf{k}\,\boldsymbol{\xi})^2}{|\mathbf{k}|^2} - \Xi\right)\chi(\mathbf{k}, \boldsymbol{\xi}) = (\boldsymbol{\xi}^2 - \Xi)\chi(\mathbf{k}, \boldsymbol{\xi}) = 0 \tag{7.4.13}$$

respectively. Next, let us calculate the Lorentz transformations of $\chi(\mathbf{k}, \xi)$. For this purpose we calculate UFU^{-1} and $\frac{1}{i}U\left(\frac{\partial}{\partial \mathbf{k}}\right)U^{-1}$ by the use of (7.4.11), and we get

$$UFU^{-1} = \frac{1}{k_0^2}\{\mathbf{k} \times (\boldsymbol{\xi} \times \mathbf{k})\}\,, \tag{7.4.14}$$

$$\frac{1}{i}U\left(\frac{\partial}{\partial \mathbf{k}}\right)U^{-1} = \frac{1}{i}\frac{\partial}{\partial \mathbf{k}} + \frac{\hat{\mathbf{l}} \times \mathbf{k}}{\mathbf{k}^2} - \frac{(\hat{\mathbf{l}}\mathbf{k})(\mathbf{n} \times \mathbf{k})}{\mathbf{k}^2(|\mathbf{k}| + k_3)}\,. \tag{7.4.15}$$

Equation (7.4.14) has been derived from $U\boldsymbol{\xi}U^{-1}$ which is calculated by the use of (7.4.11) to be

$$U\boldsymbol{\xi}U^{-1} = \boldsymbol{\xi}\cos\theta + (1-\cos\theta)\frac{\{(\mathbf{k} \times \mathbf{n})\boldsymbol{\xi}\}}{|\mathbf{k} \times \mathbf{n}|^2}(\mathbf{k} \times \mathbf{n}) + \{(\mathbf{k} \times \mathbf{n}) \times \boldsymbol{\xi}\}\frac{\sin\theta}{|\mathbf{k} \times \mathbf{n}|}\,,$$

namely

$$U\xi_1U^{-1} = \frac{\xi_1 k_3 - \xi_3 k_1}{|\mathbf{k}|} + \frac{k_2(\xi_1 k_2 - \xi_2 k_1)}{|\mathbf{k}|(|\mathbf{k}| + k_3)}\,,$$

$$U\xi_2U^{-1} = \frac{\xi_2 k_3 - \xi_3 k_2}{|\mathbf{k}|} - \frac{k_1(\xi_1 k_2 - \xi_2 k_1)}{|\mathbf{k}|(|\mathbf{k}| + k_3)}\,,$$

$$U\xi_3U^{-1} = \frac{\mathbf{k}\boldsymbol{\xi}}{|\mathbf{k}|}\,.$$

Equation (7.4.15) has been derived in the following manner. Putting $B = \frac{1}{i}\frac{\partial}{\partial \mathbf{k}}$, and $A = i\theta(\hat{l}_1 \sin\phi - \hat{l}_2 \cos\phi)$ in (7.4.11) we have

$$\begin{aligned}\frac{1}{i}U\left(\frac{\partial}{\partial \mathbf{k}}\right)U^{-1} &= \frac{1}{i}\frac{\partial}{\partial \mathbf{k}} - \frac{\partial\phi}{\partial \mathbf{k}}\frac{(\hat{\mathbf{l}}\mathbf{k})}{|\mathbf{k}|} + \frac{\partial\phi}{\partial \mathbf{k}}\hat{l}_3 - \frac{\partial\theta}{\partial \mathbf{k}}\frac{\{\mathbf{n}(\hat{\mathbf{l}} \times \mathbf{k})\}}{\sqrt{k_1^2 + k_2^2}} \\ &= \frac{1}{i}\frac{\partial}{\partial \mathbf{k}} - \frac{(\mathbf{n} \times \mathbf{k})(\hat{\mathbf{l}}\mathbf{k})}{|\mathbf{k}|(k_1^2 + k_2^2)} + \frac{\mathbf{n} \times \mathbf{k}}{k_1^2 + k_2^2}\hat{l}_3 \\ &\quad - \frac{\{(\mathbf{k}\mathbf{n})\mathbf{k} - \mathbf{k}^2\mathbf{n}\}\{\mathbf{n}(\hat{\mathbf{l}} \times \mathbf{k})\}}{\mathbf{k}^2(k_1^2 + k_2^2)}\,.\end{aligned} \tag{7.4.16}$$

On the other hand, in an identity

$$\{\mathbf{A}(\mathbf{C} \times \mathbf{D})\}\mathbf{B} = (\mathbf{A}\mathbf{B})(\mathbf{C} \times \mathbf{D}) - (\mathbf{B}\mathbf{C})(\mathbf{A} \times \mathbf{D}) + (\mathbf{B}\mathbf{D})(\mathbf{A} \times \mathbf{C})$$

which is derived from $\mathbf{A}\times\{\mathbf{B}\times(\mathbf{C}\times\mathbf{D})\}$, we put $\mathbf{A}=\mathbf{n}$, $\mathbf{C}=\hat{\mathbf{l}}$, $\mathbf{D}=\mathbf{k}$, and we put $(\mathbf{kn})\mathbf{k}-\mathbf{k}^2\mathbf{n}$ in place of $\mathbf{B}$. Then the numerator of the fourth term on the right-hand side of (7.4.16) becomes

$$\begin{aligned}\{(\mathbf{kn})^2-\mathbf{k}^2\}(\hat{\mathbf{l}}\times\mathbf{k})&-\{(\mathbf{kn})(\mathbf{k}\hat{\mathbf{l}})-\mathbf{k}^2(\mathbf{n}\hat{\mathbf{l}})\}(\mathbf{n}\times\mathbf{k})\\&+\{(\mathbf{kn})\mathbf{k}^2-\mathbf{k}^2(\mathbf{nk})\}(\mathbf{n}\times\hat{\mathbf{l}})\\=-(k_1^2+k_2^2)(\hat{\mathbf{l}}\times\mathbf{k})&-k_3(\mathbf{k}\hat{\mathbf{l}})(\mathbf{n}\times\mathbf{k})+\mathbf{k}^2(\mathbf{n}\times\mathbf{k})\hat{l}_3\ ,\end{aligned}$$

which leads to (7.4.15). Thus, considering (7.4.8) we find that $\chi(\mathbf{k},\boldsymbol{\xi})$ transforms as

$$\begin{aligned}\chi'(\mathbf{k},\boldsymbol{\xi})&=U(1+i\mathbf{J}\boldsymbol{\theta})U^{-1}\chi(\mathbf{k},\boldsymbol{\xi})\\&=\left\{1+i\boldsymbol{\theta}\left(\frac{1}{i}\mathbf{k}\times\frac{\partial}{\partial\mathbf{k}}+\hat{\mathbf{l}}\right)\right\}\chi(\mathbf{k},\boldsymbol{\xi})\end{aligned}\tag{7.4.17}$$

for the θ-transformation, and

$$\begin{aligned}\chi'(\mathbf{k},\boldsymbol{\xi})&=U(1-i\boldsymbol{\tau}\mathbf{K})U^{-1}\chi(\mathbf{k},\boldsymbol{\xi})\\&=\left\{1+ik_0\boldsymbol{\tau}\left(\frac{1}{i}\frac{\partial}{\partial\mathbf{k}}+\frac{\hat{\mathbf{l}}\times\mathbf{k}}{\mathbf{k}^2}+\frac{\mathbf{k}\times(\boldsymbol{\xi}\times\mathbf{k})}{(\mathbf{k}^2)^2}\right)\right\}\chi(\mathbf{k},\boldsymbol{\xi})\end{aligned}\tag{7.4.18}$$

for the τ-transformation.

According to (7.4.12) and (7.4.13) we can write

$$\chi(\mathbf{k},\boldsymbol{\xi})=\delta(\boldsymbol{\xi}^2-\Xi)\delta\left(\frac{\mathbf{k}\boldsymbol{\xi}}{|\mathbf{k}|}\right)\tilde{\chi}(\mathbf{k},\boldsymbol{\xi})\ .\tag{7.4.19}$$

Since $\delta(\boldsymbol{\xi}^2-\Xi)\delta(\mathbf{k}\boldsymbol{\xi}/|\mathbf{k}|)$ is $U\delta(\xi_1^2+\xi_2^2-\Xi)\delta(\xi_3)U^{-1}$ and since $\delta(\xi^2+\xi^2-\Xi)\delta(\xi_3)$ is commutable with $\mathbf{J}$ and $\mathbf{K}$, the part of δ-functions in (7.4.19) is invariant under the Lorentz transformations (7.4.17) and (7.4.18), and hence the above transformations become

$$\begin{aligned}&\delta(\boldsymbol{\xi}^2-\Xi)\delta\left(\frac{\mathbf{k}\boldsymbol{\xi}}{|\mathbf{k}|}\right)\tilde{\chi}'(\mathbf{k},\boldsymbol{\xi})\\&\quad=\delta(\boldsymbol{\xi}^2-\Xi)\delta\left(\frac{\mathbf{k}\boldsymbol{\xi}}{|\mathbf{k}|}\right)\{1+i\boldsymbol{\theta}\left(\frac{1}{i}\mathbf{k}\times\frac{\partial}{\partial\mathbf{k}}+\hat{\mathbf{l}}\right)\}\tilde{\chi}(\mathbf{k},\boldsymbol{\xi})\\&\qquad(\theta\text{-transformation})\ ,\end{aligned}\tag{7.4.20}$$

$$\delta(\boldsymbol{\xi}^2 - \Xi)\delta\Big(\frac{\mathbf{k}\boldsymbol{\xi}}{|\mathbf{k}|}\Big)\tilde{\chi}'(\mathbf{k},\boldsymbol{\xi})$$
$$= \delta(\boldsymbol{\xi}^2 - \Xi)\delta\Big(\frac{\mathbf{k}\boldsymbol{\xi}}{|\mathbf{k}|}\Big)\Big\{1 + ik_0\tau\Big(\frac{1}{i}\frac{\partial}{\partial\mathbf{k}} + \frac{\hat{\mathbf{l}}\times\mathbf{k}}{\mathbf{k}^2} + \frac{\mathbf{k}\times(\boldsymbol{\xi}\times\mathbf{k})}{(\mathbf{k}^2)^2}\Big)\Big\}\tilde{\chi}(\mathbf{k},\boldsymbol{\xi})$$
$$= \delta(\boldsymbol{\xi}^2 - \Xi)\delta\Big(\frac{\mathbf{k}\boldsymbol{\xi}}{|\mathbf{k}|}\Big)\Big\{1 + k_0\tau\Big(\frac{\partial}{\partial\mathbf{k}} + \frac{(\boldsymbol{\xi}\mathbf{k})\frac{\partial}{\partial\boldsymbol{\xi}} - \boldsymbol{\xi}(\mathbf{k}\frac{\partial}{\partial\boldsymbol{\xi}})}{\mathbf{k}^2}$$
$$+ i\frac{\mathbf{k}^2\boldsymbol{\xi} - (\mathbf{k}\boldsymbol{\xi})\mathbf{k}}{(\mathbf{k}^2)^2}\Big)\Big\}\tilde{\chi}(\mathbf{k},\boldsymbol{\xi})$$
$$= \delta(\boldsymbol{\xi}^2 - \Xi)\delta\Big(\frac{\mathbf{k}\boldsymbol{\xi}}{|\mathbf{k}|}\Big)\Big\{1 + k_0\tau\Big(\frac{\partial}{\partial\mathbf{k}} - \frac{\boldsymbol{\xi}}{\mathbf{k}^2}\Big(\mathbf{k}\frac{\partial}{\partial\boldsymbol{\xi}}\Big) + i\frac{\boldsymbol{\xi}}{\mathbf{k}^2}\Big)\Big\}\tilde{\chi}(\mathbf{k},\boldsymbol{\xi})$$
$$(\tau\text{-transformation}) \ . \tag{7.4.21}$$

Consequently we can use

$$\tilde{\chi}'(\mathbf{k},\boldsymbol{\xi}) = \Big\{1 + i\boldsymbol{\theta}\Big(\frac{1}{i}\mathbf{k}\times\frac{\partial}{\partial\mathbf{k}} + \hat{\mathbf{l}}\Big)\Big\}\tilde{\chi}(\mathbf{k},\boldsymbol{\xi})$$
$$(\theta\text{-transformation}) \ , \tag{7.4.22}$$
$$\tilde{\chi}'(\mathbf{k},\boldsymbol{\xi}) = \Big\{1 + k_0\tau\Big(\frac{\partial}{\partial\mathbf{k}} - \frac{\boldsymbol{\xi}}{\mathbf{k}^2}\Big(\mathbf{k}\frac{\partial}{\partial\mathbf{k}}\Big) + i\frac{\boldsymbol{\xi}}{\mathbf{k}^2}\Big)\Big\}\tilde{\chi}(\mathbf{k},\boldsymbol{\xi})$$
$$(\tau\text{-transformation}) \ , \tag{7.4.23}$$

for the Lorentz transformations of $\tilde{\chi}(\mathbf{k},\boldsymbol{\xi})$. If we rewrite the inner product (5.2.37) in terms of $\tilde{\chi}(\mathbf{k},\boldsymbol{\xi})$ we have

$$\langle 1|2\rangle = \langle 1,+|2,+\rangle + \langle 1,-|2,-\rangle$$
$$= \int\frac{d\mathbf{k}}{|\mathbf{k}|}d\boldsymbol{\xi}\delta(\boldsymbol{\xi}^2 - \Xi)\delta\Big(\frac{\mathbf{k}\boldsymbol{\xi}}{|\mathbf{k}|}\Big)\tilde{\chi}^*(\mathbf{k},\boldsymbol{\xi})_1\tilde{\chi}(\mathbf{k},\boldsymbol{\xi})_2$$
$$= \int d\mathbf{k}\,d\boldsymbol{\xi}\delta(\boldsymbol{\xi}^2 - \Xi)\delta(\mathbf{k}\boldsymbol{\xi})\tilde{\chi}^*(\mathbf{k},\boldsymbol{\xi})_1\tilde{\chi}(\mathbf{k},\boldsymbol{\xi})_2 \ , \tag{7.4.24}$$

where we have used the fact that the expression given in the right-hand side of (7.4.24) with $\tilde{\chi}^{(\pm)}(\mathbf{k},\boldsymbol{\xi})_1$ and $\tilde{\chi}^{(\mp)}(\mathbf{k},\boldsymbol{\xi})_2$ for $\tilde{\chi}(\mathbf{k},\boldsymbol{\xi})_1$ and $\tilde{\chi}(\mathbf{k},\boldsymbol{\xi})_2$ respectively vanishes like (5.1.26), because, by the argument after (5.1.19), $\tilde{\chi}^{(\pm)}(\mathbf{k},\boldsymbol{\xi})$ in $\tilde{\chi}(\mathbf{k},\boldsymbol{\xi}) = \tilde{\chi}^{(+)}(\mathbf{k},\boldsymbol{\xi}) + \tilde{\chi}^{(-)}(\mathbf{k},\boldsymbol{\xi})$ reads $\theta(\pm k_0)\tilde{\chi}^{(\pm)}(\mathbf{k},\boldsymbol{\xi})$.

Since the transformation coefficients in (7.4.22) and (7.4.23) depend on $\mathbf{k}$, and since $\boldsymbol{\xi}$ is a three-dimensional vector, these expressions have not yet been converted into covariant forms. Thus, introducing a 4-dimensional vector η_μ in place of $\boldsymbol{\xi}$ we shall consider a scalar function $\psi(\mathbf{k},\eta_\mu)$ whose Lorentz transformations are

$$\tilde{\psi}'(\mathbf{k},\eta_\mu) = \Big\{1 + \boldsymbol{\theta}\Big(\mathbf{k} \times \frac{\partial}{\partial \mathbf{k}} + \boldsymbol{\eta} \times \frac{\partial}{\partial \boldsymbol{\eta}}\Big)\Big\}\tilde{\psi}(\mathbf{k},\eta_\mu)$$

$$(\theta\text{-transformation}) \; , \tag{7.4.25}$$

$$\tilde{\psi}'(\mathbf{k},\eta_\mu) = \Big\{1 + \boldsymbol{\tau}\Big(k_0 \frac{\partial}{\partial \mathbf{k}} + \eta_0 \frac{\partial}{\partial \boldsymbol{\eta}} + \boldsymbol{\eta}\frac{\partial}{\partial \eta_0}\Big)\Big\}\tilde{\psi}(\mathbf{k},\eta_\mu)$$

$$(\tau\text{-transformation}) \; . \tag{7.4.26}$$

However, one of the four components of η_μ in $\tilde{\psi}(\mathbf{k},\eta_\mu)$ must be made ineffective in order that η_μ correspond to $\boldsymbol{\xi}$. We assume here that $\psi(\mathbf{k},\eta_\mu)$ satisfies

$$\begin{aligned}\tilde{\psi}(\mathbf{k},\eta_\mu) &= e^{i\eta_0/k_0}\tilde{\psi}\left(\mathbf{k},\eta_\mu - \frac{\eta_0}{k_0}k_\mu\right) \\ &= e^{i\eta_0/k_0}\tilde{\psi}\left(\mathbf{k},\boldsymbol{\eta} - \frac{\eta_0}{k_0}\mathbf{k}\right) \; ,\end{aligned} \tag{7.4.27}$$

where we have used a brief notation in the right-hand side of (7.4.27) because the fourth component of $\eta_\mu - \frac{\eta_0}{k_0}k_\mu$, $(\mu = 4)$, vanishes. According to (7.4.27) we have

$$\begin{aligned}\tilde{\psi}(\mathbf{k},\eta_\mu + \alpha k_\mu) &= e^{i\{(\eta_0/k_0)+\alpha\}}\tilde{\psi}\left(\mathbf{k},\boldsymbol{\eta} + \alpha\,\mathbf{k} - \frac{\eta_0 + \alpha k_0}{k_0}\mathbf{k}\right) \\ &= e^{i\{(\eta_0/k_0)+\alpha\}}\tilde{\psi}\left(\mathbf{k},\boldsymbol{\eta} - \frac{\eta_0}{k_0}\mathbf{k}\right)\end{aligned}$$

with an arbitrary real number α, and hence we obtain

$$\tilde{\psi}(\mathbf{k},\eta_\mu + \alpha k_\mu) = e^{i\alpha}\tilde{\psi}(\mathbf{k},\eta_\mu) \; . \tag{7.4.28}$$

Equation (7.4.27) will be derived from this equation with $\alpha = -\eta_0/k_0$. Equation (7.4.28) shows clearly that one of the 4 freedoms of η_μ is frozen. That is, the replacement of η_μ by $\eta_\mu + \alpha k_\mu$ (a gauge transformation for η_μ) causes only the phase factor $e^{i\alpha}$ to appear. Further, if we differentiate (7.4.28) with respect to α and put $\alpha = 0$, we find that $\tilde{\psi}(\mathbf{k},\eta_\mu)$ satisfies

$$(k_\nu \frac{\partial}{\partial \eta_\nu} - i)\tilde{\psi}(\mathbf{k},\eta_\mu) = 0 \; . \tag{7.4.29}$$

Furthermore, since the right-hand side of (7.4.28) can be obtained, by the use of (7.4.29), from the Taylor expansion of $\tilde{\psi}(\mathbf{k},\eta_\mu + \alpha k_\mu)$ with respect to α, we find that the solutions for this equation always satisfy (7.4.28). In other words, Eqs. (7.4.27), (7.4.28) and (7.4.29) are all equivalent conditions for $\tilde{\psi}(\mathbf{k}, \eta_\mu)$.

Using (7.4.27) we can calculate the Lorentz transformations of $\tilde{\psi}(\mathbf{k}, \boldsymbol{\eta} - \frac{\eta_0}{k_0}\mathbf{k})$, and from (7.4.25) and (7.4.26) we have

$$\tilde{\psi}'(\mathbf{k}, \boldsymbol{\eta} - \frac{\eta_0}{k_0}\mathbf{k}) = \left\{1 + \boldsymbol{\theta}\left(\mathbf{k} \times \frac{\partial}{\partial \mathbf{k}} + \boldsymbol{\xi} \times \frac{\partial}{\partial \boldsymbol{\xi}}\right)\right\} \tilde{\psi}(\mathbf{k}, \boldsymbol{\xi})\Big|_{\boldsymbol{\xi} = \boldsymbol{\eta} - \frac{\eta_0}{k_0}\mathbf{k}}$$
$$(\theta\text{-transformation})\,, \tag{7.4.30}$$

$$\tilde{\psi}'\left((\mathbf{k}, \boldsymbol{\eta} - \frac{\eta_0}{k_0}\mathbf{k}\right) = \left\{1 + k_0 \boldsymbol{\tau}\left(\frac{\partial}{\partial \mathbf{k}} - \frac{\boldsymbol{\xi}}{\mathbf{k}^2}\left(\mathbf{k}\frac{\partial}{\partial \boldsymbol{\xi}}\right) + i\frac{\boldsymbol{\xi}}{\mathbf{k}^2}\right)\right\} \tilde{\psi}(\mathbf{k}, \boldsymbol{\xi})\Big|_{\boldsymbol{\xi} = \boldsymbol{\eta} - \frac{\eta_0}{k_0}\mathbf{k}}$$
$$(\tau\text{-transformation})\,, \tag{7.4.31}$$

where we have used $\partial k_0/\partial \mathbf{k} = \mathbf{k}/k_0$. Let us compare these equations with (7.4.22) and (7.4.23), then $\tilde{\chi}(\mathbf{k}, \boldsymbol{\xi})$ is written as

$$\begin{aligned} \tilde{\chi}(\mathbf{k}, \boldsymbol{\xi}) &= \int d\boldsymbol{\eta} \delta\left(\boldsymbol{\eta} - \frac{\eta_0}{k_0}\mathbf{k} - \boldsymbol{\xi}\right) \tilde{\psi}\left(\mathbf{k}, \boldsymbol{\eta} - \frac{\eta_0}{k_0}\mathbf{k}\right) \\ &= e^{-i\eta_0/k_0} \int d\boldsymbol{\eta} \delta\left(\boldsymbol{\eta} - \frac{\eta_0}{k_0}\mathbf{k} - \boldsymbol{\xi}\right) \tilde{\psi}(\mathbf{k}, \eta_\mu)\,. \end{aligned} \tag{7.4.32}$$

Conversely from this equation

$$\tilde{\psi}(\mathbf{k}, \eta_\mu) = e^{i\eta_0/k_0} \tilde{\chi}\left(\mathbf{k}, \boldsymbol{\eta} - \frac{\eta_0}{k_0}\mathbf{k}\right) \tag{7.4.33}$$

is derived. We can therefore use $\tilde{\psi}(\mathbf{k}, \eta_\mu)$ satisfying covariant conditions (7.4.25), (7.4.26) and (7.4.29) in place of $\tilde{\chi}(\mathbf{k}, \boldsymbol{\xi})$.

Substituting (7.4.32) into the inner product (7.4.24) we have

$$\begin{aligned} \langle 1 | 2 \rangle &= e^{i(\eta_0 - \eta_0')/k_0} \int d\mathbf{k} d\boldsymbol{\xi} d\boldsymbol{\eta} d\boldsymbol{\eta}' \delta(\mathbf{k}\,\boldsymbol{\xi}) \delta(\boldsymbol{\xi}^2 - \Xi) \delta\left(\boldsymbol{\eta} - \frac{\eta_0}{k_0}\mathbf{k} - \boldsymbol{\xi}\right) \\ &\quad \times \delta\left(\boldsymbol{\eta}' - \frac{\eta_0'}{k_0}\mathbf{k} - \boldsymbol{\xi}\right) \tilde{\psi}^*(\mathbf{k}, \eta_\mu)_1 \tilde{\psi}(\mathbf{k}, \eta_\mu')_2 \\ &= e^{i(\eta_0 - \eta_0')/k_0} \int d\mathbf{k} d\boldsymbol{\eta} \delta(k_\mu \eta_\mu) \delta(\eta_\mu^2 - \Xi) \tilde{\psi}^*(\mathbf{k}, \eta_\mu)_1 \\ &\quad \times \tilde{\psi}\left(\mathbf{k}, \eta_\mu - \frac{\eta_0 - \eta_0'}{k_0} k_\mu\right)_2 \\ &= \int d\mathbf{k} d\boldsymbol{\eta} \delta(k_\mu \eta_\mu) \delta(\eta_\mu^2 - \Xi) \tilde{\psi}^*(\mathbf{k}, \eta_\mu)_1 \tilde{\psi}(\mathbf{k}, \eta_\mu)_2\,. \end{aligned} \tag{7.4.34}$$

This expression is nothing but the inner product written with $\tilde{\psi}(\mathbf{k},\eta_\mu)$.

Now, let

$$\psi(\mathbf{k},\eta_\mu)=\delta(\eta_\mu^2-\Xi)\delta(k_\mu\eta_\mu)\tilde{\psi}(\mathbf{k},\eta_\mu)\ . \tag{7.4.35}$$

Since

$$\delta(\boldsymbol{\xi}^2-\Xi)\delta\left(\frac{\mathbf{k}\boldsymbol{\xi}}{|\mathbf{k}|}\right)\Big|_{\boldsymbol{\xi}=\boldsymbol{\eta}-\frac{\eta_0}{k_0}\mathbf{k}}=|\mathbf{k}|\delta(\eta_\mu^2-\Xi)\delta(k_\mu\eta_\mu)\ , \tag{7.4.36}$$

from (7.4.19), (7.4.33), (7.4.35) and (7.4.9) we get

$$\begin{aligned}\psi(\mathbf{k},\eta_\mu)&=e^{i\eta_0/k_0}|\mathbf{k}|^{-1}\chi(\mathbf{k},\boldsymbol{\xi})\Big|_{\boldsymbol{\xi}=\boldsymbol{\eta}-\frac{\eta_0}{k_0}\mathbf{k}}\\&=|\mathbf{k}|^{-1}\exp\left\{i\theta(\hat{l}_1\sin\phi-\hat{l}_2\cos\phi)+\frac{i\eta_0}{k_0}\right\}\\&\qquad\times\phi(\mathbf{k},\boldsymbol{\xi})\Big|_{\boldsymbol{\xi}=\boldsymbol{\eta}-\frac{\eta_0}{k_0}\mathbf{k}}\ .\end{aligned} \tag{7.4.37}$$

Multiplying (7.4.32) by $U^{-1}\delta(\boldsymbol{\xi}^2-\Xi)\delta(\mathbf{k}\boldsymbol{\xi}/|\mathbf{k}|)$ we obtain, by the use of (7.4.19) and (7.4.9),

$$\begin{aligned}\phi(\mathbf{k},\boldsymbol{\xi})&=|\mathbf{k}|\exp\left\{-i\theta(\hat{l}_1\sin\phi-\hat{l}_2\cos\phi)-\frac{i\eta_0}{k_0}\right\}\\&\qquad\times\int d\boldsymbol{\eta}\,\delta\left(\boldsymbol{\eta}-\frac{\eta_0}{k_0}\mathbf{k}-\boldsymbol{\xi}\right)\psi(\mathbf{k},\eta_\mu)\ .\end{aligned} \tag{7.4.38}$$

These equations give a relation between the state vector $\phi(\mathbf{k},\boldsymbol{\xi})$ in a representation space of the Poincaré group and the covariant amplitude $\psi(\mathbf{k},\eta_\mu)$.

The equations satisfied by $\psi(\mathbf{k},\eta_\mu)$ are, from (7.4.35),

$$k_\mu\eta_\mu\psi(\mathbf{k},\eta_\mu)=0\ , \tag{7.4.39}$$

$$(\eta_\mu^2-\Xi)\psi(\mathbf{k},\eta_\mu)=0\ , \tag{7.4.40}$$

and, from (7.4.29),

$$\left(k_\mu\frac{\partial}{\partial\eta_\mu}-i\right)\psi(\mathbf{k},\eta_\mu)=0\ , \tag{7.4.41}$$

where the following relations have been used in deriving this equation:

$$\left[\left(k_\mu\frac{\partial}{\partial\eta_\mu}-i\right),\delta(\eta_\mu^2-\Xi)\right]=2k_\mu\eta_\mu\delta'(\eta_\mu^2-\Xi)\ ,$$

$$\left[\left(k_\mu\frac{\partial}{\partial\eta_\mu}-i\right),\delta(k_\mu\eta_\mu)\right]=k_\mu^2\delta'(k_\mu\eta_\mu)=0\ .$$

We can transfer to the x-representation by

$$\psi(x_\mu, \eta_\nu) = \frac{1}{2\pi} \int \frac{d\mathbf{k}}{|\mathbf{k}|} e^{i(\mathbf{kx}-k_0 t)} \psi(\mathbf{k}, \eta_\nu) , \tag{7.4.42}$$

and we immediately obtain

$$\frac{\partial^2}{\partial x_\mu^2} \psi(x_\mu, \eta_\nu) = 0 , \tag{7.4.43}$$

$$\eta_\mu \frac{\partial}{\partial x_\mu} \psi(x_\mu, \eta_\nu) = 0 , \tag{7.4.44}$$

$$(\eta_\mu^2 - \Xi) \psi(x_\mu, \eta_\nu) = 0 , \tag{7.4.45}$$

$$\left(\frac{\partial^2}{\partial x_\mu \partial \eta_\mu} + 1 \right) \psi(x_\mu, \eta_\nu) = 0 . \tag{7.4.46}$$

As is clear from (7.4.25) and (7.4.26), the Lorentz transformations are given by

$$\psi'(x_\mu, \eta_\nu) = \psi(\Lambda_{\mu\lambda}^{-1} x_\lambda, \Lambda_{\nu\rho}^{-1} \eta_\rho) . \tag{7.4.47}$$

Further the covariant inner product is

$$\begin{aligned} \langle\langle 1 | 2 \rangle\rangle &= \langle 1 | 2 \rangle \\ &= \int d\mathbf{x} d\boldsymbol{\eta} \delta(\mathbf{x}\boldsymbol{\eta}) \delta(\boldsymbol{\eta}^2 - \Xi) \frac{\partial \tilde{\psi}^*(x_\mu, \boldsymbol{\eta})_1}{\partial t} \frac{\partial \tilde{\psi}(x_\mu, \boldsymbol{\eta})_2}{\partial t} , \end{aligned} \tag{7.4.48}$$

where $\tilde{\psi}(x_\mu, \boldsymbol{\eta})$ is given by

$$\tilde{\psi}(x_\mu, \boldsymbol{\eta}) = \frac{\sqrt{\Xi}}{2\pi} \int \frac{d\mathbf{k}}{|\mathbf{k}|} e^{i(\mathbf{kx}-k_0 t)} \tilde{\psi}(\mathbf{k}, \eta_\nu) \delta(\boldsymbol{\eta}\mathbf{k})|_{\eta_0 = 0} . \tag{7.4.49}$$

This can be proved in the following way.

The expression (7.4.34) is independent of the value of η_0 as is seen in the form of the left-hand side. Thus, put $\eta_0 = 0$, then (7.4.34) becomes

$$\langle 1 | 2 \rangle = \int d\boldsymbol{\eta} d\mathbf{k} \delta(\mathbf{k}\boldsymbol{\eta}) \delta(\boldsymbol{\eta}^2 - \Xi) \tilde{\psi}^*(\mathbf{k}\boldsymbol{\eta})_1 \tilde{\psi}(\mathbf{k}, \boldsymbol{\eta})_2 . \tag{7.4.50}$$

Now we shall calculate

$$I = \int d\mathbf{k} \delta(\mathbf{k}\boldsymbol{\eta}) \tilde{\psi}^*(\mathbf{k}, \boldsymbol{\eta})_1 \tilde{\psi}(\mathbf{k}, \boldsymbol{\eta})_2 |_{\boldsymbol{\eta}^2 = \Xi} . \tag{7.4.51}$$

Denoting $\tilde{\psi}(\mathbf{k},\boldsymbol{\eta})$ as $\tilde{\psi}(k_1,k_2,k_3;\eta_1,\eta_2,\eta_3)$ and choosing the direction of $\boldsymbol{\eta}$ in the direction of the third axis we have

$$I = \frac{1}{\sqrt{\Xi}}\int dk_1 dk_2 \tilde{\psi}^*(k_1,k_2,0;0,0,\sqrt{\Xi})_1 \tilde{\psi}(k_1,k_2,0;0,0,\sqrt{\Xi})_2 \,. \tag{7.4.52}$$

On the other hand, if we choose $\boldsymbol{\eta}$ in the direction of the third axis also in (7.4.49) and put $x_3 = 0$, $\boldsymbol{\eta}^2 = \Xi$, we obtain

$$\begin{aligned}\tilde{\psi}(x_1,x_2,0,t;0,0,\sqrt{\Xi}) = \frac{1}{\sqrt{2\pi}}\int \frac{dk_1 dk_2}{\sqrt{k_1^2+k_2^2}} e^{i(k_1x_1+k_2x_2-k_0t)} \\ \times\, \tilde{\psi}(k_1,k_2,0;0,0,\sqrt{\Xi})\,,\end{aligned} \tag{7.4.53}$$

where $k_0^2 = k_1^2 + k_2^2$. Taking the inverse Fourier transform of (7.4.53) to get $\tilde{\psi}(k_1,k_2,0;0,0,\sqrt{\Xi})$ and substituting it into (7.4.52) we have

$$\begin{aligned} I &= \frac{1}{(2\pi)^2\sqrt{\Xi}}\int dx_1 dx_2 dx_1' dx_2' \tilde{\psi}^*(x_1,x_2,0,t;0,0,\sqrt{\Xi})_1 \\ &\quad \times \tilde{\psi}(x_1',x_2',0,t;0,0,\sqrt{\Xi})_2 e^{i(k_1(x_1-x_1')+k_2(x_2-x_2'))} k_0^2 dk_1 dk_2 \\ &= \frac{1}{\sqrt{\Xi}}\int dx_1 dx_2 \frac{\partial\tilde{\psi}^*(x_1,x_2,0,t;0,0,\sqrt{\Xi})_1}{\partial t}\frac{\partial\tilde{\psi}(x_1,x_2,0,t;0,0,\sqrt{\Xi})_2}{\partial t} \\ &= \int d\mathbf{x}\,\delta(\mathbf{x}\boldsymbol{\eta})\frac{\partial\tilde{\psi}^*(x_\mu,\boldsymbol{\eta})_1}{\partial t}\frac{\partial\tilde{\psi}(x_\mu,\boldsymbol{\eta})_2}{\partial t}\bigg|_{\boldsymbol{\eta}^2=\Xi} \,. \end{aligned} \tag{7.4.54}$$

The right-hand side is independent of the direction of $\boldsymbol{\eta}$ because I is invariant under a spatial rotation. If we substitute this expression into (7.4.50) we obtain (7.4.48).

Equations (7.4.43)–(7.4.49) give a completely covariant description for a continuous spin particle belonging to a single-valued representation with mass 0. Equation (7.4.48) shows that the covariant norm always takes a positive value independent of the sign of the frequency. This implies quite a different nature of the continuous spin from the cases of particles belonging to the single-valued representations with finite spin freedom mentioned previously.

Incidentally, the discrete transformations for $\psi(x_\mu,\eta_\nu)$ are as follows:

Space reflection $\psi'(x_\mu,\eta_\nu) = e^{i\delta}\psi(-\mathbf{x},t,-\boldsymbol{\eta},\eta_0)$,

Time reversal $\psi''(x_\mu,\eta_\nu) = e^{i\delta'}\psi^*(\mathbf{x},-t,\boldsymbol{\eta},-\eta_0)$,

Charge conjugation $\psi^c(x_\mu,\eta_\nu) = e^{i\delta''}\psi^*(x_\mu,\eta_\nu)$,

where $e^{i\delta}$, $e^{i\delta'}$ and $e^{i\delta''}$ are arbitrary phase factors.

b) *Particles in a Double-Valued Representation*

This kind of particles is defined by (5.2.38)–(5.2.42). Since the representations belonging to $S = \frac{1}{2}$ and $-\frac{1}{2}$ are unitary equivalent to each other as has been mentioned, we shall use the representation belonging to $S = \frac{1}{2}$ for the positive frequency and that belonging to $S = -\frac{1}{2}$ for the negative frequency in order to make use of the argument of §7.1. Similarly to (7.1.11) we put

$$\varphi^{(+)}(\mathbf{k}, \xi_1, \xi_2) = \begin{pmatrix} \phi^{(+)}_{1/2}(\mathbf{k}, \xi_1, \xi_2) \\ 0 \end{pmatrix} ,$$

$$\varphi^{(-)}(\mathbf{k}, \xi_1, \xi_2) = \begin{pmatrix} 0 \\ \phi^{(-)}_{-1/2}(\mathbf{k}, \xi_1, \xi_2) \end{pmatrix} ,$$

$$\varphi(\mathbf{k}, \xi_1, \xi_2) = \varphi^{(+)}(\mathbf{k}, \xi_1, \xi_2) + \varphi^{(-)}(\mathbf{k}, \xi_1, \xi_2) , \qquad (7.4.55)$$

and further we introduce $\varphi(\mathbf{k}, \boldsymbol{\xi}) = \delta(\xi_3)\varphi(\mathbf{k}, \xi_1, \xi_2)$ which satisfies

$$\xi_3\varphi(\mathbf{k}, \boldsymbol{\xi}) = 0 , \qquad (7.4.56)$$

$$k_0\varphi(\mathbf{k}, \boldsymbol{\xi}) = |\mathbf{k}|\sigma_3\varphi(\mathbf{k}, \boldsymbol{\xi}) . \qquad (7.4.57)$$

The generators of Lorentz transformations for $\varphi(\mathbf{k}, \boldsymbol{\xi})$ are given, using (5.2.38) and (5.2.39), by

$$\mathbf{J} = \mathbf{k} \times \left\{\frac{1}{i}\frac{\partial}{\partial \mathbf{k}} + \left(\hat{l}_3 + \frac{\sigma_3}{2}\right)\frac{\mathbf{n} \times \mathbf{k}}{|\mathbf{k}|(|\mathbf{k}| + k_3)}\right\} + \left(\hat{l}_3 + \frac{\sigma_3}{2}\right)\frac{\mathbf{k}}{|\mathbf{k}|} , \qquad (7.4.58)$$

$$\mathbf{K} = -k_0\left\{\frac{1}{i}\frac{\partial}{\partial \mathbf{k}} + \left(\hat{l}_3 + \frac{\sigma_3}{2}\right)\frac{\mathbf{n} \times \mathbf{k}}{|\mathbf{k}|(|\mathbf{k}| + k_3)}\right\} - \mathbf{F}\frac{k_0}{\mathbf{k}^2} . \qquad (7.4.59)$$

These equations show that the spin part of $\varphi(\mathbf{k}, \boldsymbol{\xi})$ is nothing but the direct product of the spin part of $\phi(\mathbf{k}, \boldsymbol{\xi})$ for a single-valued continuous spin (7.4.2) and the spin part of $\varphi^R(\mathbf{k})$ given in (7.1.13). Therefore we can get a covariant formalism of this kind of particles by combining the arguments of the above two cases. The calculation has been performed already and is not repeated here.

If we suppose that $\Psi(\mathbf{k}, \eta_\mu)$ transforms under Lorentz transformations as

$$\Psi'(\mathbf{k}, \eta_\mu) = \left\{1 + i\boldsymbol{\theta}\left(\frac{1}{i}\mathbf{k} \times \frac{\partial}{\partial \mathbf{k}} + \frac{1}{i}\boldsymbol{\eta} \times \frac{\partial}{\partial \boldsymbol{\eta}} + \frac{\boldsymbol{\sigma}}{2}\right)\right\}\Psi(\mathbf{k}, \eta_\mu)$$

$$(\theta\text{-transformation}) , \qquad (7.4.60)$$

$$\Psi'(\mathbf{k}, \eta_\mu) = \left\{1 + \boldsymbol{\tau}\left(k_0\frac{\partial}{\partial \mathbf{k}} + \eta_0\frac{\partial}{\partial \boldsymbol{\eta}} + \boldsymbol{\eta}\frac{\partial}{\partial \eta_0} - \frac{\boldsymbol{\sigma}}{2}\right)\right\}\Psi(\mathbf{k}, \eta_\mu)$$

$$(\tau\text{-transformation}) , \qquad (7.4.61)$$

and suppose that it satisfies

$$k_0\Psi(\mathbf{k},\eta_\mu) = \mathbf{k}\boldsymbol{\sigma}\Psi(\mathbf{k},\eta_\mu) \ , \tag{7.4.62}$$

$$k_\mu\eta_\mu\Psi(\mathbf{k},\eta_\mu) = 0 \ , \tag{7.4.63}$$

$$(\eta_\mu^2 - \Xi)\Psi(\mathbf{k},\eta_\mu) = 0 \ , \tag{7.4.64}$$

$$\left(k_\mu\frac{\partial}{\partial\eta_\mu} - i\right)\Psi(\mathbf{k},\eta_\mu) = 0 \ , \tag{7.4.65}$$

then we have relations between $\Psi(\mathbf{k},\eta_\mu)$ and $\varphi(\mathbf{k},\boldsymbol{\xi})$:

$$\Psi(\mathbf{k},\eta_\mu) = |\,\mathbf{k}\,|^{-1/2}\exp\left[i\theta\left\{\left(\hat{l}_1 + \frac{\sigma_1}{2}\right)\sin\phi - \left(\hat{l}_2 + \frac{\sigma_2}{2}\right)\cos\phi\right\}\right.$$
$$\left. + i\frac{\eta_0}{k_0}\right]\varphi(\mathbf{k},\boldsymbol{\xi})\Big|_{\boldsymbol{\xi}=\boldsymbol{\eta}-\frac{\eta_0}{k_0}\mathbf{k}} \ , \tag{7.4.66}$$

$$\varphi(\mathbf{k},\boldsymbol{\xi}) = |\,\mathbf{k}\,|^{1/2}\exp\left[-i\theta\left\{\left(\hat{l}_1 + \frac{\sigma_1}{2}\right)\sin\phi - \left(\hat{l}_2 + \frac{\sigma_2}{2}\right)\cos\phi\right\} - i\frac{\eta_0}{k_0}\right]$$
$$\times\int d\boldsymbol{\eta}\,\delta\left(\boldsymbol{\eta} - \frac{\eta_0}{k_0}\mathbf{k} - \boldsymbol{\xi}\right)\Psi(\mathbf{k},\eta_\mu) \ . \tag{7.4.67}$$

Since we have, from (7.4.62) and (7.4.63),

$$\Psi(\mathbf{k},\eta_\mu) = \delta(\eta_\mu^2 - \Xi)\delta(k_\mu\eta_\mu)\tilde{\Psi}(\mathbf{k},\eta_\mu) \ , \tag{7.4.68}$$

using $\tilde{\Psi}(\mathbf{k},\eta_\mu)$ we can express the inner product (5.2.42) as

$$\langle\,1,\pm\,|\,2,\pm\,\rangle = \int\frac{d\mathbf{k}}{|\,\mathbf{k}\,|}d\boldsymbol{\eta}\,\delta(k_\mu\eta_\mu)\delta(\eta_\mu^2 - \Xi)\tilde{\Psi}^{(\pm)*}(\mathbf{k},\eta_\mu)_1\tilde{\Psi}^{(\pm)}(\mathbf{k},\eta_\mu)_2$$
$$= \pm\int\frac{k_0}{|\,\mathbf{k}\,|^2}d\mathbf{k}\,d\boldsymbol{\eta}\,\delta(k_\mu\eta_\mu)\delta(\eta_\mu^2 - \Xi)\tilde{\Psi}^{(\pm)*}(\mathbf{k},\eta_\mu)_1\tilde{\Psi}^{(\pm)}(\mathbf{k},\eta_\mu)_2 \ . \tag{7.4.69}$$

We define here the amplitude in the x-representation, similarly to (7.4.42), by

$$\Psi(x_\mu,\eta_\nu) = \frac{1}{2\pi}\int\frac{d\mathbf{k}}{|\,\mathbf{k}\,|}e^{i(\mathbf{k}\mathbf{x}-k_0t)}\Psi(\mathbf{k},\eta_\nu) \ . \tag{7.4.70}$$

Then the covariant inner product is derived by a calculation similar to that in the case of single-valued representation, and is given by

$$\langle\langle\,1\,|\,2\,\rangle\rangle = \frac{1}{2i}\int d\mathbf{x}\,d\boldsymbol{\eta}\,\delta(\mathbf{x}\boldsymbol{\eta})\delta(\boldsymbol{\eta}^2 - \Xi)\left\{\frac{\partial\tilde{\Psi}^*(x_\mu,\boldsymbol{\eta})_1}{\partial t}\tilde{\Psi}(x_\mu,\boldsymbol{\eta})_2\right.$$
$$\left. - \tilde{\Psi}^*(x_\mu,\boldsymbol{\eta})_1\frac{\partial\tilde{\Psi}(x_\mu,\boldsymbol{\eta})_2}{\partial t}\right\} \ , \tag{7.4.71}$$

$$\tilde{\Psi}(x_\mu,\boldsymbol{\eta}) = \frac{\sqrt{\Xi}}{2\pi}\int\frac{d\mathbf{k}}{|\,\mathbf{k}\,|}e^{i(\mathbf{k}\mathbf{x}-k_0t)}\tilde{\Psi}(\mathbf{k},\eta_\nu)\delta(\boldsymbol{\eta}\mathbf{k})\Big|_{\eta_0=0} \ . \tag{7.4.72}$$

The covariant inner product (7.4.71) is connected to (7.4.69) as

$$\langle\langle 1,\pm\,|\,2,\pm\rangle\rangle = \pm\langle 1,\pm\,|\,2,\pm\rangle\,, \tag{7.4.73}$$

and, in contrast to the case of single-valued representation, it has a negative value for the negative frequency. This property is also quite different from that of particles in a double-valued representation with finite spin freedom.

The equations for $\psi(x_\mu,\eta_\nu)$ are, from (7.4.62)–(7.4.64) and (7.4.70),

$$\Big(\frac{\partial}{\partial t}+\boldsymbol{\sigma}\nabla\Big)\Psi(x_\mu,\eta_\nu)=0\,, \tag{7.4.74}$$

$$\eta_\mu\frac{\partial}{\partial x_\mu}\Psi(x_\mu,\eta_\nu)=0\,, \tag{7.4.75}$$

$$(\eta_\mu^2-\Xi)\Psi(x_\mu,\eta_\nu)=0\,, \tag{7.4.76}$$

$$\Big(\frac{\partial^2}{\partial x_\mu\partial\eta_\mu}+1\Big)\Psi(x_\mu,\eta_\nu)=0\,. \tag{7.4.77}$$

The Lorentz transformations are obtained from (7.4.60) and (7.4.61) as

$$\Psi'(x_\mu,\eta_\nu)=\Big(1+\frac{i}{2}\boldsymbol{\theta\sigma}\Big)\Psi(\Lambda(\theta)^{-1}_{\mu\lambda}x_\lambda,\Lambda(\theta)^{-1}_{\nu\rho}\eta_\rho) \quad (\theta\text{-transformation})\,, \tag{7.4.78}$$

$$\Psi'(x_\mu,\eta_\nu)=\Big(1-\frac{1}{2}\boldsymbol{\tau\sigma}\Big)\Psi(\Lambda(\tau)^{-1}_{\mu\lambda}x_\lambda,\Lambda(\tau)^{-1}_{\nu\rho}\eta_\rho) \quad (\tau\text{-transformation})\,. \tag{7.4.79}$$

The covariant formalism for a particle in a double-valued representation with continuous spin is completely expressed by (7.4.71) and (7.4.74)–(7.4.79).

We began with $\varphi(\mathbf{k},\xi_1,\xi_2)$ in (7.4.55) and have proceeded our discussion. Such an amplitude was chosen in order to make use of the analogy with (7.1.11), while we can, of course, use another amplitude similar to (7.1.12) which has two components with $S=-\frac{1}{2}$ for the positive frequency and $S=\frac{1}{2}$ for the negative frequency. It will be easily seen from the argument for a spin $\frac{1}{2}$ particle in §7.1 that the covariant formalism in this case is obtained from (7.4.74) and (7.4.79) with changing the signs of $\boldsymbol{\sigma}\nabla$ and $\boldsymbol{\tau\sigma}/2$. Needless to say, this formalism is equivalent to the covariant formalism explicitly given above. Indeed, if $\widehat{\Psi}(x_\mu,\eta_\nu)=\Xi^{-1/2}(\sigma_\mu\eta_\mu)\Psi(x_\mu,\eta_\nu)$, $\widehat{\Psi}(x_\mu,\eta_\nu)$ satisfies the equations derived in the above manner with the sign change, where σ_μ has been given in (2.2.1). This shows, according to the argument of §7.2, that a unitary operator of space reflection can be defined in an irreducible representation space of the Poincaré group. This point is also different from the case of massless particle with discrete spin ($|S| \geq 1/2$). Consequently we have the following discrete transformations for $\Psi(x_\mu, \eta_\nu)$:

Space reflection

$$\Psi'(x_\mu, \eta_\nu) = e^{i\delta}\Xi^{-1/2}(\sigma_\mu^\dagger \eta_\mu)\Psi(-\mathbf{x}, t, -\boldsymbol{\eta}, \eta_0) \ , \tag{7.4.80}$$

Time reversal

$$\Psi''(x_\mu, \eta_\nu) = e^{i\delta'}\sigma_2\Psi^*(\mathbf{x}, -t, \boldsymbol{\eta}, -\eta_0) \ , \tag{7.4.81}$$

Charge conjugation

$$\Psi^C(x_\mu, \eta_\nu) = e^{i\delta''}\Xi^{-1/2}(\sigma_\mu^\dagger \eta_\mu)\sigma_2\Psi^*(x_\mu, \eta_\nu) \ . \tag{7.4.82}$$

The proof that the theory is invariant under these transformations is left as a reader's exercise. It should be noted that the positive and negative frequencies are exchanged by the charge conjugation.

Chapter 8

Quantized Fields

§8.1 Quantum Theory of Matter Waves

We have studied the covariant descriptions of the theory in Chapters 6 and 7. This is not necessary, however, for the treatment of free particles. The irreducible representations are sufficient for the purpose. Of course, there appear generally the irreducible representations corresponding to negative frequency states, which have difficulties in their interpretation, but they may be abandoned for reason of the lack of physical meaning.*) This treatment indeed cannot be incorrect as far as free particles are concerned, but it will become complicated for a system where some external interaction acts on the particle. The interaction causes the exchanges of energy and momentum between the interior and the exterior of the system, and consequently the energy-momentum k_μ of the particle become not to be the constants of motion and k_μ^2 is no longer a constant in time. In this case, the Wigner rotation itself becomes meaningless. That is, the wave functions in an irreducible representation of the Poincaré group cannot be connected smoothly to that of the system with an interaction, and any effort along this line has no good theoretical prospect even if some complicated technique might be possible.

*)In this case a particle with imaginary mass has no reason to exist because the Lorentz invariant separation of the positive and negative frequencies is impossible for such a particle. Therefore, authors who wish to save this kind of particles are looking for a new interpretation of the negative frequency. See, for example, G. Feinberg: *Phys. Rev.* **159** (1967) 1089 and references therein.

Concerning this point the covariant formalism makes the discussion about the Lorentz transformations easier because the theory has no explicit dependence on k_μ. But, also in this case, the separation of the positive and negative frequencies is not possible when there is an interaction, and there can be a process where a particle in a positive frequency state interacting during an appropriate time interval becomes free in a negative frequency state. In the covariant formalism, since the inner product for an integral spin has a modified definition in which the expectation values of k_0 are always positive, if we identify these positive values with the observed values of energy we might avoid considering unphysical quantities like the negative energy. But we cannot avoid the difficulty that the norm of a negative frequency state is negative. On the other hand, in the case of half-integral spin, the norm does not become negative independent of whether the frequency is negative or not. As a substitute for it, however, the expectation value of energy in the negative frequency state becomes negative.

In addition it is doubtful whether the one-particle picture which we have studied until now is applicable to a system with an interaction. As is well known, photons are absorbed or emitted by an atom and their number decreases or increases during the interaction. A similar process occurs also in the case of particles with half-integral spin. For example, the β-decay process annihilates a neutron and creates a proton, an electron and a neutrino.

From this consideration we can conclude that the framework of the theory must be extended so that it can describe creation and annihilation of particles in order that it is connected smoothly to that of a system with an interaction. Further we want to avoid the above difficulty concerning the negative frequency as a consequence. These requirements are satisfied by the idea of field quantization which will be mentioned in this chapter. Of course, also in this case the picture of a free particle is fundamental and necessary in connection to the particle picture in quantum field theory, because it describes the system before and after the interaction where our method of irreducible representations of the Poincaré group is applicable. In this theory we can introduce a new interpretation of the negative frequency which will overcome the difficulty concerning it.

This book does not discuss, however, the details of the general theory of quantized fields with interactions, because on one hand it constitutes itself a great field in particle physics and cannot be presented in limited pages, and on the other hand there is no difficulty in learning it since many excellent text books on this subject have been published already. In this book we restrict ourselves to the fundamental discussion about the connection between irreducible representations of the Poincaré group and the quantization of a free field (which is not interacting). The basic idea of quantization of relativistic fields can be explained to some extent even within the limited pages.

In classical theory there is no particle picture of photon, and the phenomena of the light are described by the electromagnetic fields satisfying the Maxwell equations. The fields are the functions of the coordinates of space and time, and the variables of the fields at any point of space-time x_μ are determined by them, where x_μ are the scales of coordinates and are no more than parameters. The idea of probability amplitude in quantum theory is applicable neither to the electric field $\mathbf{E}(x)$ nor to the magnetic field $\mathbf{H}(x)$ which are field quantities. These waves existing in the real space-time are quite different from the imaginary waves in a space formed by quantum state vectors, and they are real waves which carry energy and which can be used as signals travelling between two separated points. These waves existing in the real space-time are generally called matter waves and are expressed by fields which are functions of space-time points.

Now, if we suppose that there is some connection between the electromagnetic fields as the matter waves and the existence of photons in quantum theory, we may consider that there exist matter waves, namely fields, corresponding to other particles. Although the electromagnetic fields are known experimentally to satisfy the Maxwell equations, it is not evident what kinds of equations are satisfied by other fields. However, the fact that the covariant amplitudes describing a photon satisfy the Maxwell equations when the probability amplitudes are rewritten into the covariant forms, as has been seen in paragraph a) of §7.3, suggests the forms of equations satisfied by other matter waves. Although the matter waves are originally different from the covariant amplitudes derived by the deformations of the probability waves, we shall here assume that the fields describing the former also satisfy the same equations which are satisfied by the covariant amplitudes. This way of introducing the assumption is never logical, but it will be justified by the following argument in this chapter.

For simplicity we shall restrict ourselves to the discussion of particles with finite mass.*) We shall consider the field $U(x)$ satisfying the Klein-Gordon equation for a spin 0 particle and the field $\psi_{(\dots)_n}(x)$ satisfying the Bargmann-Wigner equations for a spin $n/2$ ($n \geq 1$) particle. It must be noted that $U(x)$ and $\psi_{(\dots)_n}(x)$ represent the quantities of fields in this chapter although they

*) Massless particles can be considered similarly, but the argument will be somewhat complicated.

are expressed by the same symbols as those of the covariant amplitudes. As the covariant amplitudes were written by various expressions, the fields are also expressed in some different ways, but needless to say, they can be converted into one another, and any of them can be used. The Lorentz transformations of $U(x)$ and $\psi_{(\dots)_n}(x)$ are given by (6.1.10) and (6.3.26), (6.3.27) respectively.

By the assumption on the equations for the fields we have

$$(\partial_\mu^2 - m^2)U(x) = 0 \tag{8.1.1}$$

for a spin 0 particle, and

$$(\gamma_\mu^{(i)}\partial_\mu + m)\psi_{(\dots)_n}(x) = 0 \quad (i = 1, 2, \dots n) \tag{8.1.2}$$

for a spin $n/2$ particle. The quantization of fields regards $U(x)$ and $\psi_{(\dots)_n}(x)$ satisfying the above equations as quantum operators neither merely real nor complex numbers.*) We shall study step by step what kind of conditions determine the operators. Analogously to the ordinary quantum mechanics which is obtained by the quantization of a classical particle, the quantum field theory may be regarded as the quantum theory of a kind of continuous media. The parameter describing the motion of the system is the time in the former case, while the space-time coordinates x_μ are the parameters in the latter case.

Let $\mathcal{H}$ be the total energy of the field considered as continuous medium. The energy $\mathcal{H}$ is the Hamiltonian of the system which is, of course, Hermitian and is a function of the field. For example, in classical electromagnetism it is well known that $\mathcal{H}$ is proportional to $\int d\mathbf{x}(\mathbf{E}(x)^2 + \mathbf{H}(x)^2)$. While, in quantum field theory $\mathcal{H}$ is an operator such that its eigenvalues are the values of energy of the system. According to the principle of quantum mechanics we suppose that the time derivative of $U(x)$ satisfies the Heisenberg equation of motion

$$i\frac{\partial U(x)}{\partial t} = [U(x), \mathcal{H}]\,, \tag{8.1.3}$$

and $\psi_{(\dots)_n}(x)$ also satisfies a similar equation. Of course, these equations should not contradict the field equations, for example, Eqs. (8.1.1) and (8.1.2) when interactions are absent.

---

*)The corresponding quantities to the complex conjugates of the amplitudes $U(x)$ and $\psi_{(\dots)_n}(x)$ must be expressed as $U^\dagger(x)$ and $\psi^\dagger_{(\dots)_n}(x)$ in quantum field theory.

Denote the energy contained in an infinitesimal volume $d\mathbf{x}(= dx_1 dx_2 dx_3)$ as $\mathcal{H}(x)d\mathbf{x}$ where $\mathcal{H}(x)$ is called an energy density. Needless to say, H is equal to $\int d\mathbf{x}\mathcal{H}(x)$, which is the integral of $\mathcal{H}(x)$ over the whole volume. When two points x_μ and y_μ are separated spatially, i.e., $(x_\mu - y_\mu)^2 > 0$, from the relativistic requirement that the signal never travels faster than the light, we can put

$$[U(x), \mathcal{H}(y)] = 0 \quad ((x_\mu - y_\mu)^2 > 0) . \tag{8.1.4}$$

That is, the field $U(x)$ at a point x_μ, concerning its variation with time, is not affected by the energy density $\mathcal{H}(x)$ at y_μ such that $(x_\mu - y_\mu)^2 > 0$. This is nothing but the causality in a relativistic sense. Of course, a similar equation also holds for $\psi_{(\dots)_n}(x)$. The condition (8.1.4) is based on the idea that the local energy density has a physical meaning, and we shall call the equation the locality condition.*)

The Hamiltonian H is a function of the field, but we have not yet any idea as to what form it has. By a physical reason we shall now put the following condition on H. That is, the eigenvalues of H have the lower limit, which can be adjusted to 0 by adding an appropriate constant. Under this condition the energy of the system never becomes negative. The state of the lowest energy is called the vacuum, and it is assumed not to be degenerate. Namely, the vacuum must be given uniquely as the state of the lowest eigenvalue of H.

Under the above conditions we must determine the equation characterizing the field as an operator, the form of H and, at the same time, the Hilbert space on which these operators act. If this is done, it will become possible to avoid the difficulty concerning the negative frequency mentioned in the first half of the section. That is, the negative frequency is merely one of the modes of the matter wave and it does not mean that the energy of the system is negative. As a matter of fact, the energy is given by the eigenvalue of H which never takes on negative value by the assumption. Since the state vectors are defined in the Hilbert space on which the field and H act, if the norms in this space are positive, there must not exist any trouble mentioned previously.

In order to make the argument explicit we shall consider the case of free field $U(x)$. The field $U(x)$ satisfying (8.1.1) is written as

$$U(x) = U^{(+)}(x) + U^{(-)}(x) , \tag{8.1.5}$$

---

*) A weaker condition $[\mathcal{H}(x), \mathcal{H}(y)] = 0$, $(x_\mu - y_\mu)^2 > 0$, instead of (8.1.4), is sufficient to proceed with the following discussion, but we have here adopted (8.1.4) for simplicity.

$$U^{(+)}(x) = \frac{1}{(2\pi)^{3/2}} \int \frac{d\mathbf{k}}{\sqrt{2\omega_k}} e^{i(\mathbf{kx}-\omega_k t)} A(\mathbf{k}) \ , \tag{8.1.6}$$

$$U^{(-)}(x) = \frac{1}{(2\pi)^{3/2}} \int \frac{d\mathbf{k}}{\sqrt{2\omega_k}} e^{-i(\mathbf{kx}-\omega_k t)} B^\dagger(\mathbf{k}) \ , \tag{8.1.7}$$

where $A(\mathbf{k})$ and $B^\dagger(-\mathbf{k})$ correspond to $\phi^{(+)}(\mathbf{k})$ and $\phi^{(-)}(\mathbf{k})$ in §6.1 respectively and are operators here. The additional factors $1/\sqrt{2}$'s on the right-hand sides of (8.1.6) and (8.1.7) have been introduced merely for convenience. The Lorentz transformations of $A(\mathbf{k})$ and $B^\dagger(-\mathbf{k})$ have the same forms as those of $\phi^{(\pm)}(\mathbf{k})$, and from (6.1.1) and (6.1.2) we have

$$\left.\begin{aligned} A'(\mathbf{k}) &= A(\Lambda^{-1}\mathbf{k}) \ , \\ B'(\mathbf{k}) &= B(\Lambda^{-1}\mathbf{k}) \ , \end{aligned}\right\} \tag{8.1.8}$$

where $\Lambda^{-1}\mathbf{k}$ denotes the space components of $k_\mu = (\mathbf{k}, i\omega_k)$ multiplied by Λ^{-1}.

Equation (8.1.3) is valid for any x_μ, and $\mathcal{H}$ is a constant of motion, namely $d\mathcal{H}/dt = 0$. Therefore, from the equation (8.1.3) substituted by (8.1.5) and its time-derivative we get

$$\left.\begin{aligned} \omega_k A(\mathbf{k}) &= [A(\mathbf{k}), \mathcal{H}] \ , \\ \omega_k B(\mathbf{k}) &= [B(\mathbf{k}), \mathcal{H}] \ . \end{aligned}\right\} \tag{8.1.9}$$

Let $|\,E\,\rangle$ be a state belonging to the eigenvalue E of $\mathcal{H}$ and let (8.1.9) act on it. As will be seen easily, both $A(\mathbf{k})|\,E\,\rangle$ and $B(\mathbf{k})|\,E\,\rangle$ are the eigenstates of $\mathcal{H}$, and their eigenvalues are both $(E - \omega_k)$. Furthermore, an analogous argument with the Hermitian conjugate equation of (8.1.9) shows that both $A^\dagger(\mathbf{k})|\,E\,\rangle$ and $B^\dagger(\mathbf{k})|\,E\,\rangle$ are also the eigenstates of $\mathcal{H}$ whose eigenvalues are both $(E + \omega_k)$. Namely, $A^\dagger(\mathbf{k})$, $B^\dagger(\mathbf{k})$ and $A(\mathbf{k})$, $B(\mathbf{k})$ increase and decrease the eigenvalues of $\mathcal{H}$ by ω_k respectively. In this sense $A^\dagger(\mathbf{k})$ and $B^\dagger(\mathbf{k})$ are called creation operators, and $A(\mathbf{k})$ and $B(\mathbf{k})$ are called annihilation (or destruction) operators. If $|\,0\,\rangle$ denotes the vacuum, the conditions

$$A(\mathbf{k})|\,0\,\rangle = B(\mathbf{k})|\,0\,\rangle = 0 \tag{8.1.10}$$

must hold by definition. The significance of $A(\mathbf{k})$ and $B(\mathbf{k})$ will be mentioned in detail in §8.3. In short, what we must do is to determine the relations among $A(\mathbf{k})$, $B(\mathbf{k})$ and their Hermitian conjugates, and at the same time, the form of $\mathcal{H}$ satisfying the above conditions. We shall make a preparation for it in the next section. In the case where $U(x) = U^\dagger(x)$ we have $A(\mathbf{k}) = B(\mathbf{k})$, and we call $U(x)$ a real scalar field. On the other hand, $U(x)$ for which $U(x) \neq U^\dagger(x)$ is called a complex scalar field. We have discussed about $U(x)$ so far, and $\psi_{(\dots)_n}(x)$ will be considered in §8.4.

§8.2 Harmonic Oscillators

The quantization of $U(x)$ is to solve (8.1.9) which describes a system of infinite number of harmonic oscillators, as will be revealed in the following. For $\psi_{(\dots)_n}(x)$ a similar equation to (8.1.9) will also be derived (§8.4).

We shall first begin with the consideration of the quantum theory of a simplest 1-dimensional harmonic oscillator

$$\ddot{x} = -\omega^2 x \,. \tag{8.2.1}$$

Needless to say, $\omega/2\pi(>0)$ is a frequency and the dot denotes a time-derivative. The operator x is assumed to be Hermitian. Dividing (8.2.1) into two equations we get

$$\left.\begin{aligned} \dot{x} &= \omega p \,, \\ \dot{p} &= -\omega x \,. \end{aligned}\right\} \tag{8.2.2}$$

With the assumption of the existence of Hamiltonian $\mathcal{H}$, applying the Heisenberg equation $i\dot{x} = [x, \mathcal{H}]$, $i\dot{p} = [p, \mathcal{H}]$, we rewrite the above set of equations in the form

$$\left.\begin{aligned} i\omega p &= [x, \mathcal{H}] \,, \\ -i\omega x &= [p, \mathcal{H}] \,. \end{aligned}\right\} \tag{8.2.3}$$

If we use

$$a = \frac{x + ip}{\sqrt{2}}\,, \quad a^\dagger = \frac{x - ip}{\sqrt{2}} \tag{8.2.4}$$

in place of x and p, we can derive immediately

$$\omega a = [a, \mathcal{H}] \tag{8.2.5}$$

and its Hermitian conjugate equation. Equation (8.2.5) is evidently equivalent to (8.2.1). As is easily seen, a corresponds to $A(\mathbf{k})$ and $B(\mathbf{k})$ in (8.1.9). Therefore $A(\mathbf{k})$ and $B(\mathbf{k})$ describe harmonic oscillators for each $\mathbf{k}$. Since the number of possible values of $\mathbf{k}$ is infinite, (8.1.9) represents a system of infinite number of harmonic oscillators.

As far as we consider such a system of harmonic oscillators which is a basis of quantum field theory, Eq. (8.2.5) is sufficient for our present discussion and it should be noted that the canonical commutation relation between x and

p, i.e., $[x,p] = i$, is not necessary to be assumed at the beginning. That is, p is no more than a variable introduced in order to convert the second order differential equation (8.2.1) with respect to time into the first order equations. The practical image of the momentum is not necessary to be given to p at this stage. In this sense we shall hereafter investigate the properties of the harmonic oscillator (8.2.5) without the assumption of the canonical commutation relation.

In order to solve (8.2.5) we need the explicit form of $\mathcal{H}$. Since (8.2.5) and the Heisenberg equation lead to $\dot{a} = -i\omega a$, $\dot{a}^\dagger = i\omega a^\dagger$, the constants of motion of this system, including $\mathcal{H}$, should generally be functions of $a^{\dagger n}a^n$ and $a^m a^{\dagger m}$ $(n, m = 1, 2, \ldots)$. The equations for a and $a^\dagger$ are derived from (8.2.5) with such $\mathcal{H}$. Of course, a and $a^\dagger$ must be defined as operators by these equations with, if necessary, some conditions consistent with them. Further the existence of the non-degenerate lowest eigenvalue of $\mathcal{H}$ must be confirmed at the same time. If this is possible, a always gives the harmonic oscillator (8.2.1).

The argument is, however, too general to proceed further. We shall consider the simple cases where $\mathcal{H}$ has the following bilinear form of $a^\dagger$ and a:

$$\mathcal{H} = \mathcal{H}_A = \frac{\omega}{2}[a^\dagger, a] \tag{8.2.6}$$

or

$$\mathcal{H} = \mathcal{H}_S = \frac{\omega}{2}\{a^\dagger, a\}\ . \tag{8.2.7}$$

As will be seen in the following there are many solutions even in these restricted cases. The factors $\omega/2$'s on the right-hand sides are written merely for convenience, and they may be written as $\omega/2c$ $(c > 0)$ for another choice of the scale of a.*)

Let us consider first the case of $\mathcal{H}_A$. From (8.2.5) we have the equations for a and $a^\dagger$:

$$\left.\begin{aligned} 2a &= [a, [a^\dagger, a]]\ , \\ -2a^\dagger &= [a^\dagger, [a^\dagger, a]]\ . \end{aligned}\right\} \tag{8.2.8}$$

Note that these equations are the same as those satisfied by the generators of the rotation group. Namely we have the following correspondence between

---

*)The case $c < 0$ is not considered. In fact, if we use $-\omega/2$ instead of $\omega/2$, we will find that the eigenvalues of energy have no lower limit for (8.2.7). Although (8.2.7) itself leads to no inconsistency, but if we consider a system consisting of many such harmonic oscillators in order to connect it to a field theory, the negative norm inevitably appears in the Hilbert space (Y. Ohnuki and M. Yamada and S. Kamefuchi: *Phys. Lett.* **36B** (1971), 51).

$a, a^\dagger$ and $\mathbf{S}$ in (4.1.3), (4.1.6) or $S^{(\pm)}, S_3$:

$$S^{(+)} \leftrightarrow a^\dagger\,, \quad S^{(-)} \leftrightarrow a\,, \quad S_3 \leftrightarrow \frac{1}{2}[a^\dagger, a]\,. \tag{8.2.9}$$

Therefore the matrix elements of a are easily determined in terms of the irreducible representations of the rotation group. Now let us consider the irreducible representation with "angular momentum" $r/2$ ($r = 0, 1, 2, \ldots$). Then the lowest eigenvalue of $\mathcal{H}_A$ is the lowest eigenvalue of S_3 multiplied by ω. If $|0\rangle$ denotes the corresponding normalized eigenstate, we have

$$a\,|0\rangle = 0\,, \tag{8.2.10}$$

$$\mathcal{H}_A|0\rangle = -\frac{\omega}{2}r\,|0\rangle\,. \tag{8.2.11}$$

From the argument for the angular momentum it is clear that all eigenstates of $\mathcal{H}$ are obtained by applying the operator $a^\dagger$, which increases the energy eigenvalue by ω, on $|0\rangle$ successively. If $|n\rangle$ is a state given by normalizing $a^{\dagger n}|0\rangle$, from (4.1.7) and the correspondence $\xi \leftrightarrow n - r/2$, $S \leftrightarrow r/2$ we obtain

$$\left.\begin{aligned} a^\dagger|\,n\,\rangle &= \sqrt{(r-n)(n+1)}\,|\,n+1\,\rangle\,, \\ a\,|\,n\,\rangle &= \sqrt{n(r-n+1)}\,|\,n-1\,\rangle\,. \end{aligned}\right\} \tag{8.2.12}$$

That is, when a non-negative integer r is given, the corresponding a and $a^\dagger$ are uniquely determined. Now (8.2.12) gives $a^\dagger|\,r\,\rangle = 0$. This fact shows that r pieces of energy ω can be created on $|0\rangle$ by applying $a^\dagger$ r times, but the creation of more than r pieces is impossible. In the special case $r = 1$ the number of the ω created on the vacuum is at most one. In this case the harmonic oscillator is said to obey the 1-dimensional Fermi statistics and it satisfies the well known relations $\{a, a^\dagger\} = 1$, $a^{\dagger 2} = a^2 = 0$ which are derived from (8.2.12). The case $r = 0$ corresponds to the 1-dimensional representation of the rotation group and we have $a|\,0\,\rangle = a^\dagger|\,0\,\rangle = 0$, which leads to $a = a^\dagger = 0$. Therefore the solution of $r = 0$ is not interesting physically since the oscillator itself disappears in this case.

Let us next consider the case of $\mathcal{H}_S$. As is seen in (8.2.7), $\mathcal{H}_S$ has the lowest eigenvalue, which is not negative. Let it be $\omega r'/2$. Then we can write

$$\begin{aligned} a|\,0\,\rangle &= 0\,, \\ \mathcal{H}_S\,|\,0\,\rangle &= \frac{\omega}{2}r'|\,0\,\rangle \quad (r' \geq 0)\,. \end{aligned} \tag{8.2.13}$$

From this equation we get

$$aa^\dagger|\,0\,\rangle = r'|\,0\,\rangle\ , \tag{8.2.14}$$

which shows the uniqueness of the vacuum. When $r' = 0$, from (8.2.14) we find that the norm of $a^\dagger|\,0\,\rangle$ vanishes, and hence $a^\dagger|\,0\,\rangle = a|\,0\,\rangle = 0$, which gives $a = a^\dagger = 0$. Since this solution is trivial, we shall consider the case $r' > 0$ hereafter. The equations for a and $a^\dagger$ are, from (8.2.5) and (8.2.7),

$$\left.\begin{aligned} 2a &= \left[a, \{a^\dagger, a\}\right]\ , \\ -2a^\dagger &= \left[a^\dagger, \{a^\dagger, a\}\right]\ . \end{aligned}\right\} \tag{8.2.15}$$

If we use $aa^{\dagger 2} = a^{\dagger 2}a + 2a^\dagger$ which is derived from the above equations, according to the first equation of (8.2.13) and Eq. (8.2.14) we can always write the eigenstate of $\mathcal{H}_S$ as $(a^\dagger)^n|\,0\,\rangle$ whose eigenvalue is $\omega(n + r'/2)$. In order to determine a and $a^\dagger$ we shall here consider operators $a^{\dagger 2}/2$, $a^2/2$ and $\{a^\dagger, a\}/2$. According to (8.2.15) we have the following commutation relations between these operators:

$$\left.\begin{aligned} &\left[\frac{\{a^\dagger, a\}}{4}, \frac{a^{\dagger 2}}{2}\right] = \frac{a^{\dagger 2}}{2}\ , \quad \left[\frac{\{a^\dagger, a\}}{4}, \frac{a^2}{2}\right] = -\frac{a^2}{2}\ , \\ &\left[\frac{a^{\dagger 2}}{2}, \frac{a^2}{2}\right] = -\frac{1}{2}\{a^\dagger, a\}\ . \end{aligned}\right\} \tag{8.2.16}$$

Comparing these relations with (4.4.11) and (4.4.12) we find that these three operators are the generators of the 3-dimensional Lorentz group. Namely, we have the following correspondence between these operators and $H^{(\pm)}$, H_0 in §4.4:

$$H^{(+)} \leftrightarrow \frac{a^{\dagger 2}}{2}\ , \quad H^{(-)} \leftrightarrow \frac{a^2}{2}\ , \quad H_0 \leftrightarrow \frac{\{a^\dagger, a\}}{4}\ . \tag{8.2.17}$$

This correspondence holds, however, only for the case where the eigenvalues of H_0 have the lowest limit, since the eigenvalues of $\{a^\dagger, a\}/4$ are not negative. If $|\,n\,\rangle$ is a state vector defined by normalizing $(a^\dagger)^n|\,0\rangle$ similarly to the previous case, two states giving the lowest eigenvalue exist in the Hilbert space spanned by $|\,n\,\rangle$, and they are $|\,0\,\rangle$ and $|\,1\,\rangle$. Indeed, we immediately find that these states multiplied by $a^2/2$ vanish. Successive multiplications of $a^{\dagger 2}/2$ on $|\,0\,\rangle$ and $|\,1\,\rangle$ generate two unitary irreducible representation spaces of the 3-dimensional Lorentz group. That is, the state vector $|\,n\,\rangle$ belongs to either the above two irreducible representation spaces corresponding to whether n

is even or odd. The operators a and $a^\dagger$ play a role that connects the above two irreducible representation spaces.

On the basis of the above argument we can easily determine a and $a^\dagger$. First, from $\mathcal{H}_S\,|n\rangle = \omega(n+r'/2)|n\rangle$ we have

$$\frac{1}{4}\{a^\dagger, a\}|0\rangle = \frac{r'}{4}|0\rangle\,, \tag{8.2.18}$$

$$\frac{1}{4}\{a^\dagger, a\}|1\rangle = \left(\frac{1}{2}+\frac{r'}{4}\right)|1\rangle\,, \tag{8.2.19}$$

and hence the values of μ_0 specifying the irreducible representations are $r'/4$ and $\frac{1}{2}+\frac{r'}{4}$ corresponding to even n and odd n respectively. These values are not necessarily confined into integers or half-integers. We have now no reason for that the representations must be confined to the single-valued or the double-valued representations, although we have confined them in §4.4 by reason of the representations of the little group of the Poincaré group. We can use, however, the argument of §4.4 for $\mu_0 > 0$. That is, we can use $D^{(+)}\mu_0$ with $\mu_0 > 0$ in Table 4.1. Consequently we obtain

$$\frac{a^{\dagger 2}}{2}|2n\rangle = \sqrt{(n+\frac{r'}{2})(n+1)}|2n+2\rangle\,, \tag{8.2.20}$$

$$\frac{a^{\dagger 2}}{2}|2n+1\rangle = \sqrt{(n+1+\frac{r'}{2})(n+1)}|2n+3\rangle\,. \tag{8.2.21}$$

The correspondence between these expressions and those of Table 4.1 is as follows: $\mu_0 = r'/4$, $\mu = n/2 + r'/4$ for even n of $|n\rangle$, and $\mu_0 = \frac{1}{2}+r'/4$, $\mu = n/2+r'/4$ for odd n. To get $a^\dagger$ and a we shall put here

$$a^\dagger|2n\rangle = \alpha_n|2n+1\rangle\,,\quad a^\dagger|2n+1\rangle = \beta_n|2n+2\rangle\,. \tag{8.2.22}$$

Then from (8.2.20) and (8.2.21) we get

$$\alpha_n\beta_n = 2\sqrt{(n+\frac{r'}{2})(n+1)}\,, \tag{8.2.23}$$

$$\alpha_{n+1}\beta_n = 2\sqrt{(n+1+\frac{r'}{2})(n+1)}\,, \tag{8.2.24}$$

of which ratio is

$$\frac{\alpha_n}{\alpha_{n-1}} = \sqrt{\frac{2n+r'}{2(n-1)+r'}}\,,$$

and hence we obtain

$$\alpha_n = \alpha_0 \sqrt{\frac{2n+r'}{r'}} = \sqrt{2n+r'} \ , \tag{8.2.25}$$

where we have used $a^\dagger|\,0\,\rangle = \sqrt{r'}|\,1\,\rangle$, namely $\alpha_0 = \sqrt{r'}$. If the above equation is substituted into (8.2.23), β_n will be obtained. Combining these results we have

$$a^\dagger|\,n\,\rangle = \begin{cases} \sqrt{n+r'}\,|n+1\,\rangle & (n = \text{even}) \ , \\ \sqrt{n+1}\,|n+1\,\rangle & (n = \text{odd}) \ , \end{cases} \tag{8.2.26}$$

and

$$a|\,n\,\rangle = \begin{cases} \sqrt{n}\,|n-1\,\rangle & (n = \text{even}) \ , \\ \sqrt{n-1+r'}\,|\,n-1\,\rangle & (n = \text{odd}) \ . \end{cases} \tag{8.2.27}$$

In this way, we have found in the case of $\mathcal{H}_S$ that for any real number $r'(>0)$, a and $a^\dagger$ are uniquely determined. Since

$$[a, a^\dagger]|\,n\,\rangle = \begin{cases} r'\,|\,n\,\rangle & (n = \text{even}) \ , \\ (2-r')|\,n\,\rangle & (n = \text{odd}) \end{cases} \tag{8.2.28}$$

are derived from (8.2.26) and (8.2.27), we can put $[a, a^\dagger] = 1$ independent of n only when $r' = 1$. In this case the system is said to obey the 1-dimensional Bose statistics.

We have considered the 1-dimensional harmonic oscillator, and we have seen that there are various possibilities which give (8.2.1) corresponding to r and r' even when the Hamiltonian is restricted to (8.2.6) or (8.2.7). If other forms of Hamiltonian are included in the consideration the range of possibility may be more extended. What we must note is, however, as has been mentioned previously, that our argument is still confined to the quantum theory of a harmonic oscillator and it is not applicable to any other dynamical system. Therefore we cannot regard $x = (a + a^\dagger)/\sqrt{2}$ and $p = (a - a^\dagger)/\sqrt{2}$ as the position and the momentum of a particle which can be defined independently of the structure of the system. If we regard x as the postion operator and p as the momentum operator in the general sense, the relation between x and p must be invariant under the change of the origin of coordinates, namely the transformation $x \to x + c, p \to p$ with any real number c. But this is not satisfied by any operator of (8.2.8), and it is satisfied only by operators which obey (8.2.15) and the ordinary commutation relation $[x, p] = i$ (equivalent to $[a, a^\dagger] = 1$). In other words, the argument of this section is peculiar to the harmonic oscillator, and because of this the Fermi statistics can be considered as well.

We shall here proceed to the discussion of a system of many harmonic oscillators ($\omega_n > 0, n = 1, 2, \ldots$)

$$\omega_n a_n = i\dot{a}_n = [a_n, \mathcal{H}] . \tag{8.2.29}$$

Since each oscillator is independent, denoting the Hamiltonian corresponding to n as $\mathcal{H}^{(n)}$ we have

$$\omega_n \delta_{nm} a_n = [a_m, \mathcal{H}^{(n)}] , \tag{8.2.30}$$

where $\mathcal{H} = \sum_n \mathcal{H}^{(n)}$. Since $\mathcal{H}^{(n)}$ is the Hamiltonian of the n-th 1-dimensional harmonic oscillator, we can use $\mathcal{H}_A^{(n)} = \frac{\omega_n}{2}[a_n^\dagger, a_n]$ or $\mathcal{H}_S^{(n)} = \frac{\omega_n}{2}\{a_n^\dagger, a_n\}$ corresponding to (8.2.6) or (8.2.7). However, if a_n ($n = 1, 2, \ldots$) are the harmonic oscillators derived from a field, the local energy density mentioned in the previous section cannot be defined when $\mathcal{H}_A^{(n)}$ and $\mathcal{H}_S^{(m)}$ ($n \neq m$) exist together. The details of this situation will be clarified in the arguments of §8.3 and §8.4, and here we assume the total Hamiltonian for a field to be either

$$\mathcal{H}_A = \sum_n \mathcal{H}_A^{(n)} = \sum_n \frac{\omega_n}{2}[a_n^\dagger, a_n] \tag{8.2.31}$$

or

$$\mathcal{H}_S = \sum_n \mathcal{H}_S^{(n)} = \sum_n \frac{\omega_n}{2}\{a_n^\dagger, a_n\} , \tag{8.2.32}$$

which are extensions of (8.2.6) and (8.2.7). They can, of course, have additional constants. Now, we find that (8.2.30) itself is not sufficient to calculate the right-hand side of (8.1.4), because (8.1.4) requires the relations between the fields at space-time points x_μ and y_μ. That is, we need the relations of a_n such that can be rewritten into those between the fields with variables of space-time points. We shall then impose the following set of relations which is a further extension of (8.2.30):

$$\left.\begin{aligned} &[a_n, [a_m^\dagger, a_l]] = 2\delta_{nm} a_l , \\ &[a_n, [a_m^\dagger, a_l^\dagger]] = 2(\delta_{nm} a_l^\dagger - \delta_{nl} a_m^\dagger) , \\ &[a_n, [a_m, a_l]] = 0 , \end{aligned}\right\} \tag{8.2.33}$$

and their Hermitian conjugates.

By similar reasoning, in the case of $\mathcal{H}_S$ we shall assume the following set of relations for a_n:

$$\left.\begin{aligned} &[a_n, \{a_n^\dagger, a_l\}] = 2\delta_{nm} , \\ &[a_n, \{a_m^\dagger, a_l^\dagger\}] = 2(\delta_{nm} a_l^\dagger + \delta_{nl} a_m^\dagger) , \\ &[a_n, \{a_m, a_l\}] = 0 , \end{aligned}\right\} \tag{8.2.34}$$

and their Hermitian conjugates.*)

The relations (8.2.33) and (8.2.34) were proposed by Green as generalizations of the Fermi and the Bose statistics.**)

The conditions for the vacuum are given by

$$a_n|0\rangle = 0 \,, \tag{8.2.35}$$

$$a_m a_n^\dagger|0\rangle = \delta_{mn} r|0\rangle \quad (r = 0, 1, 2, \ldots) \tag{8.2.36}$$

in both cases. Equation (8.2.36) corresponds to the uniqueness of the vacuum. When the number of harmonic oscillators becomes infinity it should be noted that r takes integral values in both cases of (8.2.33) and (8.2.34) in contrast to r of (8.2.13). This property of the system of infinite number of harmonic oscillators was derived by Green and Messiah***) under the condition of uniqueness of the vacuum and the condition that the Hilbert space must have positive norms. Further it was also shown by them that if such r is given, the irreducible operators a_n and $a_n^\dagger$ are uniquely determined. Using the second equations of (8.2.33) and (8.2.34), and (8.2.35), (8.2.36) we find that any state vector is expressed by the superposition of the state vectors constructed from the vacuum multiplied by the creation operators only.

The relations of the Fermi statistics

$$\left.\begin{aligned} &\{a_n, a_m^\dagger\} = \delta_{nm} \,, \\ &\{a_n, a_m\} = \{a_n^\dagger, a_m^\dagger\} = 0 \end{aligned}\right\} \tag{8.2.37}$$

satisfy (8.2.33) and (8.2.36), and $r = 1$. Conversely, if $r = 1$ in (8.2.36), the above relations (8.2.37) are derived from (8.2.33). Although this can be verified in various ways, it is easily realized if we accept the result of Greenberg and Messiah that a_n is uniquely determined when r is given. On the other hand, (8.2.34) with $r = 1$ leads to the Bose statistics

$$\left.\begin{aligned} &[a_n, a_m^\dagger] = \delta_{nm} \,, \\ &[a_n, a_m] = [a_n^\dagger, a_m^\dagger] = 0 \,. \end{aligned}\right\} \tag{8.2.38}$$

---

*)Another method of derivation of (8.2.33) and (8.2.34) from (8.2.30) is, for example, based on the condition that the relations for a are invariant under an arbitrary unitary transformation $a_\kappa = \sum_n U_{\kappa n} a_n$ and the condition that any state vector is constructed from $|0\rangle$ multiplied by the creation operators only. (I. Bialynicki-Biula: *Nuclear Phys.* **49** (1963), 605).

)H. S. Green: *Phys. Rev.* **90 (1953) 270.

***)O. W. Greenberg and A. M. Messiah: *Phys. Rev.*, **138** (1965) B1155.

That is, the Fermi statistics and the Bose statistics are derived from (8.2.33) and (8.2.34) as special solutions with $r = 1$. For a general r, when a_n satisfy (8.2.33) the system is said to obey the para-Fermi statistics of order r, and when a_n satisfy (8.2.34) the system is said to obey the para-Bose statistics of order r. In the case $r > 0$, since the creation operators can be assumed to be neither commutable nor anti-commutable with one another, the properties of the state vectors are not simple, but the consistent quantum field theory can be constructed as in the case $r = 1$.*) However, we cannot afford to describe it here and we shall discuss mainly the case $r = 1$ hereafter. In addition, the particle which must obey the statistics of $r > 1$ have not yet been confirmed directly in experiments, and such statistics remain as a theoretical possibility. There is, however, an argument that they may be necessary for the description of the internal structures of elementary particles.

Our argument has been developed on the basis of the Hamiltonians (8.2.31) and (8.2.32). If we take other appropriate Hamiltonians, we might get other relativistic quantum field theories describing the system of infinite number of harmonic oscillators, but we have not yet discovered any examples.

§8.3 Scalar Fields

Let us begin with the quantization of a complex scalar field $U(x)$. We put $r = 1$ for simplicity, namely we consider the Bose statistics if the Hamiltonian is $\mathcal{H}_S$ and the Fermi statistics if the Hamiltonian is $\mathcal{H}_A$.

We shall first assume that $U(x)$ obey the Bose statistics. According to the argument of the previous section it is clear that $A(\mathbf{k})$ and $B(\mathbf{k})$ correspond to a_n, and $A^\dagger(\mathbf{k})$ and $B^\dagger(\mathbf{k})$ correspond to $a_n^\dagger$. As has been mentioned, the ambiguity of the scale of a_n (which depends generally on n) leads us to write the first equation of (8.2.38) in the form $[A(\mathbf{k}), A^\dagger(\mathbf{k}')] = [B(\mathbf{k}), B^\dagger(\mathbf{k}')] = f(\mathbf{k})\delta(\mathbf{k}-\mathbf{k}')$ $(f(\mathbf{k}) > 0)$. Taking the Lorentz transformation (8.1.8) and the Lorentz invariance of $\omega_k\delta(\mathbf{k}-\mathbf{k}')$ (§3.1) into account we find that $f(\mathbf{k})$ must be proportional to ω_k in order that the above commutation relations are Lorentz invariant. Thus we can write

$$\begin{aligned}[A(\mathbf{k}), A^\dagger(\mathbf{k}')] = [B(\mathbf{k}), B^\dagger(\mathbf{k}')] &= \omega_k\delta(\mathbf{k}-\mathbf{k}') \ , \\ [A(\mathbf{k}, A(\mathbf{k}')] = [B(\mathbf{k}), B(\mathbf{k}')]& \\ = [A(\mathbf{k}), B(\mathbf{k}')] &= [A(\mathbf{k}), B^\dagger(\mathbf{k}')] = 0 \ , \end{aligned} \tag{8.3.1}$$

*)See for example Y. Ohnuki and S. Kamefuchi: *Quantum Field Theory and Parastatistics*, Univ. of Tokyo Press/Springer-Verlag (1982). (Added in translation.)

and from (8.2.32) we obtain

$$\mathcal{H} = \mathcal{H}_S = \frac{1}{2}\int d\mathbf{k}(\{A^\dagger(\mathbf{k}), A(\mathbf{k})\} + \{B^\dagger(\mathbf{k}), B(\mathbf{k})\}) \ . \qquad (8.3.2)^{*)}$$

We shall rewrite these expressions in terms of $U(x)$ and $U^\dagger(x)$. Using (8.1.5)–(8.1.7) and (8.3.1) we get easily

$$\left.\begin{aligned} &[U(x), U^\dagger(y)] = i\Delta(x-y) \ , \\ &[U(x), U(y)] = 0 \ , \end{aligned}\right\} \qquad (8.3.3)$$

where $\Delta(x)$ is a Lorentz invariant odd function of x_μ ($\Delta(x) = -\Delta(-x)$) and is given with the help of

$$\varepsilon(k) = \begin{cases} 1 & (k_0 > 0) \ , \\ -1 & (k_0 < 0) \ , \end{cases} \qquad (8.3.4)$$

in the form

$$\begin{aligned} \Delta(x) &= \frac{-i}{(2\pi)^3}\int \frac{d\mathbf{k}}{2\omega_k}\left(e^{i(\mathbf{kx}-\omega_k x_0)} - e^{-i(\mathbf{kx}-\omega_k x_0)}\right) \\ &= \frac{-i}{(2\pi)^3}\int d^4k\varepsilon(k)\delta(k_\mu^2 + m^2)e^{ik_\mu x_\mu}) \ . \end{aligned} \qquad (8.3.5)$$

The expression (8.3.5) shows clearly that this function satisfies

$$(\partial_\mu^2 - m^2)\Delta(x) = 0 \ , \qquad (8.3.6)$$

$$\frac{\partial}{\partial x_0}\Delta(x)|_{x_0=0} = -\delta(\mathbf{x}) \ . \qquad (8.3.7)$$

If we put $x_0 = 0$ on the right-hand side of (8.3.5), then the integrand becomes an even function of $\mathbf{k}$ and an odd function of k_0, thus we have $\Delta(x)|_{x_0=0} = 0$. By the Lorentz invariance of $\Delta(x)$ we therefore find that for any space-like x_μ

$$\Delta(x) = 0 \quad (x_\mu^2 > 0) \ . \qquad (8.3.8)$$

---

*) An additional constant (minus infinity) is necessary on the right-hand side of (8.3.2) in order that $\mathcal{H}\,|0\rangle = 0$ is satisfied, but it will be omitted hereafter to simplify the expressions.

We can verify that there exists no Lorentz invariant function of x_μ satisfying (8.3.8) and (8.3.6) except for the functions proportional to $\Delta(x)$. Further we get from (8.1.5)–(8.1.7)

$$\left.\begin{aligned} A(k) &= \frac{1}{\sqrt{2}\,(2\pi)^{3/2}}\int d\mathbf{x}(\omega_k U(x) + i\dot{U}(x))e^{-i(\mathbf{kx}-\omega_k t)} \\ B(k) &= \frac{1}{\sqrt{2}\,(2\pi)^{3/2}}\int d\mathbf{x}(\omega_k U^\dagger(x) + i\dot{U}^\dagger(x))e^{-i(\mathbf{kx}-\omega_k t)} \end{aligned}\right\} \tag{8.3.9}$$

Substituting these expressions into (8.3.2) we shall write the Hamiltonian. In terms of the energy density

$$\mathcal{H}(x) = \frac{1}{2}(\{\nabla U^\dagger(x), \nabla U(x)\} + \left\{\frac{\partial U^\dagger(x)}{\partial t}, \frac{\partial U(x)}{\partial t}\right\} + m^2\{U^\dagger(x), U(x)\}) \tag{8.3.10}$$

we can write

$$\mathcal{H} = \int d\mathbf{x}\mathcal{H}(x)\,, \tag{8.3.11}$$

where we have used the condition that $U(x)$ tends to zero as $|\mathbf{x}| \to \infty$, namely that the surface integral vanishes when we perform the space integration by parts. Such a method of calculation will be sometimes applied hereafter. By (8.3.3) we can easily verify that the energy density (8.3.10) satisfies the locality condition (8.1.4). If we define the following covariant symmetric tensor $\theta_{\mu\nu}(k)$ of rank 2:

$$\begin{aligned} \theta_{\mu\nu}(x) = &\frac{1}{2}\left(\left\{\frac{\partial U^\dagger(x)}{\partial x_\mu}, \frac{\partial U(x)}{\partial x_\nu}\right\} + \left\{\frac{\partial U^\dagger(x)}{\partial x_\nu}, \frac{\partial U(x)}{\partial x_\mu}\right\}\right) \\ &- \frac{1}{2}\delta_{\mu\nu}\left(\left\{\frac{\partial U^\dagger(x)}{\partial x_\lambda}, \frac{\partial U(x)}{\partial x_\lambda}\right\} + m^2\{U^\dagger(x), U(x)\}\right)\,, \end{aligned} \tag{8.3.12}$$

$\mathcal{H}(x)$ is expressed as

$$\mathcal{H}(x) = \theta_{00}(x)\,. \tag{8.3.13}^{*)}$$

In general $\theta_{\mu\nu}(x)$ is called an energy-momentum tensor and satisfies

$$\partial_\mu \theta_{\mu\nu}(x) = 0 \tag{8.3.14}$$

according to (8.1.1).

---

*) To give the additional constant term of $\mathcal{H}(x)$ mentioned in the footnote of p. 185 the additional term of const. $\delta_{\mu\nu}$ is necessary in $\theta_{\mu\nu}(x)$ of (8.3.12).

We shall next consider the case where $U(x)$ is assumed to obey the Fermi statistics. In this case, from (8.2.37) and the Lorentz invariance we have

$$\left.\begin{aligned}\{A(\mathbf{k}), A^\dagger(\mathbf{k}')\} &= \{B(\mathbf{k}), B^\dagger(\mathbf{k}')\} = \omega_k\delta(\mathbf{k}-\mathbf{k}')\ ,\\ \{A(\mathbf{k}), A(\mathbf{k}')\} &= \{B(\mathbf{k}), B(\mathbf{k}')\} = \{A(\mathbf{k}), B(\mathbf{k}')\}\\ &= \{A(\mathbf{k}), B^\dagger(\mathbf{k}')\} = 0\ .\end{aligned}\right\} \tag{8.3.15}$$

Further from (8.2.31) we get

$$\mathcal{H} = \mathcal{H}_A = \frac{1}{2}\int d\mathbf{k}([A^\dagger(\mathbf{k}), A(\mathbf{k})] + [B^\dagger(\mathbf{k}), B(\mathbf{k})])\ . \tag{8.3.16}$$

Using (8.3.9) we rewrite these expressions into the x-representation, and we obtain

$$\left.\begin{aligned}\{U(x), U^\dagger(y)\} &= \Delta^{(1)}(x-y)\ ,\\ \{U(x), U(y)\} &= 0\ ,\end{aligned}\right\} \tag{8.3.17}$$

and the energy density

$$\mathcal{H}(x) = \frac{1}{2i}\left(\left[\frac{\partial U^\dagger(x)}{\partial t}, \sqrt{-\nabla^2+m^2}\,U(x)\right] - \left[\sqrt{-\nabla^2+m^2}\,U^\dagger(x), \frac{\partial U(x)}{\partial t}\right]\right)\ , \tag{8.3.18}$$

where $\Delta^{(1)}(x)$ is a Lorentz invariant even function of x_μ ($\Delta^{(1)}(x) = \Delta^{(1)}(-x)$) and is defined by

$$\begin{aligned}\Delta^{(1)}(x) &= \frac{1}{(2\pi)^3}\int \frac{d\mathbf{k}}{2\omega_k}\left(e^{i(\mathbf{kx}-\omega_k x_0)} + e^{-i(\mathbf{kx}-\omega_k x_0)}\right)\\ &= \frac{1}{(2\pi)^3}\int d^4k\,\delta(k_\mu^2+m^2)e^{ik_\mu x_\mu}\ .\end{aligned} \tag{8.3.19}$$

The function $\Delta^{(1)}(x)$ clearly satisfies the Klein-Gordon equation (8.3.6), but, in contrast to $\Delta(x)$, it does not always vanish for any space-like x_μ ($x_\mu^2 > 0$). In fact, if we perform the integration of (8.3.19) with $x_\mu^2 > 0$, we get

$$\Delta^{(1)}(x) = \frac{m}{4\pi}\frac{N_1(m\sqrt{x_\mu^2})}{\sqrt{x_\mu^2}} \quad (x_\mu^2 > 0)\ , \tag{8.3.20}$$

where N_1 is a Neumann function (the second kind Bessel function).

To investigate the locality condition we shall here calculate $[U(x), \mathcal{H}(y)]$. With the help of (8.3.17) and (8.3.18) we get

$$[U(x), \mathcal{H}(y)] = i\Big(\frac{\partial \Delta^{(1)}(x-y)}{\partial x_0}\sqrt{-\nabla_y^2 + m^2}\,U(y) + \sqrt{-\nabla_x^2 + m^2}\,\Delta^{(1)}(x-y)\frac{\partial U(y)}{\partial y_0}\Big)\,, \tag{8.3.21}$$

where ∇_x^2 and ∇_y^2 are brief accounts of $\partial^2/\partial \mathbf{x}^2$ and $\partial^2/\partial \mathbf{y}^2$ respectively. The right-hand side of (8.3.21) does not vanish when $(x_\mu - y_\mu)^2 > 0$ and except when $x_0 = y_0, \mathbf{x} \neq \mathbf{y}$. For two separated points at one instant of time the right-hand side of (8.3.21) vanishes as can be verified with the help of $\partial\Delta^{(1)}(x)/\partial x_0|_{x_0=0} = 0$ and $\sqrt{-\nabla^2 + m^2}\Delta^{(1)}(x)|_{x_0=0} = \delta(\mathbf{x})$ which are derived from (8.3.19). But the fact that (8.3.21) does not vanish for other space-like intervals contradicts evidently the requirement of relativity. This is because the energy density (8.3.18) involves noncovariant quantity $\sqrt{-\nabla^2 + m^2}U(x)$ and its Hermitian conjugate. In other words, the locality condition can be satisfied only if $\mathcal{H}(x)$ is expressed covariantly. In this sense, the covariance is a very fundamental concept in a relativistic field theory. The reason that we spent many pages for the covariant formalisms in Chapters 6 and 7 was to make a preparation for the arguments like this.

In this way, we have arrived at the conclusion that "$U(x)$ must obey the Bose statistics" in order to satisfy the requirement of relativity.

We shall now give some supplementary comments on the description of a scalar particle by $U(x)$.

Integrating $\theta_{0\mu}(x)$ of (8.3.12) over the volume we write

$$P_\mu = \int d\mathbf{x}\theta_{0\mu}(x)\,, \tag{8.3.22}$$

then P_μ transform as a 4-vector under a Lorentz transformation. This will be proved as follows. In the 4-dimensional space-time we consider a 3-dimensional surface σ on which any two points x_μ and x'_μ are located in a space-like separation $(x_\mu - x'_\mu)^2 > 0$. Such a surface σ is called a space-like surface or simply a σ-surface. Next, define infinitesimal surface elements $d\sigma_\mu$ ($\mu = 1, 2, 3, 4$) to consider an integral on this surface:

$$d\sigma_\mu = (dx_2dx_3dx_0, dx_1dx_3dx_0, dx_1dx_2dx_0, -idx_1dx_2dx_3)\,. \tag{8.3.23}$$

As is easily verified, $d\sigma_\mu$ transforms as a 4-vector. Now performing the integration on the σ-surface we put

$$P_\mu(\sigma) = \int_\sigma d\sigma_\nu\theta_{\nu\mu}(x)\,. \tag{8.3.24}$$

Under the Lorentz transformation $U(x) \to U'(x) = U(\Lambda^{-1}x)$ we have

$$P_\mu(\sigma) \to P'_\mu(\sigma) = \Lambda_{\mu\nu} P_\nu(\Lambda^{-1}\sigma) \;, \tag{8.3.25}$$

where $\Lambda^{-1}\sigma$ on the right-hand side represents a space-like surface which is formed by all points on σ transformed by Λ^{-1}. Let us consider a surface σ which is obtained from σ by an infinitesimal deformation (Fig. 8.1). Then $\int_{\sigma'} - \int_\sigma$ is a surface integral on the infinitesimal 4-dimensional volume Ω with the boundaries σ and σ', and thus it can be written by Gauss's theorem in the form

$$P_\mu(\sigma') - P_\mu(\sigma) = \int_\Omega d^4x \partial_\nu \theta_{\nu\mu}(x) = 0 \;, \tag{8.3.26}$$

where (8.3.14) has been used. Therefore $P_\mu(\sigma)$ does not depend on σ. If we now take as σ a flat surface perpendicular to the time axis, namely the ordinary 3-dimensional space, we can put $P_\mu(\sigma) = P_\mu$, which transform as a 4-vector as is seen in (8.3.25).

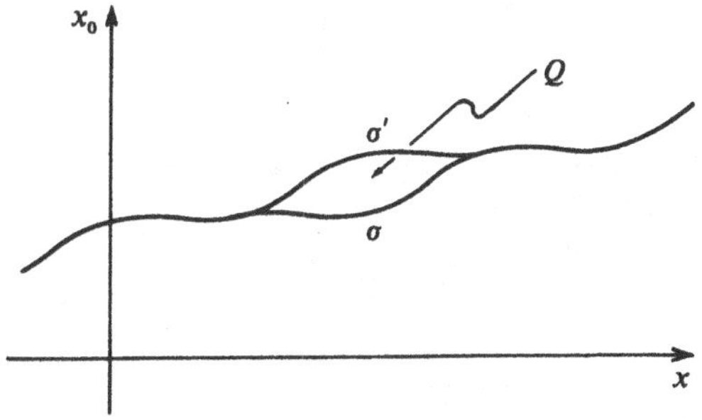

Fig. 8.1.

Since $P_0 = \mathcal{H}$ is the energy, P_i $(i = 1, 2, 3)$ which form a 4-vector together with P_0 are the operators of the momentum of the field. On the other hand, using (8.3.3) we can derive

$$[P_\mu, U(x)] = i\partial_\mu U(x) \;, \tag{8.3.27}$$

$$[P_\mu, P_\nu] = 0 \;. \tag{8.3.28}$$

Since (8.3.27) is also written in the form

$$[A(\mathbf{k}), P_\mu] = k_\mu A(\mathbf{k}) \;, \tag{8.3.29}$$

$$[B(\mathbf{k}), P_\mu] = k_\mu B(\mathbf{k}) \quad (k_\mu = (\mathbf{k}, i\omega_k)) \;, \tag{8.3.30}$$

we find that $A(\mathbf{k})$ and $B(\mathbf{k})$ are the annihilation operators of particles with momentum-energy $(\mathbf{k}, \omega_k)$, and $A^\dagger(\mathbf{k})$ and $B^\dagger(\mathbf{k})$ are the creation operators of the particles. The particle created or annihilated by $A^\dagger(\mathbf{k})$ or $A(\mathbf{k})$ is not the same as that created or annihilated by $B^\dagger(\mathbf{k})$ or $B(\mathbf{k})$. The latter is called an anti-particle.

The Hamiltonian $\mathcal{H}$ given by (8.3.11) is nothing but an operator which is derived in the following way: Rewrite the energy expectation value of a one-particle system by the covariant norm mentioned in §6.1 $\langle\langle K_0 \rangle\rangle = \int d\mathbf{k}(|\phi^{(+)}(\mathbf{k})|^2 + |\phi^{(-)}(\mathbf{k})|^2)$ in terms of the covariant amplitude $U(x)$ of (6.1.8), and regarding $U(x)$ as an operator, symmetrize the products that have appeared. A similar operation on the covariant norm (6.1.9) with $U(x)_1 = U(x)_2 = U(x)$ leads to

$$\langle\langle 1 \rangle\rangle \to \mathcal{N} = \int d\mathbf{x} \mathcal{J}_0(x) , \tag{8.3.31}$$

$$\mathcal{J}_0(x) = \frac{1}{4i}\left(\left\{\frac{\partial U^\dagger(x)}{\partial t}, U(x)\right\} - \left\{U^\dagger(x), \frac{\partial U(x)}{\partial t}\right\}\right) , \tag{8.3.32}$$

where

$$\mathcal{J}_\mu(x) = \frac{i}{4}(\{\partial_\mu U^\dagger(x), U(x)\} - \{U^\dagger(x), \partial_\mu U(x)\}) . \tag{8.3.33}$$

As will be easily derived from (8.1.1) we have

$$\partial_\mu \mathcal{J}_\mu(x) = 0 , \tag{8.3.34}$$

and thus, according to an argument similar to the previous one, we can write

$$\mathcal{N} = \int_\sigma d\sigma_\mu \mathcal{J}_\mu(x) . \tag{8.3.35}$$

The quantity $\mathcal{N}$ is a constant of motion independent of σ, that is, by (8.3.34) this quantity is invariant when σ is slightly deformed into σ', and it is a scalar under a Lorentz transformation. This is expressed in terms of $A(\mathbf{k})$ and $B(\mathbf{k})$ in the form

$$\mathcal{N} = \frac{1}{2}\int \frac{d\mathbf{k}}{\omega_k}(\{A^\dagger(\mathbf{k}), A(\mathbf{k})\} - \{B^\dagger(\mathbf{k}), B(\mathbf{k})\}) . \tag{8.3.36}$$

As is seen in this expression $\mathcal{N}$ is the difference of numbers of particles and anti-particles, and takes both positive and negative values. It is, however, an operator here, and is not the norm for a particle mentioned in §6.1. Therefore there does not now exist the problem of the negative norm mentioned previously. The operator $\mathcal{N}$ is called a number operator and $J_\mu(x)$ is called a current density.

We have seen in the above argument that of the Bose and the Fermi statistics, $U(x)$ must obey the former, and also how the problems mentioned in §8.1 is avoided. On the basis of these considerations we should treat a system with an interaction, but it is beyond the scope of this book. We shall give a comment that the above argument can be easily extended to the case of the para-statistics and we can show analogously that $U(x)$ must obey the para-Bose statistics, not the para-Fermi statistics. In that case $\theta_{\mu\nu}(x)$ and $J_\mu(x)$ also have the same forms as (8.3.12) and (8.3.33). However, the commutation relations are

$$\begin{aligned}
&[U(x), \{U^\dagger(y), U(z)\}] = 2i\Delta(x-y)U(z) , \\
&[U(x), \{U^\dagger(y), U^\dagger(z)\}] = 2i(\Delta(x-y)U^\dagger(z) + \Delta(x-z)U^\dagger(y)) , \\
&[U(x), \{U(y), U(z)\}] = 0 ,
\end{aligned} \tag{8.3.37}$$

corresponding to (8.2.34). The equations corresponding to (8.2.36) are

$$\left.\begin{aligned}
&A(\mathbf{k})A^\dagger(\mathbf{k}')|\,0\,\rangle = B(\mathbf{k})B^\dagger(\mathbf{k}')|\,0\,\rangle = r\omega_k\delta(\mathbf{k}-\mathbf{k}')|\,0\,\rangle , \\
&A(\mathbf{k})B^\dagger(\mathbf{k}')|\,0\,\rangle = B(\mathbf{k})A^\dagger(\mathbf{k}')|\,0\,\rangle = 0
\end{aligned}\right\} \tag{8.3.38}$$

for order r.

§8.4 Spin and Statistics

In order to study the quantum field theory of a spin $n/2$ $(n \geq 1)$ particle, generalizing the argument of the previous section we shall consider a free field $\psi_{(\dots)_n}(x)$ satisfying the Bargmann-Wigner equations (8.1.2). The Lorentz transformations of the field are given by (6.3.26) and (6.3.27). The Heisenberg equation of motion

$$i\frac{\partial\psi_{(\dots)_n}(x)}{\partial t} = [\psi_{(\dots)_n}(x), \mathcal{H}] \tag{8.4.1}$$

plays a fundamental role for the quantization of field also in this case. We must first extract independent harmonic oscillators from $\psi_{(\dots)_n}(x)$. This can be carried out by tracing the argument of §6.3 conversely. That is, $\psi_{(\dots)_n}(\mathbf{k})$ $(= \psi^{(+)}_{(\dots)_n}(\mathbf{k}) + \psi^{(-)}_{(\dots)_n}(\mathbf{k}))$ is derived from (6.3.24), and (6.3.17), (6.3.5) and (6.3.6) lead to the forms of $\phi^{(\pm)}_\xi(\mathbf{k})$ $(\xi = 1, 2, \dots, n+1)$. Of course, they are all operators in this case. To use a systematic notation in this chapter we shall write

$$\left.\begin{aligned} A_\xi(\mathbf{k}) &= \phi_\xi^{(+)}(\mathbf{k}) , \\ B_\xi^\dagger(-\mathbf{k}) &= \phi_\xi^{(-)}(\mathbf{k}) , \end{aligned}\right\} \tag{8.4.2}$$

then we obtain the Lorentz transformations of $A_\xi(\mathbf{k})$ and $B_\xi(\mathbf{k})$ from (5.1.16), (5.1.17) and (5.1.6):

$$\left.\begin{aligned} A_\xi(\mathbf{k}) &\longrightarrow \sum_{\xi'} Q^{(+)}(\lambda_k, l)_{\xi\xi'} A_{\xi'}(\Lambda^{-1}\mathbf{k}) , \\ B_\xi(\mathbf{k}) &\longrightarrow \sum_{\xi'} [Q^{(+)}(\lambda_k, l)_{\xi\xi'}]^* B_{\xi'}(\Lambda^{-1}\mathbf{k}) , \end{aligned}\right\} \tag{8.4.3}$$

where $Q^{(+)}(\lambda_k, l)$ is the Wigner rotation for the positive frequency of a particle with mass m.*) The operators $A_\xi(\mathbf{k})$ and $B_\xi(\mathbf{k})$ are independent operators and they form harmonic oscillators since (8.4.1) leads to

$$\left.\begin{aligned} \omega_k A_\xi(\mathbf{k}) &= [A_\xi(\mathbf{k}), \mathcal{H}] , \\ \omega_k B_\xi(\mathbf{k}) &= [B_\xi(\mathbf{k}), \mathcal{H}] . \end{aligned}\right\} \tag{8.4.4}$$

Hence, according to the argument of §8.2 and the Lorentz invariance, the commutation relations are written as

$$\begin{aligned} [A_\xi(\mathbf{k}), A_{\xi'}^\dagger(\mathbf{k}')]_\mp &= [B_\xi(\mathbf{k}), B_{\xi'}^\dagger(\mathbf{k}')]_\mp = \delta_{\xi\xi'}\omega_k\delta(\mathbf{k}-\mathbf{k}') , \\ [A_\xi(\mathbf{k}), A_{\xi'}(\mathbf{k}')]_\mp &= [B_\xi(\mathbf{k}), B_{\xi'}(\mathbf{k}')]_\mp = [A_\xi(\mathbf{k}), B_{\xi'}(\mathbf{k}')]_\mp \\ &= [A_\xi(\mathbf{k}), B_{\xi'}^\dagger(\mathbf{k}')]_\mp = 0 , \end{aligned} \tag{8.4.5}$$

and the corresponding Hamiltonian is given by

$$\mathcal{H} = \frac{1}{2}\sum_\xi \int d\mathbf{k}([A_\xi^\dagger(\mathbf{k}), A_\xi(\mathbf{k})]_\pm + [B_\xi^\dagger(\mathbf{k}), B_\xi(\mathbf{k})]_\pm) , \tag{8.4.6}$$

where $[\,,\,]_-$ means the minus type commutation relation $[\,,\,]$, and $[\,,\,]_+$ means the plus type commutation relation $\{\,,\,\}$. That is, if we restrict our argument into the Bose or the Fermi statistics, among the subscripts of $[\,,\,]$ the uppers

---

*)The operators $[Q^{(+)}(\lambda_k, l)]^*$ and $Q^{(+)}(\lambda_k, l)$ are unitary equivalent in a representation of the little group. That is, if we take the representation of (4.1.5)–(4.1.7) for spin matrices of the little group, from the form of $Q^{(+)}(\lambda_k, l)$ for an infinitesimal transformation (§5.1) we find that $[Q^{(+)}(\lambda_k, l)]^* = e^{-i\pi S_2} Q^{(+)}(\lambda_k, l) e^{i\pi S_2}$.

represent the Bose statistics and the lowers represent the Fermi statistics. This notation will be used in every expression hereafter.

We are now going to derive the relation between spin and statistics by rewriting these equations in terms of $\psi_{(\dots)_n}(x)$. To do this we shall first rewrite $\phi_\xi^{(\pm)}(\mathbf{k})$ in terms of $\varphi^{(\pm)}_{(\dots)_n}(\mathbf{k})$ with the help of (6.3.5) and (6.3.6). Taking account of (8.4.2) we put

$$\left.\begin{aligned} A_{(\dots)_n}(\mathbf{k}) &= \varphi^{(+)}_{(\dots)_n}(\mathbf{k}) \ , \\ B^\dagger_{(\dots)_n}(-\mathbf{k}) &= \varphi^{(-)}_{(\dots)_n}(\mathbf{k}) \ , \end{aligned}\right\} \tag{8.4.7}$$

then we obtain the following result from (8.4.5) and (8.4.6):

$$[A_{(a_1\dots a_n)}(\mathbf{k}), A^\dagger_{(a'_1\dots a'_n)}(\mathbf{k}')]_\mp = \frac{\omega_k\delta(\mathbf{k}-\mathbf{k}')}{2^n n!}|N|^2 \sum_{\substack{\text{all perm of}\\ a_1\dots a_n}} \prod_{i=1}^n (1+\beta)_{a_i a'_i} \ ,$$

$$[B_{(a_1\dots a_n)}(\mathbf{k}), B^\dagger_{(a'_1\dots a'_n)}(\mathbf{k}')]_\mp = \frac{\omega_k\delta(\mathbf{k}-\mathbf{k}')}{2^n n!}|N|^2 \sum_{\substack{\text{all perm of}\\ a_1\dots a_n}} \prod_{i=1}^n (1-\beta)_{a_i a'_i} \ ,$$

$$\begin{aligned} &[A_{(a_1\dots a_n)}(\mathbf{k}), A_{(a'_1\dots a'_n)}(\mathbf{k}')]_\mp = [B_{(a_1\dots a_n)}(\mathbf{k}), B_{(a'_1\dots a'_n)}(\mathbf{k}')]_\mp \\ &\quad = [A_{(a_1\dots a_n)}(\mathbf{k}), B_{(a'_1\dots a'_n)}(\mathbf{k}')]_\mp = [A_{(a_1\dots a_n)}(\mathbf{k}), B^\dagger_{(a'_1\dots a'_n)}(\mathbf{k}')]_\mp \\ &\quad = 0 \ , \end{aligned} \tag{8.4.8}$$

where $N(\neq 0)$ is a normalization constant in (6.3.5) and (6.3.6), and can take a suitable value for the calculation in each problem.

The sum $\sum_{\substack{\text{all perm of}\\ a_1\dots a_n}}$ means a summation over all permutations of the indices $a_1, \dots, a_n$. Further the Hamiltonian for (8.4.5) is given by

$$\begin{aligned} \mathcal{H} = \frac{1}{2|N|^2} \sum_{a_1\dots a_n} \int d\mathbf{k}([A^\dagger_{(a_1\dots a_n)}(\mathbf{k}), A_{(a_1\dots a_n)}(\mathbf{k})]_\pm \\ + [B^\dagger_{(a_1\dots a_n)}(\mathbf{k}), B_{(a_1\dots a_n)}(\mathbf{k})]_\pm) \ . \end{aligned} \tag{8.4.9}$$

The field $\psi_{(\dots)_n}(x)$ can be written with the help of (8.4.7), (6.3.7), (6.3.17) and (6.3.24) in the form

$$\begin{aligned} \psi_{(\dots)_n}(x) = \frac{1}{(2\pi)^{3/2}} \int \frac{d\mathbf{k}}{\omega_k} \omega_k^{n/2} \prod_{i=1}^n U_F^{(i)}(\mathbf{k})(A_{(\dots)_n}(\mathbf{k})e^{i(\mathbf{kx}-\omega_k t)} \\ + B^\dagger_{(\dots)_n}(-\mathbf{k})e^{i(\mathbf{kx}+\omega_k t)}) \ . \end{aligned} \tag{8.4.10}$$

Thus, the corresponding commutation relations are, by the use of (8.4.8), given by

$$[\psi_{(a_1\ldots a_n)}(x),\quad \psi_{(a'_1\ldots a'_n)}(y)]_{\mp} = 0\,. \tag{8.4.11}$$

If ε is -1 for the Bose statistics and 1 for the Fermi statistics, we have

$$\begin{aligned}
[\psi_{(a_1\ldots a_n)}(x),\ \psi^{\dagger}_{(a'_1\ldots a'_n)}(y)]_{\mp} &= \frac{|N|^2}{2^{n-1}n!(2\pi)^3} \\
&\times \sum_{\substack{\text{all perm. of}\\ a_1\ldots a_n}} \int \frac{d\mathbf{k}}{2\omega_k}\left(\prod_{i=1}^{n}[U_F(\mathbf{k})(\omega_k+\beta\omega_k)U_F(\mathbf{k})^{-1}]_{a_ia'_i}e^{i\{\mathbf{k}(\mathbf{x}-\mathbf{y})-\omega_k(x_0-y_0)\}}\right. \\
&\left. +\varepsilon\prod_{i=1}^{n}[U_F(-\mathbf{k})(\omega_k-\beta\omega_k)U_F(-\mathbf{k})^{-1}]_{a_ia'_i}e^{-i\{\mathbf{k}(\mathbf{x}-\mathbf{y})-\omega_k(x_0-y_0)\}}\right) \\
&= \frac{|N|^2}{2^{n-1}n!(2\pi)^3} \\
&\times \sum_{\substack{\text{all perm. of}\\ a_1\ldots a_n}} \int \frac{d\mathbf{k}}{2\omega_k}\left(\prod_{i=1}^{n}[(\omega_k\beta-\beta\boldsymbol{\alpha}\mathbf{k}+m)\beta]_{a_ia'_i}e^{i\{\mathbf{k}(\mathbf{x}-\mathbf{y})-\omega_k(x_0-y_0)\}}\right. \\
&\left. +\varepsilon\prod_{i=1}^{n}[(\omega_k\beta-\beta\boldsymbol{\alpha}\mathbf{k}-m)\beta]_{a_ia'_i}e^{-i\{\mathbf{k}(\mathbf{x}-\mathbf{y})-\omega_k(x_0-y_0)\}}\right) \\
&= \frac{|N|^2}{2^{n-1}n!(2\pi)^3} \\
&\times \sum_{\substack{\text{all perm. of}\\ a_1\ldots a_n}} \prod_{i=1}^{n}\left[\left(m-\gamma_\mu\frac{\partial}{\partial x_\mu}\right)\beta\right]_{a_ia'_i}\int\frac{d\mathbf{k}}{2\omega_k}\left(e^{i\{\mathbf{k}(\mathbf{x}-\mathbf{y})-\omega_k(x_0-y_0)\}}\right. \\
&\left. +(-1)^n\varepsilon e^{-i\{\mathbf{k}(\mathbf{x}-\mathbf{y})-\omega_k(x_0-y_0)\}}\right)\,.
\end{aligned} \tag{8.4.12}$$

In deriving this expression we have used (6.2.15) and (6.2.42). Therefore, from (8.4.12),
(i) in the case n = even number (integral spin) we have

$$\begin{aligned}
&[\psi_{(a_1\ldots a_n)}(x), \bar{\psi}_{(a'_1\ldots a'_n)}(y)] \\
&= \frac{|N|^2 i}{2^{n-1}n!}\sum_{\substack{\text{all perm. of}\\ a_1\ldots a_n}}\prod_{i=1}^{n}\left(m-\gamma_\mu\frac{\partial}{\partial x_\mu}\right)_{a_ia'_i}\Delta(x-y)
\end{aligned} \tag{8.4.13}$$

for the Bose statistics $((-1)^{n-1}\varepsilon = 1)$, and

$$\{\psi_{(a_1\dots a_n)}(x), \bar{\psi}_{(a'_1\dots a'_n)}(y)\}$$
$$= \frac{|N|^2 i}{2^{n-1}n!} \sum_{\substack{\text{all perm. of} \\ a_1\dots a_n}} \prod_{i=1}^{n} \left(m - \gamma_\mu \frac{\partial}{\partial x_\mu}\right)_{a_i a'_i} \Delta^{(1)}(x-y) \tag{8.4.13'}$$

for the Fermi statistics $((-1)^n \varepsilon = 1)$,

(ii) in the case $n =$ odd number (half-integral spin) we have

$$[\psi_{(a_1\dots a_n)}(x), \bar{\psi}_{(a'_1\dots a'_n)}(y)]$$
$$= \frac{|N|^2}{2^{n-1}n!} \sum_{\substack{\text{all perm. of} \\ a_1\dots a_n}} \prod_{i=1}^{n} \left(m - \gamma_\mu \frac{\partial}{\partial x_\mu}\right)_{a_i a'_i} \Delta^{(1)}(x-y) \tag{8.4.14}$$

for the Bose statistics $((-1)^n \varepsilon = 1)$, and

$$\{\psi_{(a_1\dots a_n)}(x), \bar{\psi}_{(a'_1\dots a'_n)}(y)\}$$
$$= \frac{|N|^2 i}{2^{n-1}n!} \sum_{\substack{\text{all perm. of} \\ a_1\dots a_n}} \prod_{i=1}^{n} \left(m - \gamma_\mu \frac{\partial}{\partial x_\mu}\right)_{a_i a'_i} \Delta(x-y) \tag{8.4.14'}$$

for the Fermi statistics $((-1)^{n-1}\varepsilon = 1)$.

From the above argument we find that the commutation relations vanish for space-like $x_\mu - y_\mu$ in the case $(-1)^{n-1}\varepsilon = 1$, namely in the Bose statistics for an integral spin and in the Fermi statistics for a half-integral spin. Only in these cases the locality condition is satisfied as is realized by the fact that $[\psi(x), \mathcal{H}(y)]$ is expressed, with the help of the form of $\mathcal{H}$ given later, by a linear combination of the products of the finite derivatives of $\Delta(x-y)$ and the field. As a matter of fact, if we conversely supposed to take the Fermi statistics for an integral spin and the Bose statistics for a half-integral spin, $[\psi(x), \mathcal{H}(y)]$ will involve $\Delta^{(1)}(x-y)$ and its derivatives which lead to the breaking of locality.

Before constructing the explicit form of $\mathcal{H}(x)$ we shall consider the number operator

$$\mathcal{N} = \sum_\xi \int \frac{d\mathbf{k}}{2\omega_k}([A^\dagger_\xi(\mathbf{k}), A_\xi(\mathbf{k})]_\pm - [B^\dagger_\xi(\mathbf{k}), B_\xi(\mathbf{k})]_\pm)$$
$$= \frac{1}{2|N|^2} \sum_{a_1\dots a_n} \int \frac{d\mathbf{k}}{\omega_k}([A^\dagger_{(a_1\dots a_n)}(\mathbf{k}), A_{(a_1\dots a_n)}(\mathbf{k})]_\pm$$
$$+ \varepsilon[B_{(a_1\dots a_n)}(\mathbf{k}), B^\dagger_{(a_1\dots a_n)}(\mathbf{k})]_\pm) \,. \tag{8.4.15}$$

Similarly to the case of spin 0, $A_\xi(\mathbf{x})$ and $B_\xi(\mathbf{k})$ are also the annihilation operators of a particle and an anti-particle respectively. According to the argument of (8.4.7) and (6.3.29)–(6.3.37) we have

$$\mathcal{N} = \frac{1}{2n|N|^2m^{n-1}}\sum_{j=1}^{n}\int d\mathbf{x}([\overline{\psi^{(+)}_{(\dots)_n}}(x), \gamma_4^{(j)}\psi^{(+)}_{(\dots)_n}(x)]_\pm$$
$$+(-1)^{n-1}\varepsilon[\overline{\psi^{(-)}_{(\dots)_n}}(x), \gamma_4^{(j)}\psi^{(-)}_{(\dots)_n}(x)]_\pm) \ . \tag{8.4.16}$$

Of course $\psi^{(+)}_{(\dots)_n}(x)$ and $\psi^{(-)}_{(\dots)_n}(x)$ represent the positive and negative frequency parts of $\psi_{(\dots)_n}(x)$:

$$\psi_{(\dots)_n}(x) = \psi^{(+)}_{(\dots)_n}(x) + \psi^{(-)}_{(\dots)_n}(x) \ . \tag{8.4.17}$$

Now, in order to write (8.4.16) covariantly we need the condition $(-1)^{n-1}\varepsilon = 1$. That is, if we write

$$\mathcal{N} = \int_\sigma d\sigma_\mu J_\mu(x) \ , \tag{8.4.18}$$

by the use of $(-1)^{n-1}\varepsilon = 1$ we get

$$J_\mu(x) = \begin{cases} \frac{i}{n|N|^2m^{n-1}}\sum_{i=1}^{n}\sum_{a_1\dots a_n}\frac{1}{2}\{\bar{\psi}_{(a_1\dots a_n)}(x), \gamma_\mu^{(i)}\psi_{(a_1\dots a_n)}(x)\} \\ \qquad (n = \text{even}) \\ \frac{i}{n|N|^2m^{n-1}}\sum_{i=1}^{n}\sum_{a_1\dots a_n}\frac{1}{2}[\bar{\psi}_{(a_1\dots a_n)}(x), \gamma_\mu^{(i)}\psi_{(a_1\dots a_n)}(x)] \\ \qquad (n = \text{odd}) \ . \end{cases} \tag{8.4.19}$$

Since the current operator $J_\mu(x)$ satisfies

$$\partial_\mu J_\mu(x) = 0 \tag{8.4.20}$$

by the Bargmann-Wigner equations, (8.4.18) is a scalar quantity independent of σ. On the other hand, if we put $(-1)^{n-1}\varepsilon = -1$ we cannot express $\mathcal{N}$ in a covariant form as has been evident from the argument of §6.3. The explicit proof of it can be done as follows. Differentiate each side of (8.4.17), and solve it combined with (8.4.17) for $\psi^{(\pm)}_{(\dots)_n}(x)$, then

$$\psi^{(\pm)}_{(\dots)_n}(x) = \frac{1}{2}(\psi_{(\dots)_n}(x) \pm i(-\nabla^2 + m^2)^{-\frac{1}{2}}\dot{\psi}_{(\dots)_n}(x)) \ , \tag{8.4.21}$$

where $(-\nabla^2 + m^2)^{-\frac{1}{2}}\dot{\psi}_{(\dots)_n}(x)$ means $\int d\mathbf{k}(\mathbf{k}^2 + m^2)^{-\frac{1}{2}}\varphi_{(\dots)_n}(\mathbf{k})$ when $\dot{\psi}_{(\dots)_n}(x) = \int d\mathbf{k}e^{-i\mathbf{kx}}\varphi_{(\dots)_n}(\mathbf{k})$. Equation (8.4.21) shows that when $\psi^{(\pm)}_{(\dots)_n}(x)$ is expressed in terms of the field quantities at the same instant of time non-covariant differential operators like $(-\nabla^2+m^2)^{-\frac{1}{2}}$ appear inevitably. Therefore, in order that (8.4.16) with the above $\psi^{(\pm)}_{(\dots)_n}(x)$ substituted is written in a covariant form, $(-\nabla^2 + m^2)^{-\frac{1}{2}}$ must disappear. But it is impossible in the case $(-1)^{n-1}\varepsilon = -1$ as will be easily verified by a direct calculation. An analogous argument is applicable to $\mathcal{H}(x)$ which will be considered next. In addition, $\mathcal{J}_0(x)$ given by (8.4.19) is what we obtain from the integrand in the norm of (6.3.39) by symmetrizing or anti-symmetrizing it with respect to $\bar{\psi}_{(\dots)_n}(x)$ and $\psi_{(\dots)_n}(x)$ regarded as operators.

Let us next consider $\mathcal{H}(x)$. Since (8.4.9) is written as

$$\mathcal{H} = \frac{1}{2|N|^2}\sum_{a_1\dots a_n}\int \frac{d\mathbf{k}}{\omega_k}(\omega_k[A^\dagger_{(a_1\dots a_n)}(\mathbf{k}), A_{(a_1\dots a_n)}(\mathbf{k})]_\pm + (-\omega_k)\varepsilon[B_{(a_1\dots a_n)}(\mathbf{k}), B^\dagger_{(a_1\dots a_n)}(\mathbf{k})]_\pm)\,, \tag{8.4.22}$$

in a similar manner when we derived (8.4.16) from (8.4.15) we obtain

$$\begin{aligned}\mathcal{H} &= \frac{1}{2n|N|^2m^{n-1}}\sum_{j=1}^{n}\sum_{a_1\dots a_n}\int d\mathbf{x}([\overline{\psi^{(+)}_{(a_1\dots a_n)}}(x), i\gamma_4^{(j)}\frac{\partial}{\partial x_0}\psi^{(+)}_{(a_1\dots a_n)}(x)]_\pm \\ &\quad + (-1)^{n-1}\varepsilon[\overline{\psi^{(-)}_{(a_1\dots a_n)}}(x), i\gamma_4^{(j)}\frac{\partial}{\partial x_0}\psi^{(-)}_{(a_1\dots a_n)}(x)]_\pm) \\ &= \frac{1}{4n|N|^2m^{n-1}}\sum_{j=1}^{n}\sum_{a_1\dots a_n}\int d\mathbf{x}([\overline{\psi^{(+)}_{(a_1\dots a_n)}}(x), \frac{\overleftrightarrow{\partial}}{\partial x_4}\gamma_4^{(j)}\psi^{(+)}_{(a_1\dots a_n)}(x)]_\pm \\ &\quad + (-1)^{n-1}\varepsilon[\overline{\psi^{(-)}_{(a_1\dots a_n)}}(x), \frac{\overleftrightarrow{\partial}}{\partial x_4}\gamma_4^{(j)}\psi^{(-)}_{(a_1\dots a_n)}(x)]_\pm)\,,\end{aligned} \tag{8.4.23}$$

where $\overleftrightarrow{\partial}/\partial x_4$ is generally defined by

$$\Big[F(x), \frac{\overleftrightarrow{\partial}}{\partial x_\mu}G(x)\Big]_\pm = \Big[\frac{\partial}{\partial x_\mu}F(x), G(x)\Big]_\pm - \Big[F(x), \frac{\partial}{\partial x_\mu}G(x)\Big]_\pm\,. \tag{8.4.24}$$

In order to write (8.4.23) covariantly we need again $(-1)^{n-1}\varepsilon = 1$, and if we define the energy-momentum tensor $\theta_{\mu\nu}(x)$ by

$$\theta_{\mu\nu}(x) = \begin{cases} \frac{-1}{8n|N|^2 m^{n-1}} \sum_{j=1}^{n} \left(\left\{ \bar{\psi}_{(\dots)_n}(x), \frac{\overleftrightarrow{\partial}}{\partial x_\mu} \gamma_\nu^{(j)} \psi_{(\dots)_n}(x) \right\} \right. \\ \qquad \left. + \left\{ \bar{\psi}_{(\dots)_n}(x), \frac{\overleftrightarrow{\partial}}{\partial x_\nu} \gamma_\mu^{(j)} \psi_{(\dots)_n}(x) \right\} \right) \qquad (n = \text{even}) , \\ \frac{-1}{8n|N|^2 m^{n-1}} \sum_{j=1}^{n} \left(\left[\bar{\psi}_{(\dots)_n}(x), \frac{\overleftrightarrow{\partial}}{\partial x_\mu} \gamma_\nu^{(j)} \psi_{(\dots)_n}(x) \right] \right. \\ \qquad \left. + \left[\bar{\psi}_{(\dots)_n}(x), \frac{\overleftrightarrow{\partial}}{\partial x_\nu} \gamma_\mu^{(j)} \psi_{(\dots)_n}(x) \right] \right) , \qquad (n = \text{odd}) , \end{cases}$$

(8.4.25)*)

by the equation of motion (8.1.2) we get

$$\partial_\mu \theta_{\mu\nu}(x) = 0 , \tag{8.4.26}$$

and hence we can write

$$\mathcal{H} = \int_\sigma d\sigma_\mu \theta_{\mu 0}(x) = \int d\mathbf{x} \theta_{00}(x) . \tag{8.4.27}$$

Needless to say, the energy density is $\theta_{00}(x)$ and is the 0-0 component of the covariant tensor $\theta_{\mu\nu}(x)$. It should be noted that the right-hand side of (8.4.27) is what we obtain by symmetrizing or anit-symmetrizing $\bar{\psi}_{(\dots)_n}(x)$ and $\psi_{(\dots)_n}(x)$ in the expectation value of k_0 by the covariant norm in §6.3, $\langle\langle k_0 \rangle\rangle$.

So far we have investigated the relation between spin and statistics of a particle with finite mass in the relativistic quantum field theory. Our basic assumptions were (1) the theory of harmonic oscillator studied in §6.2, and (2) the locality condition. The theory (1) has the contents that the minimum eigenvalue of Hamiltonian can be adjusted to zero by an appropriate additional constant, and the corresponding state (the vacuum) is unique, and that the norms of Hilbert space are positive. As is seen in our argument so far developed, we can use the following condition instead of (2). That is, (2′) the commutation relations (or anti-commutation relations) between the fields at space-like separated two points always vanish, or equivalently, since the condition $(-1)^{n-1}\varepsilon = 1$ which is needed in that case is the condition for

*) Here, of course, the sums over n indices of Dirac spinor have been taken, but the explicit expressions have been omitted.

covariance, (2″) $\mathcal{H}(x)$ must have a covariant form. They are natural conditions, and consequently we have derived the conclusion that *"If the statistics of particles are restricted to either the Bose statistics or the Fermi statistics, the particles with integral spin must obey the Bose statistics and those with half-integral spin must obey the Fermi statistics"*. This result can also be extended to massless particles with discrete spin. Combining (1) with the one of (2), (2′) and (2″) appropriately we might have other sets of conditions, but we do not enter into the details of them here. This beautiful theorem on spin and statistics was first derived by Pauli.*) The way of his argument was slightly different from ours but the content is the same essentially. In this book the theorem was studied in connection to the irreducible representations of the Poincaré group, and consequently the explicit forms of commutation relations, $\theta_{\mu\nu}(x)$ and $\mathcal{J}_\mu(x)$ could be determined at the same time.

In addition, as can be seen in the above argument the property of the covariant inner product in a one-particle system is closely connected to the relation between spin and statistics. This is related to the covariance of $\mathcal{H}(x)$ (and $\mathcal{J}_0(x)$ at the same time) as has been seen already, and in a one-particle system the particle obeys the Bose statistics if $\langle 1, \pm \,|\, 2, \pm \rangle = \pm \langle\langle 1, \pm \,|\, 2, \pm \rangle\rangle$, and it obeys the Fermi statistics if $\langle 1, \pm \,|\, 2, \pm \rangle = \langle\langle 1, \pm \,|\, 2, \pm \rangle\rangle$. In this sense, in the case of particle with infinite spin freedom mentioned in §7.4, since the covariant inner products have an inverse property, the relation between spin and statistics might be the Fermi statistics for a single-valued representation of the Poincaré group and the Bose statistics for a double-valued representation. It should be noted that this is an essentially different point from the case of a particle with finite spin freedom.

§8.5 Poincaré Group and Free Fields

Let us continue the discussion of the case of finite mass. Similarly to (8.3.24) we define P_μ as

$$\begin{aligned} P_\mu &= \int_\sigma d\sigma_\nu \theta_{\nu\mu}(x) \\ &= \frac{1}{2} \sum_\xi \int d\mathbf{k} \frac{k_\mu}{\omega_k} ([A_\xi^\dagger(\mathbf{k}), A_\xi(\mathbf{k})]_\pm + [B_\xi^\dagger(\mathbf{k}), B_\xi(\mathbf{k})]_\pm) \\ &\qquad\qquad (k_\mu = (\mathbf{k}, i\omega_k)) \,, \end{aligned} \tag{8.5.1}$$

then P_μ form a 4-vector giving the energy-momentum of the field. According to the relation between spin and statistics, $[\,,\,]_+$ on the right-hand side of the

*) W. Pauli: ***Phys. Rev.*** **58** (1940) 716.

above equation corresponds to an integral spin and $[\,,\,]_-$ corresponds to a half-integral spin. The symbols $[\,,\,]_\pm$ will always be used in this sense hereafter. It is evident from the right-hand side of (8.5.1) that P_μ are commutable with one another. The quantity $\theta_{00}(x)$ represents the energy density and $\theta_{0i}(x)$ ($i = 1,2,3$) may be considered to represent the momentum densities. Therefore $x_i\theta_{0j}(x) - x_j\theta_{0i}(x)$ can be regarded as the angular momentum density for a rotation in the plane spanned by i- and j-axes. Generalizing this expression we shall put

$$M_{\rho[\mu\nu]}(x) = x_\mu \theta_{\rho\nu}(x) - x_\nu \theta_{\rho\mu}(x) \ . \tag{8.5.2}$$

According to (8.4.26) and $\theta_{\mu\nu}(x) = \theta_{\nu\mu}(x)$ we have

$$\partial_\rho M_{\rho[\mu\nu]}(x) = 0 \ , \tag{8.5.3}$$

and hence the anti-symmetric tensor as a constant of motion

$$M_{[\mu\nu]} = \int_\sigma d\sigma_\rho M_{\rho[\mu\nu]}(x) = \int d\mathbf{x} M_{0[\mu\nu]}(x) \tag{8.5.4}$$

is expected to form, together with P_μ, the generators of the Poincaré group in quantum field theory. As a matter of fact, this expectation can be verified by the relations

$$\left.\begin{aligned} [M_{[\mu\nu]}, P_\lambda] &= i(\delta_{\mu\lambda}P_\nu - \delta_{\nu\lambda}P_\mu) \ , \\ [M_{[\mu\nu]}, M_{[\lambda\rho]}] &= i(\delta_{\mu\lambda}M_{[\nu\rho]} + \delta_{\mu\rho}M_{[\lambda\nu]} + \delta_{\nu\rho}M_{[\mu\lambda]} + \delta_{\nu\lambda}M_{[\rho\mu]}) \ . \end{aligned}\right\} \tag{8.5.5}$$

To derive (8.5.5) we may use the commutation relations, i.e., (8.4.13) for an integral spin or (8.4.14′) for a half-integral spin, but the calculation will become somewhat complicated. It is rather easy to use $M_{[\mu\nu]}$ expressed in terms of $A_\xi(\mathbf{k})$ and $B_\xi(\mathbf{k})$. For this purpose we change the time-derivatives in $\theta_{\mu\nu}(x)$ into the space-derivatives by the use of the field equations (8.1.2), perform the integration by parts, and operate it on $\psi_{(\dots)_n}(x)$. Further, after transforming into the momentum representation by a Fourier transformation we trace the argument of §6.3 conversely, where we use the Dirac representation for γ matrices. The result of such a calculation is

$$\begin{aligned} M_{[\mu\nu]} = \frac{1}{2}\sum_{\xi,\xi'}\int \frac{d\mathbf{k}}{\omega_k}&([A^\dagger_\xi(\mathbf{k}), (J_{[\mu\nu]})_{\xi\xi'}A_{\xi'}(\mathbf{k})]_\pm \\ &+ (-1)^{n-1}[B_\xi(-\mathbf{k}), (J_{[\mu\nu]})_{\xi\xi'}B^\dagger_{\xi'}(-\mathbf{k})]_\pm) \ , \end{aligned} \tag{8.5.6}$$

where $J_{[\mu\nu]}$ has been derived by (2.3.14) with $\mathbf{J}$ and $\mathbf{K}$ of (5.1.21) and (5.1.22). It should be noted that k_0 in $\mathbf{K}$ is ω_k when it acts on $A_\xi(\mathbf{k})$, and $-\omega_k$ when it acts on $B_\xi^\dagger(-\mathbf{k})$. Taking account of the relation between spin and statistics and using (8.4.5) and (8.5.6) we obtain

$$\left.\begin{aligned} [A_\xi(\mathbf{k}), M_{[\mu\nu]}] &= \sum_{\xi'} (J_{[\mu\nu]})_{\xi\xi'} A_{\xi'}(\mathbf{k}) \, , \\ [B_\xi^\dagger(-\mathbf{k}), M_{[\mu\nu]}] &= \sum_{\xi'} (J_{[\mu\nu]})_{\xi\xi'} B_{\xi'}^\dagger(-\mathbf{k}) \, . \end{aligned}\right\} \tag{8.5.7}$$

The second equation of (8.5.5) can be easily derived from these equations, their Hermitian conjugates and (2.3.14), and the first equation of (8.5.5) can be derived with the help of the right-hand side of (8.5.1).

Since $M_{[ij]}$, $M_{[i0]}$ $(i, j = 1, 2, 3)$ are Hermitian operators, $1 + \frac{i}{2} M_{[\mu\nu]}\omega_{[\mu\nu]}$ is a unitary operator and gives an infinitesimal Lorentz transformation, where $\omega_{[\mu\nu]}$ are infinitesimal parameters defined by (2.3.16). Thus the transformation of $A_\xi(\mathbf{k})$ by this operator is obtained from (8.5.7) in the form

$$\begin{aligned} A_\xi(\mathbf{k}) \longrightarrow & \left(1 - \frac{i}{2} M_{[\mu\nu]}\omega_{[\mu\nu]}\right) A_\xi(\mathbf{k}) \left(1 + \frac{i}{2} M_{[\mu\nu]}\omega_{[\mu\nu]}\right) \\ & = A_\xi(\mathbf{k}) + \frac{i}{2}\omega_{[\mu\nu]} \sum_{\xi'} (J_{[\mu\nu]})_{\xi\xi'} A_{\xi'}(\mathbf{k}) \, . \end{aligned} \tag{8.5.8}$$

If transformations like this are applied successively, the unitary transformation operator $\mathcal{L}(\Lambda)$ for a finite Λ can be built up, and it will give

$$\mathcal{L}(\Lambda)^{-1} A_\xi(\mathbf{k}) \mathcal{L}(\Lambda) = \sum_{\xi'} Q_{\xi\xi'}^{(+)}(\lambda_k, l) A_{\xi'}(\Lambda^{-1}\mathbf{k}) \, , \tag{8.5.9}$$

and similarly

$$\mathcal{L}(\Lambda)^{-1} B_\xi(\mathbf{k}) \mathcal{L}(\Lambda) = \sum_{\xi'} [Q_{\xi\xi'}^{(+)}(\lambda_k, l)]^* B_{\xi'}(\Lambda^{-1}\mathbf{k}) \, . \tag{8.5.10}$$

Therefore we can conclude that the transformation of (8.4.3) is given by the above equation.

When the operators transform in the above manner, the state vectors do not change. This situation is just the same as that in the Heisenberg picture where only operators vary with time and state vectors do not change. Thus, if the operators do not change under a Lorentz transformation, the state vector $|\ \rangle$ must transform into $\mathcal{L}(\Lambda)|\ \rangle$. In order to see it explicitly we shall consider, for example, the case where the state vector $|\ \rangle$ represents a one-particle state $A^\dagger_\xi(\mathbf{k})|0\rangle$. Here we may impose the condition

$$\mathcal{L}(\Lambda)|0\rangle = |0\rangle \tag{8.5.11}$$

by virtue of the assumption of the unique vaccum. Then we obtain by the use of (8.5.9) with Λ replaced by Λ^{-1}

$$\begin{aligned}\mathcal{L}(\Lambda)A^\dagger_\xi(\mathbf{k})|0\rangle &= \mathcal{L}(\Lambda)A^\dagger_\xi(\mathbf{k})\mathcal{L}(\Lambda)^{-1}|0\rangle \\ &= \sum_{\xi'}[Q^{(+)}_{\xi\xi'}(\tilde{\lambda}_k, l)]^* A^\dagger_{\xi'}(\Lambda\mathbf{k})|0\rangle\ .\end{aligned} \tag{8.5.12}$$

Since $\tilde{\lambda}_k$ is an element of the little group for Λ^{-1}, we get, from (3.3.4), $\tilde{\lambda}_k = \alpha_k^{-1}\Lambda^{-1}\alpha_{\Lambda k}$ which is equal to $(\lambda_{\Lambda k})^{-1}$. Thus we obtain $Q^{(+)}(\tilde{\lambda}_k, l) = Q^{(+)}(\lambda_{\Lambda k}, l)^{-1}$ from (3.3.16), and hence $[Q^{(+)}_{\xi\xi'}(\tilde{\lambda}_k, l)]^* = Q^{(+)}_{\xi'\xi}(\lambda_{\Lambda k}, l)$, where we have used the fact that $Q^{(+)}(\lambda_{\Lambda k}, l)$ is a unitary matrix. As a result, the transformation of the state $A^\dagger_\xi(\mathbf{k})|0\rangle$ becomes

$$\mathcal{L}(\Lambda)A^\dagger_\xi(\mathbf{k})|0\rangle = \sum_{\xi'} Q^{(+)}_{\xi\xi'}(\lambda_{\Lambda k}, l)A^\dagger_{\xi'}(\Lambda\mathbf{k})|0\rangle\ , \tag{8.5.13}$$

which is just the same as the transformation of a one-particle (positive frequency) state $|\bar{k}, \bar{\xi}\rangle$, $L(\Lambda)|\bar{k}, \bar{\xi}\rangle$, mentioned in §3.2, because we have

$$\begin{aligned}L(\Lambda)|\bar{k}, \bar{\xi}\rangle &= Q^{(+)}(\lambda_k, l)|\Lambda\bar{k}, \bar{\xi}\rangle = Q^{(+)}(\lambda_{\Lambda\bar{k}}, l)|\Lambda\bar{k}, \bar{\xi}\rangle \\ &= \sum_{\xi'} Q^{(+)}_{\bar{\xi}\bar{\xi}'}(\lambda_{\Lambda\bar{k}}, l)|\Lambda\bar{k}, \bar{\xi}'\rangle\ .\end{aligned} \tag{8.5.14}$$

It may be clear that we can similarly discuss the problem of a one-particle state of anti-particle $B^\dagger(\mathbf{k})|0\rangle$.*) In this way, in place of k_μ and $L(\Lambda)$ in a one-particle system mentioned in Chap. 3, in the quantum field theory P_μ and $\mathcal{L}(\Lambda)$ are used as operators playing the same role. In the later case, these operators can give not only the transformation of a one-particle state vector but also that of a multi-particle state vector, and we have a further advantage that state vectors have non-negative energy.**)

---

*)In this case, however, $[Q^{(+)}(\lambda_{\Lambda k}, l)]^*$ appears in place of $Q^{(+)}(\lambda_{\Lambda k}, l)$. They are unitary equivalent to each other as has been noted in the footnote after (8.4.3).

**)We have discussed here $\mathcal{L}(\Lambda)$ in the case of spin $n/2$ $(n \geq 1)$, and our method is applicable also to constructing $M_{\rho[\mu\nu]}$ and $\mathcal{L}(\Lambda)$ from $\theta_{\mu\nu}$ in the case of spin 0 mentioned in the previous section.

The unitary operators giving the space reflection (6.6.1) and the charge conjugation (6.6.27), i.e., P and C satisfying

$$P^{-1}\psi_{(\dots)_n}(\mathbf{x},t)P = e^{i\delta}\prod_{j=1}^{n}(i\gamma_4^{(j)})\psi_{(\dots)_n}(-\mathbf{x},t)\,, \tag{8.5.15}$$

$$C^{-1}\psi_{(\dots)_n}(x)C = \psi^c_{(\dots)_n}(x)\,, \tag{8.5.16}$$

can be also expressed in terms of $\psi_{(\dots)_n}(x)$ in the field theory. The derivation of these operators is left as a reader's exercise. Using the above P and C we can obtain the following relations:

$$\left.\begin{array}{ll} P^{-1}P_iP = -P_i\,, & P^{-1}HP = H\,, \\ P^{-1}M_{[ij]}P = M_{[ij]}\,, & P^{-1}M_{[i0]}P = -M_{[i0]}\,, \\ P^{-1}\mathcal{N}P = \mathcal{N}\,, & \end{array}\right\} \tag{8.5.17}$$

and

$$C^{-1}P_\mu C = P_\mu\,, \quad C^{-1}M_{[\mu\nu]}C = M_{[\mu\nu]}\,, \quad C^{-1}\mathcal{N}C = -\mathcal{N}\,. \tag{8.5.18}$$

It should be noted particularly that (8.5.18) can be derived with the help of the relation between spin and statics. This point must be compared with (6.6.35) where $\psi_{(\dots)_n}(x)$ were considered as mere complex numbers.

The time reversal cannot be realized by any unitary transformation also in the field theory, and it needs an anti-unitary transformation. We took the complex conjugate of a wave function to represent this transformation in §6.6, but in the field theory we cannot generally write a state vector in terms of the wave function. Thus we first begin with the following preparation.

Corresponding to each state vector $|\ \rangle$ in the Hilbert space h we define $|\tilde{\ }\rangle$ as follows. That is, if

$$|1\rangle \longleftrightarrow |\tilde{1}\rangle, \quad |2\rangle \longleftrightarrow |\tilde{2}\rangle, \quad |3\rangle \longleftrightarrow |\tilde{3}\rangle, \dots$$

then

$$\langle\tilde{1}|\tilde{2}\rangle = \langle 2|1\rangle\,, \tag{8.5.19}$$

$$c_1|1\rangle + c_2|2\rangle \longleftrightarrow c_1^*|\tilde{1}\rangle + c_2^*|\tilde{2}\rangle\,, \tag{8.5.20}$$

where c_1 and c_2 are complex numbers. We denote the Hilbert space formed by $|\tilde{\ }\rangle$ as $\tilde{h}$ and call it the dual space of h. Corresponding to an arbitrary operator F in h, an operator $\tilde{F}$ in $\tilde{h}$ is defined by

$$F|\ \rangle \longleftrightarrow \widetilde{F|\ \rangle} = \tilde{F}|\tilde{\ }\rangle\,. \tag{8.5.21}$$

Then, if

$$F_3 = c_1 F_1 + c_2 F_2 \,, \tag{8.5.22}$$

(8.5.20) and (8.5.21) lead to

$$\tilde{F}_3 = c_1^* \tilde{F}_1 + c_2^* \tilde{F}_2 \,, \tag{8.5.23}$$

$$(\widetilde{F_1 F_2}) = \tilde{F}_1 \tilde{F}_2 \,, \tag{8.5.24}$$

and, from (8.5.19) and (8.5.21) we obtain

$$\langle \tilde{1} | \tilde{F} | \tilde{2} \rangle = \langle \tilde{1} | (\widetilde{F|2\rangle}) \rangle = (\langle 2 | F^\dagger) | 1 \rangle = \langle 2 | F^\dagger | 1 \rangle \,. \tag{8.5.25}$$

As a result, we get $\tilde{F}^\dagger = \widetilde{F^\dagger}$, which is denoted as F^t. Namely, if

$$F^t = \tilde{F}^\dagger = \widetilde{F^\dagger} \,, \tag{8.5.26}$$

from (8.5.22)–(8.5.25) we obtain

$$F_3^t = c_1 F_1^t + c_2 F_2^t \,, \tag{8.5.27}$$

$$(F_1 F_2)^t = F_2^t F_1^t \,, \tag{8.5.28}$$

$$\langle \tilde{1} | F^t | \tilde{2} \rangle = \langle 2 | F | 1 \rangle \,, \tag{8.5.29}$$

$$F^{t\dagger} = F^{\dagger t} \,. \tag{8.5.30}$$

Needless to say, F^t is an operator in $\tilde{h}$ as well as $\tilde{F}$.

Now, on the basis of the above argument we shall define the time reversal in the quantum field theory by

$$|\ \rangle \longrightarrow |\tilde{\ }\rangle \,, \tag{8.5.31}$$

$$\psi_{(\dots)_n}(\mathbf{x}, t) \longrightarrow \psi'_{(\dots)_n}(\mathbf{x}, t) = e^{i\delta'} \prod_{j=1}^{n} (i\gamma_5^{(j)} \gamma_4^{(j)}) (\psi^c_{(\dots)_n}(\mathbf{x}, -t))^t \,. \tag{8.5.32}$$

(8.5.32)*)

Equation (8.5.32) corresponding to (6.6.28) has the superscript t in order to transform the field as an operator in h into that in the dual space. It can be easily verified that $\psi'_{(\dots)_n}(x)$ of (8.5.32) satisfy the equations of motion (8.1.2),

*) For a complex scalar field we put $U(\mathbf{x}, t) \longrightarrow U'(\mathbf{x}, t) = U^{\dagger t}(\mathbf{x}, -t) = \tilde{U}(\mathbf{x}, -t)$ in place of the above expression. In this case the space reflection is $P^{-1} U(\mathbf{x}, t) P = e^{i\delta} U(-\mathbf{x}, t)$ and the charge conjugation is $C^{-1} U(x) C = U^\dagger(x)$.

and satisfy the commutation relations (8.4.13) or (8.4.14′) corresponding to even n or odd n. Therefore the energy-momentum in the time-reversed world can be expressed by P_μ with $\psi_{(\dots)_n}(x)$ replaced by $\psi'_{(\dots)_n}(x)$. Denoting it as $P'_\mu = (\mathbf{P}', iH')$ and using the fact that P_μ are conserved quantities and are time independent we have

$$H' = H^t\,, \quad P'_i = -P^t_i\,, \tag{8.5.33}$$

and hence by (8.5.29) the expectation values of these operators become

$$\langle\tilde{\ }|H'|\tilde{\ }\rangle = \langle\,|H|\,\rangle\,, \quad \langle\tilde{\ }|P'_i|\tilde{\ }\rangle = -\langle\,|P_i|\,\rangle\,. \tag{8.5.34}$$

That is, the energy expectation value does not vary under the time reversal, but the momentum expectation values change their signs. Similarly, replacing $\psi_{(\dots)_n}(x)$ in $M_{[\mu\nu]}$ and $\mathcal{N}$ by $\psi'_{(\dots)_n}(x)$ we can construct $M'_{[\mu\nu]}$ and $\mathcal{N}'$, and we obtain

$$\left.\begin{aligned} \langle\tilde{\ }|M'_{[ij]}|\tilde{\ }\rangle &= -\langle\,|M_{[ij]}|\,\rangle\,, \\ \langle\tilde{\ }|M'_{[i0]}|\tilde{\ }\rangle &= \langle\,|M_{[i0]}|\,\rangle\,, \\ \langle\tilde{\ }|\mathcal{N}'|\tilde{\ }\rangle &= \langle\,|\mathcal{N}|\,\rangle\,. \end{aligned}\right\} \tag{8.5.35}$$

In this way the discrete transformations have been introduced into the quantum field theory. However, further discussion of the details of them is omitted here because it needs many pages.

We have obtained the relations between $\psi_{(\dots)_n}(x)$ and $A_\xi(\mathbf{k}), B^\dagger_\xi(-\mathbf{k})$ in terms of $U_F(\mathbf{k})$. Before concluding this section we shall give an additional argument that expresses the relations in another form.

Let $\varsigma, \xi = 1, 2, \dots, n+1$, and let two sets of complex numbers $f^\varsigma_\xi(\mathbf{k})$ and $g^\varsigma_\xi(\mathbf{k})$ form complete orthogonal bases, i.e.,

$$\sum_\xi f^{\varsigma *}_\xi(\mathbf{k}) f^{\varsigma'}_\xi(\mathbf{k}) = \sum_\xi g^{\varsigma *}_\xi(\mathbf{k}) g^{\varsigma'}_\xi(\mathbf{k}) = \delta_{\varsigma\varsigma'}\,, \tag{8.5.36}$$

$$\sum_\varsigma f^\varsigma_\xi(\mathbf{k}) f^{\varsigma *}_{\xi'}(\mathbf{k}) = \sum_\varsigma g^\varsigma_\xi(\mathbf{k}) g^{\varsigma *}_{\xi'}(\mathbf{k}) = \delta_{\xi\xi'}\,. \tag{8.5.37}$$

In terms of these we shall write

$$\left.\begin{aligned} A_\xi(\mathbf{k}) &= \sum_\varsigma A_\varsigma(\mathbf{k}) f^\varsigma_\xi(\mathbf{k})\,, \\ B^\dagger_\xi(\mathbf{k}) &= \sum_\varsigma B^\dagger_\varsigma(\mathbf{k}) g^\varsigma_\xi(\mathbf{k})\,. \end{aligned}\right\} \tag{8.5.38}$$

The direction of the spin by ξ and that by ς are not necessarily the same,*) and, of course, $A_\varsigma(\mathbf{k})$ and $B_\varsigma(\mathbf{k})$ give independent harmonic oscillators as well as $A_\xi(\mathbf{k})$ and $B_\xi(\mathbf{k})$. As we did in the first part of §6.3, we introduce $f^\varsigma_{(a_1\ldots a_n)}(\mathbf{k})$ and $g^\varsigma_{(a_1\ldots a_n)}(\mathbf{k})$ from $f^\varsigma_\xi(\mathbf{k})$ and $g^\varsigma_\xi(\mathbf{k})$ by increasing the number of indices, and assume they satisfy (6.3.13) and (6.3.14), namely

$$\left.\begin{aligned} \omega_k f^\varsigma_{(\ldots)_n}(\mathbf{k}) &= \omega_k \beta^{(i)} f^\varsigma_{(\ldots)_n}(\mathbf{k})\ , \\ \omega_k g^\varsigma_{(\ldots)_n}(\mathbf{k}) &= -\omega_k \beta^{(i)} g^\varsigma_{(\ldots)_n}(\mathbf{k})\ . \end{aligned}\right\} \tag{8.5.39}$$

The functions $f^\varsigma_{(\ldots)_n}(\mathbf{k})$ and $g^\varsigma_{(\ldots)_n}(\mathbf{k})$ correspond to the positive and negative frequency amplitudes $\varphi^{(+)}_{(\ldots)_n}(\mathbf{k})$ and $\varphi^{(-)}_{(\ldots)_n}(\mathbf{k})$ of a one-particle system respectively, and by definition they satisfy

$$\sum_{a_1\ldots a_n} f^{\varsigma *}_{(a_1\ldots a_n)}(\mathbf{k}) g^{\varsigma'}_{(a_1\ldots a_n)}(\mathbf{k}) = 0\ . \tag{8.5.40}$$

Equations (8.5.36) and (8.5.37) lead to

$$\begin{aligned} \sum_{a_1\ldots a_n} f^{\varsigma *}_{(a_1\ldots a_n)}(\mathbf{k}) f^{\varsigma'}_{(a_1\ldots a_n)}(\mathbf{k}) &= \sum_{a_1\ldots a_n} g^{\varsigma *}_{(a_1\ldots a_n)}(\mathbf{k}) g^{\varsigma'}_{(a_1\ldots a_n)}(\mathbf{k}) \\ &= |N|^2 \delta_{\varsigma\varsigma'}\ , \end{aligned} \tag{8.5.41}$$

$$\sum_\varsigma f^\varsigma_{(a_1\ldots a_n)}(\mathbf{k}) f^{\varsigma *}_{(a'_1\ldots a'_n)}(\mathbf{k}) = \frac{|N|^2}{2^n n!} \sum_{\substack{\text{all perm. of} \\ a_1\ldots a_n}} \prod_{i=1}^n (1+\beta)_{a_i a'_i} \tag{8.5.42}$$

$$\sum_\varsigma g^\varsigma_{(a_1\ldots a_n)}(\mathbf{k}) g^{\varsigma *}_{(a'_1\ldots a'_n)}(\mathbf{k}) = \frac{|N|^2}{2^n n!} \sum_{\substack{\text{all perm. of} \\ a_1\ldots a_n}} \prod_{i=1}^n (1-\beta)_{a_i a'_i} \tag{8.5.43}$$

where $|N|^2$ is a normalization constant. We shall here put

$$\left.\begin{aligned} u^\varsigma_{(\ldots)_n}(\mathbf{k}) &= \omega_k^{n/2} \prod_{i=1}^n U_F^{(i)}(\mathbf{k}) f^\varsigma_{(\ldots)_n}(\mathbf{k})\ , \\ v^\varsigma_{(\ldots)_n}(\mathbf{k}) &= \omega_k^{n/2} \prod_{i=1}^n U_F^{(i)}(\mathbf{k}) g^\varsigma_{(\ldots)_n}(\mathbf{k})\ . \end{aligned}\right. \tag{8.5.44}$$

Writing

$$k_\mu = (\mathbf{k}, i\omega_k)\ , \tag{8.5.45}$$

*)Put $f^\varsigma_\xi(\mathbf{k}) = g^\varsigma_\xi(\mathbf{k}) = \delta_{\varsigma\xi}$ to make them the same.

namely $k_0 = \omega_k$, we repeat the discussion in §6.3, then we can write (8.5.39) in the form

$$\left.\begin{array}{l}(ik_\mu\gamma_\mu^{(j)} + m)u^\varsigma_{(\dots)_n}(\mathbf{k}) = 0 \;, \\ (ik_\mu\gamma_\mu^{(j)} - m)v^\varsigma_{(\dots)_n}(\mathbf{k}) = 0\end{array} \quad (j = 1, 2, \dots n) \;, \right\} \tag{8.5.46}$$

and write (8.5.40)–(8.5.43) in the form

$$\sum_{a_1\dots a_n} \bar{u}^\varsigma_{(a_1\dots a_n)}(\mathbf{k}) v^{\varsigma'}_{(a_1\dots a_n)}(\mathbf{k}) = 0 \;, \tag{8.5.47}$$

$$\sum_{a_1\dots a_n} \bar{u}^\varsigma_{(a_1\dots a_n)}(\mathbf{k}) u^{\varsigma'}_{(a_1\dots a_n)}(\mathbf{k}) = m^n |N|^2 \delta_{\varsigma\varsigma'} \;, \tag{8.5.48}$$

$$\sum_{a_1\dots a_n} \bar{v}^\varsigma_{(a_1\dots a_n)}(\mathbf{k}) v^{\varsigma'}_{(a_1\dots a_n)}(\mathbf{k}) = (-m)^n |N|^2 \delta_{\varsigma\varsigma'} \;, \tag{8.5.49}$$

$$\sum_\varsigma u^\varsigma_{(a_1\dots a_n)}(\mathbf{k}) \bar{u}^\varsigma_{(a'_1\dots a'_n)}(\mathbf{k}) = \frac{|N|^2}{2^n n!} \sum_{\substack{\text{all perm. of} \\ a_1\dots a_n}} \prod_{j=1}^n (m - ik_\mu\gamma_\mu)_{a_j a'_j} \;, \tag{8.5.50}$$

$$\sum_\varsigma v^\varsigma_{(a_1\dots a_n)}(\mathbf{k}) \bar{v}^\varsigma_{(a_1\dots a'_n)}(\mathbf{k}) = \frac{|N|^2}{2^n n!} \sum_{\substack{\text{all perm. of} \\ a_1\dots a_n}} \prod_{j=1}^n (-m - ik_\mu\gamma_\mu)_{a_j a'_j} \;. \tag{8.5.51}$$

The function $\psi_{(\dots)_n}(x)$ is given by

$$\psi_{(\dots)_n}(x) = \frac{1}{(2\pi)^{3/2}} \sum_\varsigma \int \frac{d\mathbf{k}}{\omega_k} (A_\varsigma(\mathbf{k}) u^\varsigma_{(\dots)_n}(\mathbf{k}) e^{ik_\mu x_\mu} + B^\dagger_\varsigma(\mathbf{k}) v^\varsigma_{(\dots)_n}(\mathbf{k}) e^{-ik_\mu x_\mu}) \;. \tag{8.5.52}$$

In the above discussion we introduced $u^\varsigma_{(\dots)_n}(\mathbf{k})$ and $v^\varsigma_{(\dots)_n}(\mathbf{k})$ from $f^\varsigma_\xi(\mathbf{k})$ and $g^\varsigma_\xi(\mathbf{k})$, however forgetting the way we can define $u^\varsigma_{(\dots)_n}(\mathbf{k})$ and $v^\varsigma_{(\dots)_n}(\mathbf{k})$ directly by (8.5.46)–(8.5.51). In this case $A_\varsigma(\mathbf{k})$ and $B_\varsigma(\mathbf{k})$ can be obtained from (8.5.52) with the help of (8.5.46)–(8.5.49). If we especially chose $v^\varsigma_{(\dots)_n}(\mathbf{k})$ such that

$$v^\varsigma_{(\dots)_n}(\mathbf{k}) = (u^\varsigma_{(\dots)_n}(\mathbf{k}))^c \;. \tag{8.5.53}$$

(This choice is always possible), we have

$$\left.\begin{array}{l} A_\varsigma(\mathbf{k}) = \dfrac{1}{(2\pi)^{3/2} n m^{n-1} |N|^2} \displaystyle\sum_{j=1}^n \int d\mathbf{x} e^{-ik_\mu x_\mu} \bar{u}^\varsigma_{(\dots)_n}(\mathbf{k}) \gamma_4^{(j)} \psi_{(\dots)_n}(x) \;, \\ B_\varsigma(\mathbf{k}) = \dfrac{1}{(2\pi)^{3/2} n m^{n-1} |N|^2} \displaystyle\sum_{j=1}^n \int d\mathbf{x} e^{-ik_\mu x_\mu} \bar{u}^\varsigma_{(\dots)_n}(\mathbf{k}) \gamma_4^{(j)} \psi^c_{(\dots)_n}(x) \;. \end{array}\right\} \tag{8.5.54}$$

In the derivation of these equation we have used the relation (6.3.35). Taking account of (8.5.16) we find the following relation between $A_\varsigma(\mathbf{k})$ and $B_\varsigma(\mathbf{k})$:

$$C^{-1}A_\varsigma(\mathbf{k})C = B_\varsigma(\mathbf{k}) \ . \tag{8.5.55}$$

In usual quantum field theory we frequently use

$$\left.\begin{aligned} a_\varsigma(\mathbf{k}) &= \omega_k^{-1/2}A_\varsigma(\mathbf{k}) \ , \\ b_\varsigma(\mathbf{k}) &= \omega_k^{-1/2}B_\varsigma(\mathbf{k}) \ , \end{aligned}\right\} \tag{8.5.56}$$

in place of $A_\varsigma(\mathbf{k})$ and $B_\varsigma(\mathbf{k})$. Their commutation relations are

$$\left.\begin{aligned} [a_\varsigma(\mathbf{k}), a^\dagger_{\varsigma'}(\mathbf{k}')]_\mp &= [b_\varsigma(\mathbf{k}), b^\dagger_{\varsigma'}(\mathbf{k}')]_\mp = \delta_{\varsigma\varsigma'}\delta(\mathbf{k}-\mathbf{k}') \ , \\ [a_\varsigma(\mathbf{k}), a_{\varsigma'}(\mathbf{k}')]_\mp &= [b_\varsigma(\mathbf{k}), b_{\varsigma'}(\mathbf{k}')]_\mp = [a_\varsigma(\mathbf{k}), b_{\varsigma'}(\mathbf{k}')]_\mp \\ &= [a_\varsigma(\mathbf{k}), b^\dagger_{\varsigma'}(\mathbf{k}')]_\mp = 0 \ , \end{aligned}\right\} \tag{8.5.57}$$

where it should be noted that the right-hand side of the first row is not Lorentz invariant under a transformation $\mathbf{k} \to \Lambda^{-1}\mathbf{k}$, $\mathbf{k}' \to \Lambda^{-1}\mathbf{k}'$.

In addition, we possibly consider also the case where the field satisfies the relation $\psi^c_{(\dots)_n}(x) = e^{i\delta}\psi_{(\dots)_n}(x)$. In this case, as is seen in (8.5.54), we have $A_\varsigma(\mathbf{k}) = e^{i\delta}B_\varsigma(\mathbf{k})$, which shows that a particle and its anti-particle are the same. Such a field is called a Majorana field. For a Majorana field we get $\mathcal{N} = 0$ from (8.4.15). Since the creation operators are only $a^\dagger_\varsigma(\mathbf{k})$ (or $A^\dagger_\varsigma(\mathbf{k})$), the commutation relations are simply given by

$$\left.\begin{aligned} [a_\varsigma(\mathbf{k}), a^\dagger_{\varsigma'}(\mathbf{k})]_\mp &= \delta_{\varsigma\varsigma'}\delta(\mathbf{k}-\mathbf{k}') \ , \\ [a_\varsigma(\mathbf{k}), a_{\varsigma'}(\mathbf{k}')]_\mp &= 0 \end{aligned}\right\} \tag{8.5.58}$$

in place of (8.5.57).

In the previous section as well as this, our argument was restricted to the two statistics, i.e., the Bose statistics and the Fermi statistics. It is, however, not difficult to extend the argument to the para-statistics. In that case the relation between spin and statistics for a particle with finite spin freedom will be shown, by the same reason as before, that an integral spin particle obeys the para-Bose statistics and a half-integral spin particle obeys the para-Fermi statistics. The quantities $\theta_{\mu\nu}(x)$ and $\mathcal{N}$ have the same expressions (apart from additonal constants) as we have obtained here.

References and Bibliography

This book was written to be as self-contained as possible, but the following related references may be useful for the readers.

The application of group theory in quantum mechanics is described in the following classical literature:

1. H. Weyl, *The Theory of Groups and Quantum Mechanics*, Dover (1983).
2. E. P. Wigner, *Group Theory and its Application to the Quantum Mechanics of Atomic Spectra*, Academic Press (1959).
3. B. L. van der Wearden, *Die Gruppen Theoretische Methode in der Quantum Mechanik*, Springer (1983).

For the recent literature see

4. M. Hamermesh, *Group Theory and its Application to Physical Problems*, Addison-Wesley (1962).

As for the references on the representation theory of rotation group

5. T. Yamanouchi, *The Rotation Group and its Representations*, Iwanami (1957),

is an appropriate introductory textbook.

The exact treatment of the representations of proper Lorentz group is found in

6. M. A. Naimark, *Linear Representations of the Lorentz Group*, Pergamon Press (1964).
7. I. M. Gel'fand, R. A. Minlos and Z. Y. Shapiro, *Representations of Rotation and Lorentz Groups and Their Applications*, Pergamon Press (1963).

I do not know of any good textbooks on the unitary representations of the Poincaré group. Wigner's laborious work

8. E. P. Wigner, *On Unitary Representation of the Inhomogeneous Lorentz Group*, Ann. Math., **40** (1939) 149

is a classical and basic paper. Related papers are, for example,

9. E. P. Wigner, *Relativistische Wellengleichungen*, Z. Physik, **124** (1947), 665.
10. V. Bargmann and E. P. Wigner, *Group Theoretical Discussion of Relativistic Wave Equations*, Pro. Natl. Acad. Sci. U.S., **34** (1948) 211

and this subject was reviewed in

11. Ju M. Shirokov, *A Group-Theoretical Consideration of the Basis of Relativistic Quantum Mechanics I, II, III, IV*, Soviet Phys. - JETP, **6** (1958) 664, 918, 929, ibid., **7** (1958) 493.

The irreducible decomposition of direct products of the Poincaré group and the kinematics of *S*-matrices which were not considered in the present book can be found in

12. H. Joos, *Zur Darstellungstheorie der Inhomogenen Lorentzgrouppe als Grundlage Quantenmechanische Kinematick*, Fortschritte der Phys., **10** (1962) 65.

Many old papers on relativistic wave equation are referred in

13. E. M. Corson, *Introduction to Tensors, Spinors, and Relativistic Wave-Equations*, Hafner (1953).

The basic considerations of field quantization and the general theory of its application to systems with interactions are written in

14. Y. Takahashi, *An Introduction to Field Quantization*, Pergamon Press (1968).

The examples of quantum fields applied to particle physics can be found in

15. K. Nishijima, *Field and Particles*, Benjamin (1969).
16. N. Nakanishi, *Quantum Field Theory*, Baifukan (1975).
17. H. Yukawa and Y. Katayama, *Elementary Particle Theory* (Iwanami lectures of the foundation of modern physics 11), Iwanami (1974).

Index